S0-AVW-389

New Polymeric Materials

ACS SYMPOSIUM SERIES **916**

New Polymeric Materials

Ljiljana S. Korugic-Karasz, Editor
University of Massachusetts

William J. MacKnight, Editor
University of Massachusetts

Ezio Martuscelli, Editor
CNR Italy

Sponsored by the
ACS Division of Polymer Chemistry, Inc.

American Chemical Society, Washington, DC

Library of Congress Cataloging-in-Publication Data

New polymeric materials / Ljiljana S. Korugic-Karasz, editor, William J. MacKnight, editor, Ezio Martuscelli, editor ; sponsored by the ACS Division of Polymer Chemistry, Inc.

p. cm.—(ACS symposium series ; 916)

Includes bibliographical references and index.

ISBN 0–8412–3928–2 (alk. paper)

1. Polymers—Congresses.

I. Korugic-Karasz, Ljiljana S., -1953- II. MacKnight, William J. III. Martuscelli, Ezio. IV. American Chemical Society. Division of Polymer Chemistry, Inc. V. Series.

TA455.P58N46 2005
668.9—dc22 2005041195

The paper used in this publication meets the minimum requirements of American National Standard for Information Sciences—Permanence of Paper for Printed Library Materials, ANSI Z39.48–1984.

Distributed by Oxford University Press

PRINTED IN THE UNITED STATES OF AMERICA

Foreword

The ACS Symposium Series was first published in 1974 to provide a mechanism for publishing symposia quickly in book form. The purpose of the series is to publish timely, comprehensive books developed from ACS sponsored symposia based on current scientific research. Occasionally, books are developed from symposia sponsored by other organizations when the topic is of keen interest to the chemistry audience.

Before agreeing to publish a book, the proposed table of contents is reviewed for appropriate and comprehensive coverage and for interest to the audience. Some papers may be excluded to better focus the book; others may be added to provide comprehensiveness. When appropriate, overview or introductory chapters are added. Drafts of chapters are peer-reviewed prior to final acceptance or rejection, and manuscripts are prepared in camera-ready format.

As a rule, only original research papers and original review papers are included in the volumes. Verbatim reproductions of previously published papers are not accepted.

ACS Books Department

Contents

Preface..........xi

1. **Introduction**..........1
Ljiljana S. Korugic-Karasz, William J. MacKnight, and Ezio Martuscelli

New Polymers in Optoelectronics

2. **Developments and Opportunities in Polymer-Based New Frontiers of Nanophotonics and Biophotonics**..........6
Paras N. Prasad

3. **Conferring Smart Behavior to Polyolefins through Blending with Organic Dyes and Metal Derivatives**..........18
Andrea Pucci, Vincenzo Liuzzo, Giacomo Ruggeri, and Francesco Ciardelli

4. **Functional Materials Based on Self-Assembled Comb-Shaped Supramolecules**..........34
Gerrit ten Brinke and Olli Ikkala

5. **Helical Conformations of Conjugted Polymers and Their Control of Cholesteric Order**..........47
Akio Teramoto, Kazuto Yoshiba, Shigeru Matsushima, and Naotake Nakamura

6. **Molecular Orbital Theory Predictions for Photophysical Properties of Polymers: Toward Computer-Aided Design of New Luminescent Materials**..........63
Zoltán A. Fekete

7. **Impact of Cyano-Functional Groups on Luminescence of Poly(*m*-phenylenevinylene) Derivatives: Its Dependence on Conjugation Length**.....76
Liang Liao, Yi Pang, Liming Ding, and Frank E. Karasz

New Polymers in Biomedical Application

8. **Interaction of Poly(ethylene oxide) and Poly(perfluorohexylethyl methacrylate) -containing Block Copolymers with Biological Systems**.....92
H. Hussain, K. Busse, A. Kerth, A. Blume, N. S. Melik-Nubarov, and J. Kressler

9. **Elucidating the Kinetics of β-Amyloid Fibril Formation**.....106
Nadia J. Edwin, Grigor B. Bantchev, Paul S. Russo, Robert P. Hammer, and Robin L. McCarley

10. **Synthesis of an ABA Triblock Copolymer of Poly(DL-lactide) and Poly(ethylene glycol) and Blends with Poly(ε-caprolactone) as a Promising Material for Biomedical Application**.....119
Aleksandra Porjazoska, Oksan Karal Yilmaz, Nilhan Kayaman-Apohan, Kemal Baysal, Maja Cvetkovska, and Bahattin M. Baysal

New Polymeric Materials in Nanotechnology

11. **Nanocomposites Based on Cyclic Ester Oligomers**.....138
Amiya R. Tripathy, Stephen N. Kukureka, and William J. MacKnight

12. **Selective Solvent-Induced Reversible Surface Reconstruction of Diblock Copolymer Thin Films**.....158
Ting Xu, Matthew J. Misner, Seunghyun Kim, James D. Sievert, Oleg Gang, Ben Ocko, and Thomas P. Russell

13. **Thermoplastic Molecular Sieves: New Polymeric Materials for Molecular Packaging**.....171
Giuseppe Milano, Christophe Daniel, Vincenzo Venditto, Paola Rizzo, Gaetano Guerra, Pellegrino Musto, and Giuseppe Mensitieri

14. **Modeling Transport Properties in High Free Volume Glassy Polymers**......187
Xiao-Yan Wang, Kenneth M. Lee, Ying Lu, Matthew T. Stone, I. C. Sanchez, and B. D. Freeman

15. **New Hyperbranched Urethane Acrylates**......201
Branislav Bozic, Srba Tasic, Radomir Matovic, Radomir N. Saicic, and Branko Dunjic

16. **Charge Percolation Mechanism of Ziegler–Natta Polymerization: Part II: Importance of Support Nanoparticles**......215
Branka Pilic, Dragoslav Stoiljkovic, Ivana Bakocevic, Slobodan Jovanovic, Davor Panic, and Ljiljana Korugic-Karasz

17. **Preparation of Molecularly Imprinted Cross-Linked Copolymers by Thermal Degradation of Poly(methacryl-*N,N′*-diisopropylurea-co-ethylene glycol dimethacrylate)**......229
Radivoje Vuković, Ana Erceg Kuzmić, Grozdana Bogdanić, and Dragutin Fleš

18. **Confined Macromolecules in Polymer Materials and Processes**......238
Peter Cifra and Tomas Bleha

19. **Chain Conformational Statistics and Mechanical Properties of Elastomer Blends**......252
Milenko Plavsic, Ivana Pajic-Lijakovic, and Paula Putanov

New Techniques for Polymer Characterization

20. **Polymer Dynamics and Broadband Dielectric Spectroscopy**......268
Graham Williams

21. **Viscometry and Light Scattering of Polymer Blend Solutions**......282
M. J. K. Chee, C. Kummerlöwe, and H. W. Kammer

22. **Spectroscopic Studies of the Diffusion of Water and Ammonia in Polyimide and Polyimide–Silica Hybrids**......296
P. Musto, G. Ragosta, G. Scarinzi, and G. Mensitieri

23. **Connectivity of Domains in Microphase Separated Polymer Materials: Morphological Characterization and Influence on Properties**......309
Samuel P. Gido

24. Dual Detection High-Performance Size-Exclusion Chromatography System as an Aid in Copolymer Characterization.....325
Tatjana Tomić, Marko Rogošić, Zvonimir Matusinović, and Nikola Šegudović

25. Polymer Permeability Measurements via TGA.....339
Brandi D. Holcomb and Harvey E. Bair

26. Thermal Characterization and Morphological Studies of Binary and Ternary Polymeric Blends of Polycarbonate, Brominated Polystyrene, and Poly(2,6-dimethyl-1,4-phenylene oxide).....351
A. Z. Aroguz, Z. Misirli, and B. M. Baysal

Polymers for Environmentally Sustainable Applications

27. Polymers for the Conservation of Cultural Heritage.....370
M. Cocca, L. D'Arienzo, G. Gentile, E. Martuscelli, and L. D'Orazio

28. Matrix Free Ultra-High Molecular Weight Polyethylene Fiber-Reinforced Composites: Process, Structure, Properties, and Applications.....391
Tao Xu and Richard J. Farris

Indexes

Author Index.....409

Subject Index.....411

Preface

I am greatly honored to be given the opportunity to participate and write a few words not only on my behalf but also on behalf of the Serbian Academy of Sciences and Arts on such a solemn yet joyful occasion which marks the jubilee of our esteemed colleague, Professor Frank Karasz. The title as well as the many chapters in this book could hardly be more appropriate, considering the area of polymers that Professor Karasz has covered and to which he has made and is still making so many important contributions. We have been following this area of science for the past 50 years—from the fundamental side, in which the complex phenomena appearing in polymer science were barely understood, to the practical side, when very few plastics were used until the present where we understand more every day. Our use of plastics now makes it difficult to understand how we could live without them. As everyone is now saying "When one wishes to touch wood one realizes that the world around us is composed of plastics." It is in this world that Professor Karasz made major advances. We held this meeting in this beautiful place, Capri in recognition of his achievements and in the recognition of his love for the beauties of the world.

Personally, I and this group of people have an additional reason to be pleased to greet Professor Karasz on this occasion. When my country and my nation were going through difficult times, Professor Karasz never stopped supporting our scientists in their scientific endeavors. Every one of us who knows about this support feels sincere gratitude to Professor Karasz. To show our gratitude, we elected Professor Karasz to foreign membership of the Serbian Academy of Sciences and Arts, the most that our intellectual community can do for an esteemed foreigner, who is close to our minds and hearts.

All that I can do at this moment is to wish Professor Karasz a long and productive life in the continuation of his activities for the benefit of science.

To remind him of our deep esteem, I am honored to present him on this occasion of his birthday a modest gift—a book on Serbian culture.

Paula Putanov
Serbian Academy of Sciences and Arts
Serbia and Montenegro
Europe

Chapter 1

Introduction

Ljiljana S. Korugic-Karasz[1], William J. MacKnight[2], and Ezio Martuscelli[2]

[1]Department of Polymer Science and Engineering, University of Massachusetts, Amherst, MA 01003
[2]CAMPEC serl, Via G. Porzio, CDN, Isola F4, 80143 Napoli, Italy

The chapters in this volume are representative of the contributions presented at a symposium, *New Polymeric Materials* held in Capri, Italy, October 22–25, 2003 on the occasion of the 70th birthday of Frank E. Karasz.

Frank Erwin Karasz was born in Vienna. His undergraduate education was at the Royal College of Science, part of Imperial College, University of London, where he graduated with honors in Chemistry. In his 3rd year at Imperial he was given a research project having to do with the molecular weight determination of PMMA, which, apparently, sparked his later interest in polymers. For some years however his research was in the area of the theory of small molecules. His Ph.D. thesis at the University of Washington was con-cerned with the statistical mechanics of liquid argon/helium mixtures. After a year's post-doctoral work with the well-known biopolymer theoretician Terrell Hill at the University of Oregon, he returned to small molecule behavior in collaborative research at the National Physical Laboratory, near London, with the late John Pople. Pople and Karasz published a series of papers dealing with transitions in what were then called plastic crystals, nowadays rotationally disordered crystals. Remarkably the basic Karasz-Pople theory is still somewhat in vogue after more than 40 years with commentaries and modifications appearing every now and then.

After two and a half years at the NPL, Karasz returned to the United States to take a position at the General Electric Research Laboratory in Schenectady, New York. One of his interests of that era was in the thermodynamics of the helix-coil or order-disorder transition in synthetic polypeptides. He and his collaborator, James O'Reilly, were in fact the first to directly measure the transition enthalpy in such a system, namely solubilized poly-γ-benzyl-L-glutamate. This was the time of the introduction of new higher performance polymers by G.E. and others, and, in particular, the development of Noryl, an alloy of polystyrene and poly dimethyl phenylene oxide. This mixture is not only compatible but also miscible, with a single composition dependent glass transition. Karasz thus became interested in the general problem of polymer blends. Karasz left G.E. and went to the University of Massachusetts in 1967 where he and eventually all of the undersigned collaborated in an extensive set of studies of polymer blends, which in some cases has continued to this day. A major advance in this area was made in 1982 when, with Gerrit ten Brinke, a comprehensive quantitative theory of copolymer blends was developed which had both interpretive and predictive value. This mean field theory was applied, with a long list of collaborators, to a wide variety of systems, and has been used by many others in the past twenty years.

In 1972, Karasz, with the late Roger Porter, was successful in establishing an NSF Materials Research Laboratory at the University of Massachusetts. In his capacity as co-Director of this laboratory, he became acquainted around 1980 with the work of Alan Heeger and Alan MacDiarmid, both at that time at the University of Pennsylvania, who invited Karasz and erstwhile colleague James Chien to collaborate in the rapidly developing field of π-conjugated conducting polymers. This led to a highly successful four P.I. team which produced a large number of pioneering results. Karasz and his co-workers continued to carry out research in this area and about a decade later he expanded his field of interest to include optical properties of these polymers. Electroluminescence was first observed in poly phenylene vinylene around 1991 and in 1993 Karasz was able to produce the first electroluminescent polymer to emit blue light and he has continued, again with a long list of co-workers, to be a major contributor to this field to this day.

We note that Karasz's first scientific paper was on the adsorption of water on AgI and TiO_2, published in 1956; thus he is now in his sixth decade of a prolific publishing career, with a total of about 550 papers to date. He has received substantial recognition for his scientific contributions starting with the Mettler Medal in 1975. In 1984 he and one of the undersigned (W.J. MacKnight) were co-recipients of the Ford Prize of the American Physical Society, and in 1985 he received the Research Award from the Society of Plastic Engineers. He has received numerous other awards including the University of Ferrara 600th Year Commemorative Medal, and in 2000 Polytechnic University presented him the S.E.A.M. Award. In 2002 he received the Herman Mark Medal from the Austrian Institute for Chemical Research.

In 1991 Karasz was elected to the U.S. National Academy of Engineering and since then he has been elected to membership in the Croatian, Indian and Serbian Academies of Science and Art. He is also a Chevalier of the Knights of St. John and a Fellow of the American Physical Society, the AAAS, and the

PMSE Division of the ACS, and is currently a member of the National Materials Advisory Board.

Karasz has long had a strong relationship with the European polymer community and has until recently served, for example, as President of the Scientific Council of the CNR Polymer Laboratory in Naples. He has also been a lecturer sponsored by the Accademia Lincei of Rome. He continues to be very actively involved at the frontiers of polymer research.

New Polymers in Optoelectronics

Chapter 2

Developments and Opportunities in Polymer-Based New Frontiers of Nanophotonics and Biophotonics

Paras N. Prasad

Department of Chemistry and the Institute for Lasers, Photonics, and Biophotonics, University at Buffalo, The State University of New York, Buffalo, NY 14260-3000

Abstract

This article describes some recent developments at our Institute and opportunities created by a fusion of photonics with nanoscale science and technology as well as with biomedical research. This fusion has opened up the new frontiers of Nanophotonics and Biophotonics. Nanostructured random media such as polymeric nanocomposites are showing great promises for many photonics/optoelectronics applications, the ones specifically discussed here are for photovoltaics and photorefractivity. Ordered nanostructures such as photonic crystals, discussed here, exhibit new phenomena which can be exploited for novel applications. Polymeric nanoparticles show great potential for optical diagnostics and targeted optically tracked and light-activated therapy. Some examples of these biophotonics applications for nanomedicine are also presented.

Introduction

Photonics, the science and technology utilizing photons, has already shown applications of polymeric media for new technological applications such as polymeric light emitting diodes and polymer optical fibers. The fusion of photonics with nanotechnology has lead to the field of nanophotonics which deals with light-matter interaction on a scale significantly smaller than the wavelength of light used.[1] Polymeric nanostructures offer new opportunities for nanophotonics. One can create nanocomposites consisting of nanodomains where each domain performs a separate function or can induce a function in another domain, thus producing multifunctionality. If the domains are significantly smaller than the wavelength of light, they do not significantly scatter. As a result the medium is optically transparent and thus very suitable for photonics applications. Some selected examples of these nanocomposites are presented here. An emerging class of nanophotonic materials is photonic crystals which consist of alternating domains of different refractive index, with the periodicity (alternating length scale) match with the wavelength of light.[1,2] These photonic crystals exhibit a number of new optical phenomena that confined technological applications. Polymeric media, again, offer opportunities for fabrication of a broad range of photonic crystals with varying optical properties. Examples of polymer-based photonic crystals are also presented here.

Biophotonics is another new frontier, created by the fusion of photonics with biomedical research.[3] Here light-matter interaction can be used for bioimaging, biosensing, and other types of biodetection serving as powerful tools for diagnostics. Light-matter interaction can also be used for optically tracked and light-activated therapy. Polymer nanoparticles have given a major impetus to the emerging field of nanomedicine, as multiple probes for optical diagnostics can be incorporated, together with therapeutic agents, for optically tracked light-activated therapy. An example presented here is that for photodynamic therapy which is based on light-induced activation of a photosensitizer.

Random Nanocomposites

Nanocomposite processing offers the prospect of producing a mixed phase with unique properties which may be thermodynamically not stable. An example provided here is that of formation of an inorganic oxide glass:organic polymer optical nanocomposite by sol-gel processing, a work which we did in collaboration with Professor Frank Karasz.[4] Generally, when an inorganic glass and organic polymer are mixed together, one would expect phase separation of the glass and the polymer. Nanoscale processing of a composite of such incompatible materials provides an opportunity to limit the phase separation in nanometers. Further phase separation beyond this size scale becomes kinetically too slow to be of any practical consequence. Since the resulting phase separation is on the scale of nanometers, significantly smaller than the wavelength of light, the material is optically transparent.

We utilized sol-gel processing to produce a glass-polymer composite containing up to 50% of an inorganic oxide glass (e.g. silica or vanadium oxide) and a conjugated polymer, poly-p-phenylenevinylene, often abbreviated as

PPV.[4-6] Professor Karasz and his group have done pioneering work on this polymer and its applications. PPV exhibits many interesting optical and nonlinear optical properties. It is a rigid polymer that, without any flexible side chain, is insoluble, once formed. In our approach we used predoping of the silica (or V_2O_5) precursor sol by mixing it with the PPV monomer precursor in the solution phase, which was subsequently hydrolyzed to form the silica network. Simultaneously the PPV monomer was polymerized. By controlling the processing condition, the composite was made to form a rigid structure to be kinetically too slow to phase separate beyond several nanometers. The TEM study of the composite confirmed the nanoscale phase separation.[7] The resulting nanocomposite PPV:glass even at the high level (~50%) of mixing is optically transparent and exhibits very low optical loss compared to a film of pure PPV, a feature clearly derived from the presence of an inorganic glass.[4] The composite exhibits many features derived from inorganic glass (eg. silica) such as high surface quality and easily polishable ends for end-fire coupling of light into the waveguide. It also exhibits features derived from the polymer PPV, such as mechanical strength and high optical nonlinearity.[8] One technological application demonstrated for this nanocomposite is for dense wavelength division multiplexing (DWDM) function, an important part of optical communication technology. For this function, holographic gratings were recorded in the PPV/sol-gel glass nanocomposite. Figure 1 shows the spectral characteristics of these gratings.[9] Wavelength dispersion with very high spectral resolution and extremely narrow band width of <50 pm were achieved.

Another example provided here is for an optoelectronic application using photorefractivity. The photorefractive nanocomposites described here involve quantum dots, liquid crystal nanodroplets, and a hole transporting polymer to produce the function of photorefractivity. The photorefractive effect requires a combined action of photocharge generation, charge separation and electro-optic effect, the latter property producing a change of refractive index by the application of an electric field.[10,11] Thus light absorption by photosensitizers produce charge-carriers which separate in a charge transporting medium and create a space charge field (internally generated electric field). This electric field changes the refractive index of the medium due to the presence of the electro-optic property. Both inorganic and organic photorefractive materials exist and have been widely investigated.[10,11] However, a nanocomposite consisting of both inorganic and organic domains offer numerous advantages. In our approach we utilize inorganic semiconductor quantum dots for photosensitization to generate charge carriers. The quantum dots are nanoparticles of semiconductors which exhibit size dependent optical absorption.[1] Hence by selecting their size and composition, one can judiciously select the wavelength of absorption maximum to produce photosensitization (and hence photorefractivity) at a specific wavelength. An advantage is that one can prepare photorefractive composites for communication wavelengths in the IR (1.3 μm and 1.55 μm) which is difficult to achieve with purely organic photosensitizers.[12] These quantum dots are dispersed in a hole transporting medium (in our case PMMA doped with

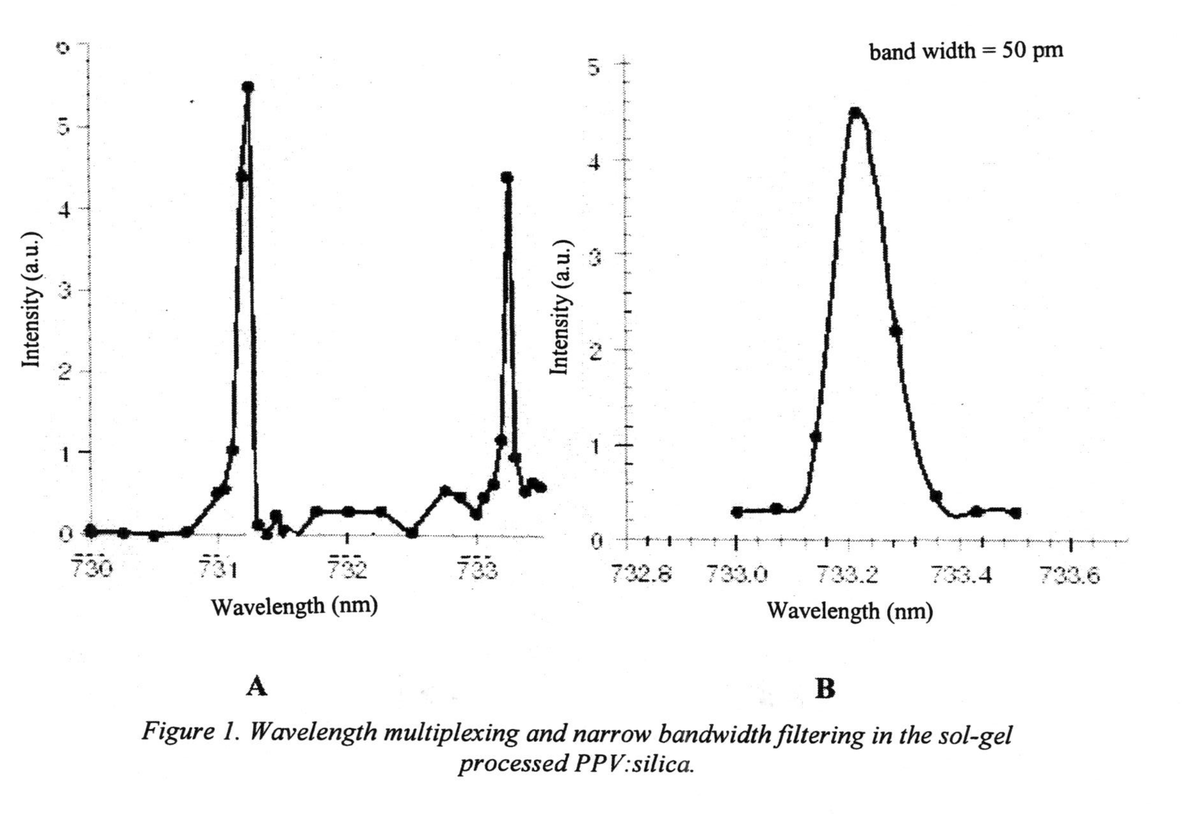

Figure 1. Wavelength multiplexing and narrow bandwidth filtering in the sol-gel processed PPV:silica.

ethylcarbazole). The liquid crystal nanodroplets in the polymer host perform the function of an electro-optic medium, by their ability to re-orient in the electric field.[13]

Figure 2 shows a schematic representation of our photorefractive nanocomposite polymer dispersed liquid crystal (PDLC) system, containing nanodroplets of liquid crystal and ~10 nm size CdS quantum dots dispersed in a PMMA polymer.[13] CdS, produced in-situ, generates charge carriers by absorbing the 532 nm laser light. The electrons remain trapped in the nanoparticle. The holes move and get trapped in the dark region created by an interference of two beams (holographic exposure). There is a resulting space-charge field which reorients the liquid crystal nanodroplet to change the effective refractive index. The change of effective refractive index, produced by the liquid crystal, can be very large, as the extraordinary index of 1.71 and ordinary index of 1.5 are widely different. One can form a grating using crossing of two optical beams, with a large refractive index change, Δn. The result is that one can achieve very high diffraction efficiency, even in a thin film. A maximum net diffraction efficiency of ~70% was achieved in our experiment. If one takes the internal optical losses into account, the diffraction efficiency is ~100%. However, the response time to reorient the nanodroplets in the plastic medium is relatively slow. The formation of the grating is mainly limited by the reorientation of these nanodroplets of liquid crystal, which is in seconds. We are conducting experiments to control the nanodroplet size and the interfacial interaction for producing a faster response.

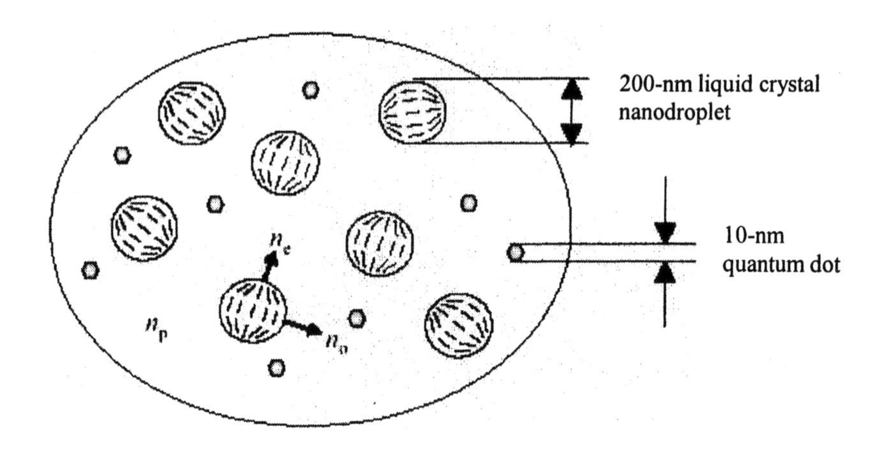

Figure 2. Schematic representation of polymer-dispersed liquid crystals containing quantum dots.

Photonic Crystals

Photonic crystals are ordered nanostructures in which two media with different dielectric constants or refractive indices are arranged in a periodic form.[1,2] Thus they form two interpenetrating domains, one domain of higher refractive index and another of lower refractive index. It is a periodic domain with periodicity on the order of wavelength of light. In analogy with an electronic crystal of a semiconductor with a periodic lattice for a certain range of energy of photons and certain wave vectors (that is, direction of propagation), light is not allowed to propagate in a photonic crystal. If we generate light inside the photonic crystal, it cannot propagate in this direction. If we send light from outside, it is reflected. Figure 3 shows typical transmission and reflection spectra of a photonic crystal produced by close packing of polystyrene spheres.
Here the alternating domains are that of polystyrene and air. The region of minimum transmission (or maximum reflection) corresponds to the not-allowed photon states and referred to as stop gap. The stop gap in a photonic crystal is determined by the Bragg diffraction condition of a periodic lattice. Hence it is dependent on the periodicity (the size of the microsphere in the present case). Polymeric materials offer opportunities to manipulate this stop gap by the choice of a wide variety of available polymers *(e.g.* polystyrene, PMMA, *etc.).* Another approach recently demonstrated by us is that of a photonic crystal alloy.[14] These photonic crystal alloys were formed by mixing of polystyrene and polymethylmethacrylate spheres of same size in different volume ratios. These photonic crystal alloys are structurally ordered, but contain refractive index disorder due to random distribution of the polystyrene and PMMA spheres. They thus provide a random variation of scattering potential. The stop-gap shows a monotonic shift in wavelength as a function of the composition, which can be fitted by assuming an effective dielectric constant for the colloidal spheres.

Photonic crystals exhibit a number of unique optical and nonlinear optical properties which can find important applications in optical communications, low-threshold lasing, frequency conversion, and sensing. These are:[1]

- *Presence of Stop gap.* The existence of these stop gaps can make them suitable for high-quality narrow-band filters. The effective periodicity can also be changed in response to adsorption (or inclusion) of an analyte, making photonic crystal media suitable for chemical and biological sensing. The position of stop-gap can be changed in a nonlinear crystal by using Kerr gate optical nonlinearity, whereby increasing intensity, the refractive index of a nonlinear periodic domain can be changed, resulting in a change of the stop gap position. This manifestation can be utilized to produce optical switching for light of frequency at the band edge.

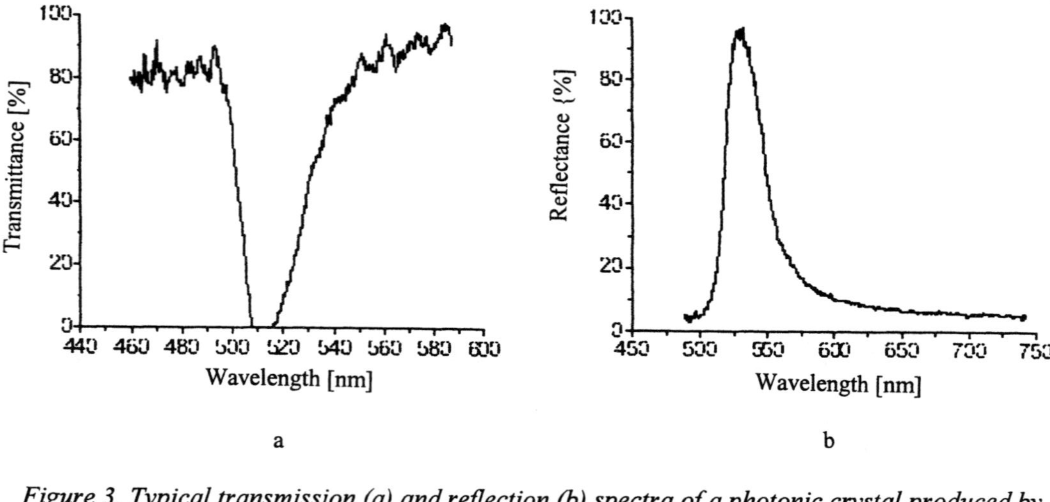

Figure 3. Typical transmission (a) and reflection (b) spectra of a photonic crystal produced by close packing of polystyrene spheres. The diameters of the polystyrene spheres are 220 nm for transmittance study and 230 nm for reflection measurement.

- *Local Field Enhancement.* Spatial distribution of an electromagnetic field can be manipulated in a photonic crystal to produce local field enhancement in one dielectric or the other. This field enhancement in a nonlinear photonic crystal can be utilized to enhance nonlinear optical effects that are strongly dependent on the local field.

- *Anomalous Group Velocity Dispersion.* A photonic band structure produces the dispersion of the allowed propagation frequencies as a function of the propagation vector **k**, and determines the group velocity with which an optical wave packet (such as a short pulse of light) propagates inside a medium. Many new phenomena related to this dispersion effect are predicted and have been verified. One is the *superprism* phenomenon, which is extraordinary angle-sensitive light propagation and dispersion due to refraction.
- *Anomalous Refractive Index Dispersion.* In nonabsorbing spectral regions, dielectric medium exhibits a steady decrease of refractive index with increasing wavelength (decreasing frequency). This is called normal refractive index dispersion. Photonic crystals exhibit anomalous dispersion of the effective refractive index near their high-frequency band edge. This anomalous dispersion can be utilized to produce phase matching for efficient generation of second or third harmonic of light from the incident fundamental wavelength. Phase-matching requirement dictates that the phase velocities (c/n) for the fundamental and the harmonic waves are the same, so that power is continuously transferred from the fundamental wave to the harmonic wave, as they travel together in phase. Our work on close-packed polystyrene spheres has established enhancement of third harmonic generation due to phase-matching derived from anomalous dispersion.[15]
- *Microcavity Effect in Photonic Crystals.* A photonic crystal also provides the prospect of designing optical micro- and nanocavities embedded into it by creating defects *(e.g.,* dislocations, holes) into it. The size and the shape of these defect sites can readily be tailored in a polymeric based photonic crystal to produce microcavities of different sizes and dimensions. These defect sites create defect associated photon states (defect modes) in the bandgap region and are analogous to the impurity (dopant) states between the conduction and the valence bands of a semiconductor. Microcavitiy resonances can produce low threshold lasing.

Polymeric Nanoparticles for Biophotonics and Nanomedicine

Polymer nanoparticles are useful for a number of biophotonics applications such as in bioimaging, optical diagnostics and nanomedicine.[1,3] Nanomedicine is a new field which utilizes nanoparticles in new methods of minimally invasive diagnostics for early detection of diseases, as well as for facilitating targeted drug delivery, enhancing effectiveness of therapy, and providing ability for real-time monitoring of drug action.[1] Nanoparticles with optical probes, light activated therapeutic agents, and specific carrier groups to direct them to the diseased cells or tissues, provide targeted drug delivery, with an opportunity for real-time monitoring of drug efficacy.[1,3]

Polymeric nanoparticles provide three different structural platforms for diagnostics and therapy: (a) An interior volume in which various probes and therapeutic agents can be encapsulated. (b) A surface which can be functionalized to attach targeting groups to carry the nanoparticles to cells or biological sites expressing appropriate receptors. In addition, the surface can bind to specific biological molecules for intracellular delivery (we have used this approach to attach DNA for gene therapy). The surface can also be functionalized to introduce a hydrophilic (polar), hydrophobic (non-polar) or amphiphilic character to enable dispersibility in a variety of fluid media. (c) Pores in the nanoparticles, which can be tailored to be of specific sizes to allow selective intake or release of biologically active molecules or to activate therapeutic agents. Thus the area of mesoporous polymers with controlled porosity is of intense current activity and interest.

Here we describe our approach of "nanoclinics" for optically tracked delivery. These nanoclinics are complex surface functionalized polymeric (silica or organically modified silica) nanoshells containing various probes for diagnostics, and drugs for targeted delivery.[16] They are produced using multistep nanochemistry and surface functionalized with known biotargeting agents. The size of these nanoclinics is small enough for them to enter the cell, in order for them to function from within the cell. Through the development of nanoclinics (functionalized nanometer-sized particles that can serve as carriers), new therapeutic approaches to disease can be accomplished from within the cell.

We present here the example of a nanoclinic used for photodynamic therapy. Photodynamic therapy utilizes light sensitive species or photosensitizers (PS) which preferentially localize in tumor tissues upon systematic administration.[3] When such photosensitizers are irradiated with an appropriate wavelength of visible or near infra-red (NIR) light, the excited molecules can transfer their energy to molecular oxygen (in its triplet ground state) present in the surrounding media. This energy transfer results in the formation of reactive oxygen species (ROS), like singlet oxygen ($^1O_2$) or free radicals.[3] ROS are responsible for oxidizing various cellular compartments, including plasma, mitochondria, lysosomal and nuclear membranes etc., resulting in irreversible damage of cells. Therefore, under appropriate conditions, photodynamic therapy offers the advantage of an effective and selective method of destroying diseased tissues, without damaging adjacent healthy ones.[17]

Most photosensitizing drugs (PS) are hydrophobic, *i.e.* poorly water soluble, and, therefore, not dispersible in biological fluids. We used a nanoparticle approach in which the photosensitizer was doped in organically modified silica-based nanoclinic nanoparticles (diameter ~30 nm), produced by controlled hydrolysis of triethoxyvinylsilane in micellar media.[18] We used a hydrophobic photosensitizer known as HPPH, which is in Phase I/II clinical trials at Roswell Park Cancer Institute, Buffalo, NY, USA.[19] The doped nanoparticles were spherical and highly monodispersed as determined by TEM and shown in Figure 4. The inherent porosity of this ceramic matrix enabled the photosensitizing drugs, entrapped within them, to interact with molecular oxygen which can diffuse through the pores. This can lead to the formation of singlet oxygen by energy transfer from the excited photosensitizer to oxygen, which can then diffuse out of the porous matrix to produce cytotoxic effect in tumor cells. The generation of singlet oxygen, after excitation of HPPH, was checked by singlet oxygen luminescence at 1270 nm. The singlet oxygen luminescence generated by HPPH, solubilized in micelles as well as entrapped in nanoparticles show similar intensity and peak positions (1270 nm), indicating similar efficiencies of singlet oxygen generation in both cases. A control study, using nanoparticles without HPPH, shows no singlet oxygen luminescence. The doped nanoparticles are actively taken up by tumor cells. Irradiation with visible light results in irreversible destruction of such impregnated cells. These observations suggest the potential of polymeric nanoparticles as carriers for photodynamic drugs.[18]

Acknowledgement

The work reported here is supported by a grant from the Directorate of Chemistry and Life Sciences of the Air Force Office of Scientific Research under the Defense University Research Initiative on Nanotechnology (DURINT) Program. Partial support from NSF Solid State Chemistry Program is also acknowledged.

References

1. P. N. Prasad, "Nanophotonics", Wiley-Interscience, New York, 2004.
2. J. D. Jodnnopoulos, R. D. Meade, and J. N. Winn, "Photonic Crystals", Princeton University Press, Princeton, N.J. 1995.
3. P. N. Prasad, "Introduction to Biophotonics", Wiley-Interscience, New York, 2003.
4. C. J. Wung, Y. Pang, P. N. Prasad and F. E. Karasz, Polymer 32, 605 (1991).
5. G. S. He, C. J. Wung, G. C. Xu, and P. N. Prasad, Appl. Opt. 30, 3810 (1991).
6. C. J. Wung, W. M. K.P. Wijekoon, and P. N. Prasad, Polymer 34, 1174 (1993).

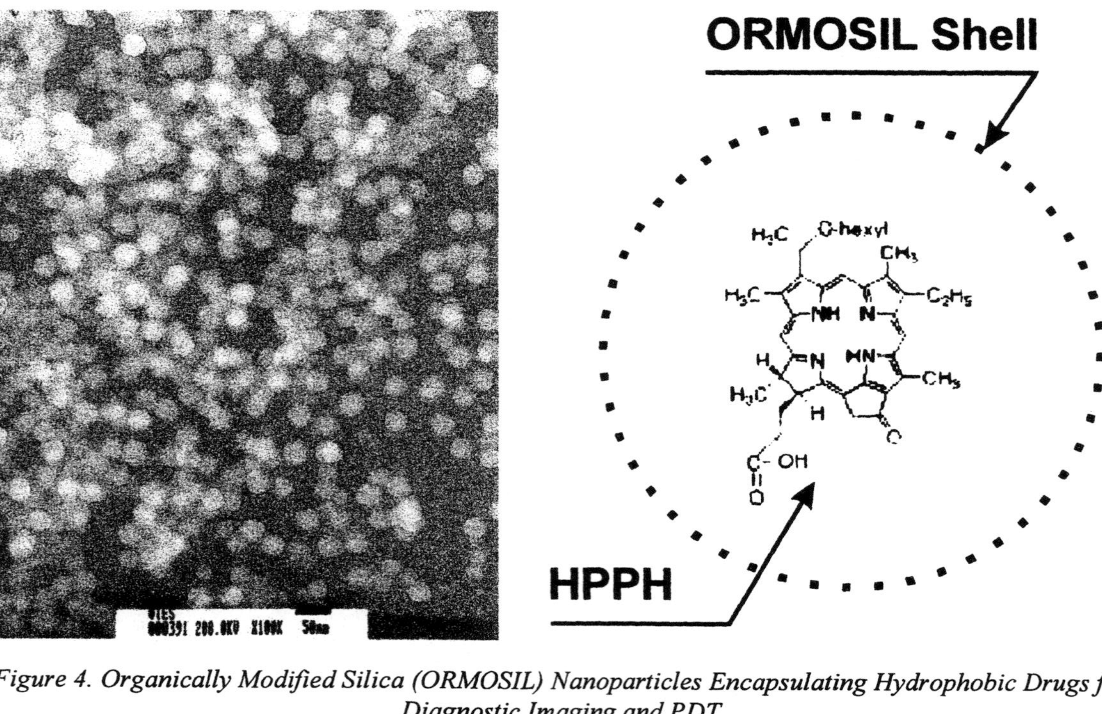

Figure 4. Organically Modified Silica (ORMOSIL) Nanoparticles Encapsulating Hydrophobic Drugs for Diagnostic Imaging and PDT

7. F. W. Embs, E. L. Thomas, C. J. Wung, and P. N. Prasad, Polymer 34, 4607 (1993).
8. Y. Pang, M. Samoc, and P. N. Prasad, J. Chem. Phys. 94, 5282 (1991).
9. D. N. Kumar and P. N. Prasad, unpublished work.
10. P. Yeh, Introduction to Photorefractive Nonlinear Optics, John Wiley & Sons, New York, 1993.
11. W. E. Moerner and S. M. Silence, Chem. Rev. 94, 127 (1994).
12. J. G. Winiarz, L. Zhang, J. Park, and P. N. Prasad, J. Phys. Chem. B 106, 967 (2002).
13. J. G. Winiarz and P. N. Prasad, Opt. Lett. 27,1330 (2002).
14. H. Tiryaki, K. Baba, P.P. Markowicz, and P. N. Prasad, Opt. Lett. 29, 1 (2004).
15. P. P. Markowicz, H. Tiryaki, H. Pudavar, P. N. Prasad, N. Lepeshkin and R. W. Boyd, Phys. Rev. Lett. 92, 083903-1 (2004).
16. L. Levy, Y. Sahoo, K.-S. Kim, E. J. Bergey, and P. N. Prasad, Chem. Mater. 14, 3715 (2002).
17. T. Hasan, A. C. E. Moor, and B. Ortel in Cancer Medicine, 5th ed., J. F. Holland, E. Frei, R. C. Bast, et al. Eds., Decker, Hamilton (2000), pp. 489.
18. I. Roy, T. Ohulchanskyy, H. E. Pudavar, E. J. Bergey, A. R. Oseroff, J. Morgan, T. J. Dougherty, and P. N. Prasad, J. Am. Chem. Soc. 125, 7860 (2003).
19. B. W. Henderson, D. A. Bellnier, W. R. Graco, A. Sharma, R. K. Pandey, L. Vaughan, K. Weishaupt, and T. J. Dougherty, Cancer Res. 57, 4000 (1997).

Chapter 3

Conferring Smart Behavior to Polyolefins through Blending with Organic Dyes and Metal Derivatives

Andrea Pucci[1,2], Vincenzo Liuzzo[1], Giacomo Ruggeri[1,2], and Francesco Ciardelli[1,3,*]

[1]Department of Chemistry and Industrial Chemistry, University of Pisa, Via Risorgimento 35, 56126 Pisa, Italy
[2]INSTM, Pisa Research Unit, Via Risorgimento 35, 56126 Pisa, Italy
[3]INFM, Pisa Research Unit, Via Risorgimento 35, 56126 Pisa, Italy

Bis(salicylaldiminate) M(II) complexes of Nickel, Copper and Cobalt have been prepared and mixed with ultra high molecular weight polyethylene (UHMWPE) in order to prepare new composite materials with interesting dispersion and optical properties by profiting of the presence of long and linear alkyl chains connected to the ligand structure. The phase dispersion behaviour of the binary polyethylene films has been studied by scanning electron microscopy (SEM) and x-ray microanalysis, whereas the optical properties of the oriented samples have been evaluated by UV-Vis spectroscopy in linearly polarized light and discussed in terms of the anisotropy induced by the mechanical orientation of the polymer matrix. The dichroism of the d π^* electronic transition of the oriented complexes can be modulated by changing the nature of the metal centre according to the different strength of the transition dipole moment.

Introduction

Polymeric films for linear polarizer applications were generally prepared by incorporating into oriented polymer matrices highly conjugated low molecular weight organic molecules (1-9).

In this contest, terthiophene based chromophores properly functionalized with push-pull organic derivatives and long alkyl lateral chains were extensively studied for the formulation of new highly oriented ultra high molecular weight polyethylene (UHMWPE) binary films (10, 11).

The modulation of the chromophore structure by introducing linear or branched alkyl chains thus providing molecules with lower degree of crystallinity and the use of polyethylene matrices with different molecular weight and density, with the possibility to take advantages from different preparation routes, allowed to reduce or even eliminate the distinct phase separation between the components (12-14). Actually, the formation of chromophore aggregates as a consequence of their separated crystallization from the polymer matrix strongly limits their alignment along the macromolecular chains providing oriented films with poor dichroism (13, 15, 16).

Analogously, the efficient strategy based on coupling the optical and dispersion properties of the organic moiety and the metal properties of the inorganic core with extremely high dispersion in polymer matrices suggested the preparation of different metal complexes from ligands bearing alkyl chains with different length, in order to provide the metal with a hydrophobic shell thus allowing a favourite dispersion into polyethylene matrix (17). Nickel and Copper molecular Schiff-Base complexes (Figure 1a) were prepared and then mixed in the polyethylene matrix providing materials characterized by a great level of homogeneity even, in some cases, at high complex concentration (~12 wt.%) (17).

In this work new Bis(salicylaldiminate) metal(II) complexes with similar structure but increased conjugation of the ligand moiety have been prepared and dispersed into polyethylene in small concentration (3 wt.%).

The results are discussed in terms of the effect of the molecular structure of the metal complexes on their dispersion into UHMWPE and on their anisotropic absorption properties induced by polymer orientation.

Experimental

Apparatus and methods

FT-IR spectra were recorded by a Perkin-Elmer Spectrum One spectrophotometer on dispersions in KBr. ^{1}H-NMR spectra were recorded with the help of a Varian-Gemini 200 on 5-10% $CDCl_3$ (Aldrich, 100.0 atom % D) solutions. NMR spectra were registered at 20°C and the chemical shifts were assigned in ppm using the solvent signal as reference.

The melting points were accomplished by a Reichert Polyvar optical microscope with crossed polarizers, equipped with a programmable Mettler FP 52 hot stage. Elementary analyses were made by microanalysis laboratory at the Faculty of Pharmacy, University of Pisa. The Scanning Electron Microscopy (SEM) analysis was performed with a Jeol 5600-LV microscope, equipped with Oxford X-rays EDS microprobe, instrument at the Chemical Engineering Department of Pisa University. Optical absorption studies were carried out in dilute ($5 \cdot 10^{-5}$ M) solutions with a Jasco 7850 UV-Vis spectrophotometer or on polymer films in polarized light with the same instrument, fitted with Sterling Optics UV linear polarizer. The films roughness was diminished, using ultra-pure silicon oil (Poly(methylphenylsiloxane), 710® fluid, Aldrich) to reduce surface scattering between the polymeric films and the quartz slides used to keep them planar. In the analysis of the absorption and emission data, the scattering contribution was corrected by the use of appropriate baselines. The fitting procedure was performed by using Origin 5.0, software by Microcal Origin®.

Materials

2,4-dihydroxybenzaldehyde, 1-bromooctadecane, *o*-Phenylene-diamine, Nickel(II) acetate tetrahydrate ($Ni(OAc)_2 \cdot 4H_2O$), Copper (II) acetate monohydrate ($Cu(OAc)_2 \cdot H_2O$) and Cobalt(II) acetate tetrahydrate ($Co(OAc)_2 \cdot 4H_2O$) (Aldrich) were used as received. Ultra High Molecular Weight Polyethylene (UHMWPE), $\overline{M}_w = 3.6 \cdot 10^6$, density: 0.928 g/cm^3 (Stamylan UH210, DSM, The Netherlands) was used as polymer matrix.

2-Hydroxy-4-(n-octadecyloxy)benzaldehyde

3.7 g (0.026 mol) of 2,4- dihydroxybenzaldehyde, 9 g (0.026 mol) of 1-bromooctadecane, 3.1 g (0.03 mol) di $KHCO_3$ and 0.2 g (0.001 mol) of KI, were refluxed for 24 h in 100 mL of acetonitrile. After cooling, the obtained precipitate was filtered off and recrystallized from heptane (7.1 g, yield 70 %); m.p: 60 °C - IR (KBr): 3444 cm^{-1} ($\nu_{O\text{-}H}$); 1671 cm^{-1} ($\nu_{C=O}$) - ^{1}H-NMR ($CDCl_3$): 0.9-1.8 (m, 35 H), 4 (t, 2 H, O-**CH$_2$**), 6.4 (s, 1H, Ar), 6.5 (d, 1H, Ar), 7.4 (d, 1H, Ar), 9.7 (s, 1H, **CHO**), 11.5 (s, 1H, **OH**) ppm.

N,N'-Bis[4-(n-octadecyloxy)-2-hydroxy-benzylidene]o-phenylenediamine (C_{18}OSalophen)

A mixture of 2-hydroxy-4-(*n*-octadecyloxy)benzaldehyde (0.6 g, 1.53 mmol) and *o*-phenylendiamine (0.09 g, 0.83 mmol) in ethanol (15 mL) was refluxed for 4 h. The yellow product obtanied was then filtered off and air-dried

(0.5 g, yield 76%); m.p.= 82-85°C. FT-IR (KBr): 1613 cm^{-1} ($\nu_{C=N}$). UV-VIS (CH_2Cl_2): λ_{max} = 330 nm; ε = 40000 $Lmol^{-1}cm^{-1}$ - ^{1}H-NMR ($CDCl_3$): 0.9-1.8 (m, 70 H), 4 (t, 4 H, O-$\mathbf{CH_2}$), 6.53 (m, 4H, Ar), 7.3 (m, 6H, Ar), 8.6 (s, 2H, C**H**N), 13.7 (sl, 2H, O**H**) ppm.

Nickel(II) complex (NiC$_{18}$OSalophen)

A mixture of C_{18}OSalophen (0.3 g, 0.35 mmol) and $Ni(OAc)_2 \cdot 4H_2O$ (0.09 g, 0.36 mmol) in ethanol (20 mL) was refluxed for 1 h. The red solid precipitated was filtered off and crystallized from heptane (0.3 g, yield 94%); m.p.= 120-122°C - FT-IR (KBr): 1605 cm^{-1} ($\nu_{C=N}$). ^{1}H-NMR ($CDCl_3$): 0.89-1.79 ppm (m, 70 H), 3.96 ppm (t, 4 H, O-$\mathbf{CH_2}$), 6.3 (d, 2H, Ar), 6.6 (s, 2H, Ar), 7.2 (m, 4H, Ar), 7.6 (d, 2H, Ar), 8.1 (s, 2H, C**H**N). Anal. Calc. $C_{56}H_{86}N_2O_4Ni$: C, 73.91%; H, 9.53%; N, 3.08%. Found: C, 69.43%; H, 9.26%; N, 2.83%.

Copper(II) complex (CuC$_{18}$OSalophen)

A mixture of C_{18}OSalophen (0.3 g, 0.35 mmol) and of $Cu(OAc)_2 \cdot 2H_2O$ (0.09 g, 0.36 mmol) in ethanol (20 mL) was refluxed for 1 h. The olive solid obtained was filtered off and crystallized from heptane (0.29 g, yield 90%). m.p.= 123-125°C. FT-IR (KBr): 1613 cm^{-1} ($\nu_{C=N}$). Anal. Calc. $C_{56}H_{86}N_2O_4Cu$: C, 73.52%; H, 9.48%; N, 3.06%. Found: C, 71.70%; H, 9.67%; N, 2.90%.

Cobalt (II) complex (CoC$_{18}$OSalophen)

$Co(OAc)_2 \cdot 4H_2O$ (0.05 g, 0.2 mmol) was dissolved in 100 mL of degassed ethanol. A solution of C_{18}OSalophen (0.17 g, 0.19 mmol) in 5 mL of chloroform was slowly added at room temperature and under nitrogen atmosphere. The mixture was stirred for 2 h. The brown precipitate was filtered off, dried and crystallized from chloroform/ethanol (0.12 g, yield 65%). m.p.= 129-131°C. FT-IR (KBr): 1613 cm^{-1} ($\nu_{C=N}$). Anal. Calc. $C_{56}H_{86}N_2O_4Co$: C, 73.89%; H, 9.52%; N, 3.08%. Found: C, 70.25%; H, 9.08%; N, 2.66%.

Film preparation by solution casting

0.5 g of UHMWPE and the appropriate amount of the metal complex were dissolved in 75 ml of *p*-xylene at 125°C and stirred until complete dissolution occurred; the solution was then cast on a glass and slowly evaporated at room temperature.

Polymer orientation

Solid state drawings of the host-guest films were performed on thermostatically controlled hot stage at 125°C. The draw ratio was determined by measuring the displacement of ink-marks printed onto the films before stretching.

Results and Discussion

Bis(salicylaldiminate) metal(II) complexes with long alkyl chains of Nickel, Copper and Cobalt, respectively $NiC_{18}OSalophen$, $CuC_{18}OSalophen$ and $CoC_{18}OSalophen$, were prepared by the incorporation of metals(II) on substituted salicylaldehydes with *o*-phenylenediamine according to literature procedures (18-21). All the prepared non-centrosymmetric complexes are known to be planar (22) and thermally stable and are widely studied as materials for nonlinear optical applications and more generally as precursors for the fabrication of conjugated materials to be employed as opto-electronic devices (23-25).

The ligand functionalization with long and linear alkyl chains was introduced to ensure the complexes good phase dispersion into polyethylene matrix (Figure 1b) (17).

The Bis(salicylaldiminate) metal complexes were successively mixed (3% by weight) with UHMWPE by solution-casting and the phase dispersion behaviour of the binary films investigated by scanning electron microscopy (SEM).

SEM micrographs taken for 3 wt.% $NiC_{18}OSalophen$ (Figure 2, left) and 3 wt.% $CuC_{18}OSalophen$ (Figure 3, left) polyethylene films revealed a good dispersion behaviour of the metal complexes into the polymer matrix. No molecular aggregates are actually detectable at the polymer film surface. A similar morphology can be proposed also for the UHMWPE film containing the Cobalt complex on the basis of the analogy with Nickel and Copper observed by optical microscopy investigations.

The x-ray energy dispersive spectra (EDS) (26) obtained in analyzing the film surface displayed at 0.85 and 7.48 keV (Figure 2, right) and at 0.93 and 8.05 keV (Figure 3, right) the strongest L_α and K_α lines of respectively Nickel and Copper indicating the presence of the metal complexes near the polymer films surface but in a well dispersed forms.

The SEM investigations were also performed on the section of the same films (Figure 4).

In all cases, a preferential presence of the metal complex close to the surface exposed to air, in the first 10 μm thick layer approximately, was observed as evidenced by the x-ray microanalysis, similar to what previously observed for terthiophene chromophores dispersed by solution casting in polyethylene films (10, 12, 14).

The binary films were successively oriented by thermo-mechanical stretching up to 30 times the original length of each tape at a temperature close to, but below, the melting temperature of the polymer (130°C).

a

R = $-CH_2(CH_2)_4CH_3$

$-CH_2(CH_2)_{16}CH_3$

M = Ni(II)
Cu(II)

b

$C_{18}H_{37}O$ $OC_{18}H_{37}$

M = Ni(II)
Cu(II)
Co(II)

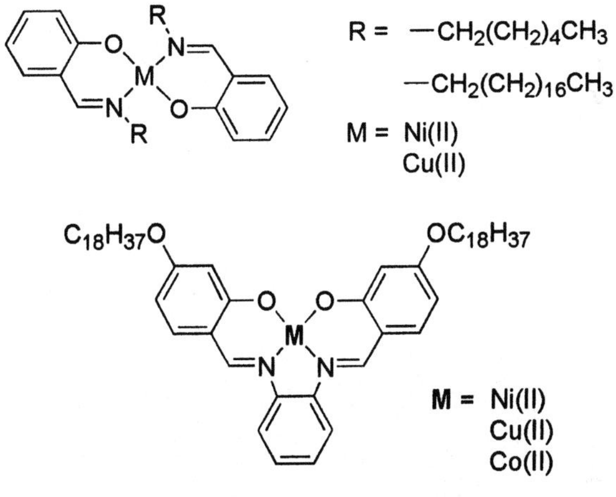

Figure 1. Previously studied Shiff-Base metal complexes (1a) and Salophen metal complexes prepared in this work (1b)

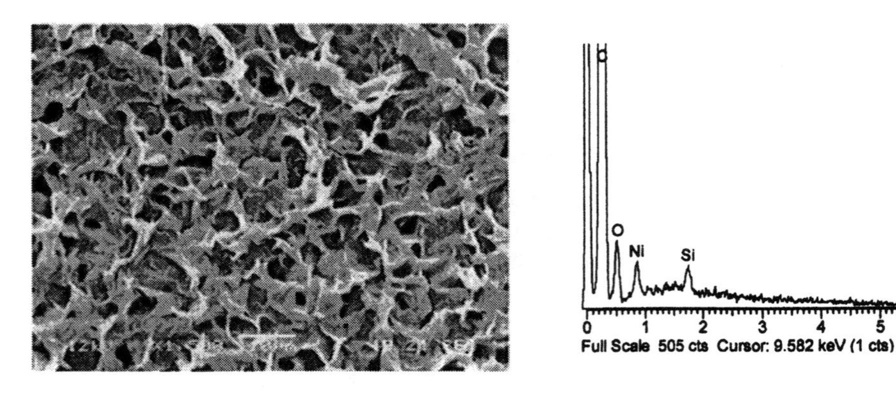

Figure 2. SEM micrograph of 3 wt.% $NiC_{18}OSalophen$/UHMWPE film (left) and energy dispersive (EDS) spectrum of the polymer surface (right).

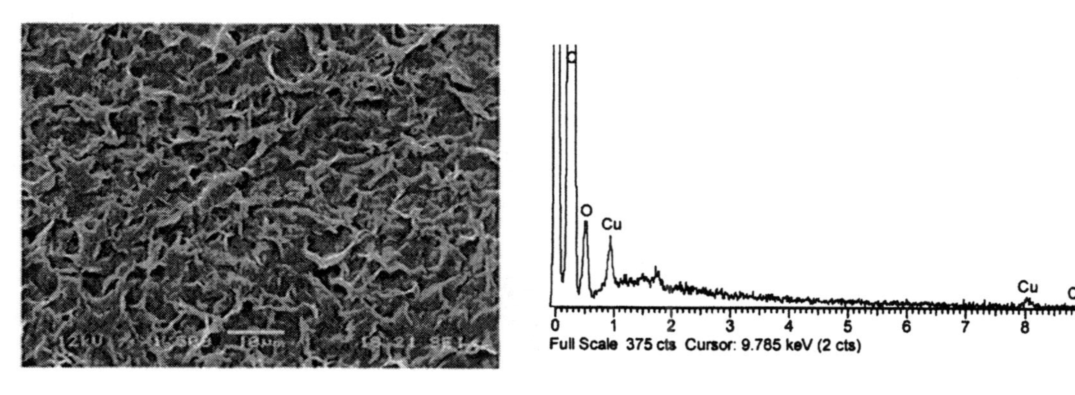

Figure 3. SEM micrograph of 3 wt.% $CuC_{18}OSalophen/UHMWPE$ film (left) and energy dispersive (EDS) spectrum of the polymer surface (right).

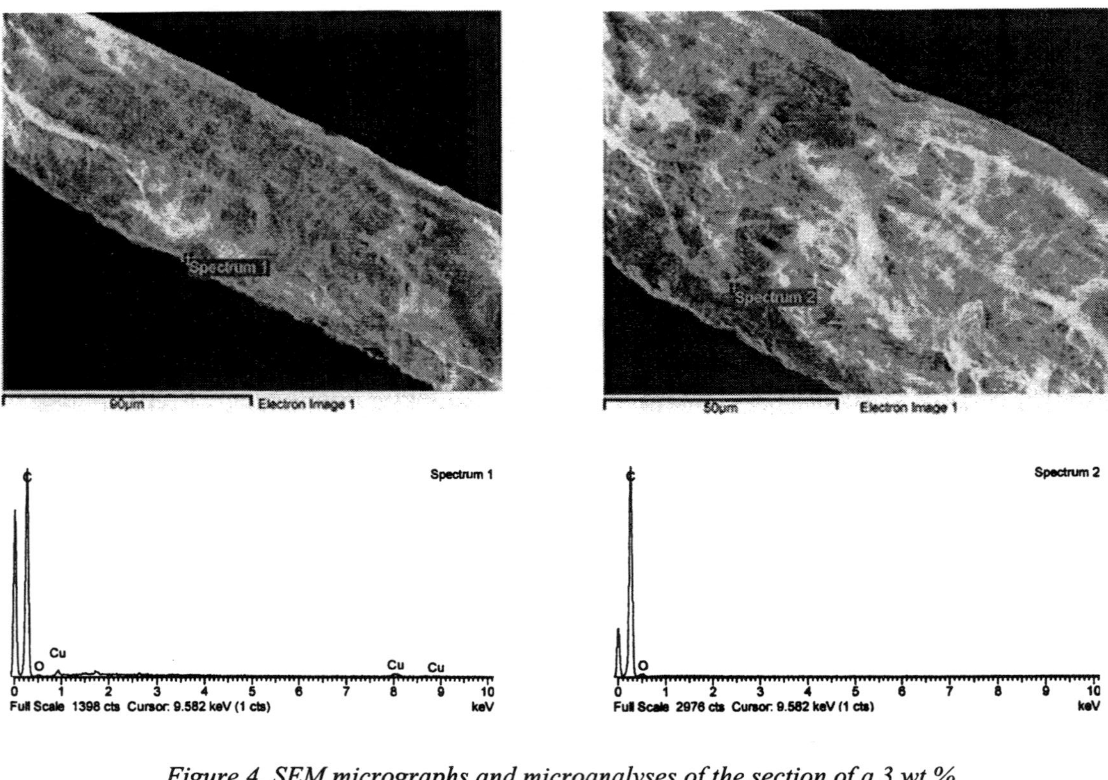

Figure 4. SEM micrographs and microanalyses of the section of a 3 wt.% $CuC_{18}OSalophen$/UHMWPE film near the surface exposed to air (left) and 10 µm deeper.

The opto-electronic properties of the metal complexes were compared to those of the pure ligand dispersed in polyethylene and studied in terms of their sensitivity to polymer matrix orientation.

The UV-Vis spectra recorded exciting an oriented UHMWPE film (draw ratio = 20) containing the 3 wt.% of the Bis(salicylaldiminate) based ligand (whose chemical structure is depicted in Figure 5) with a linearly polarized light respectively parallel (0°) and perpendicular (90°) to the drawing direction, show a broad absorption band centred at 330 nm with a shoulder at higher wavelengths attributed to the π   π* intraligand transitions (18, 19).

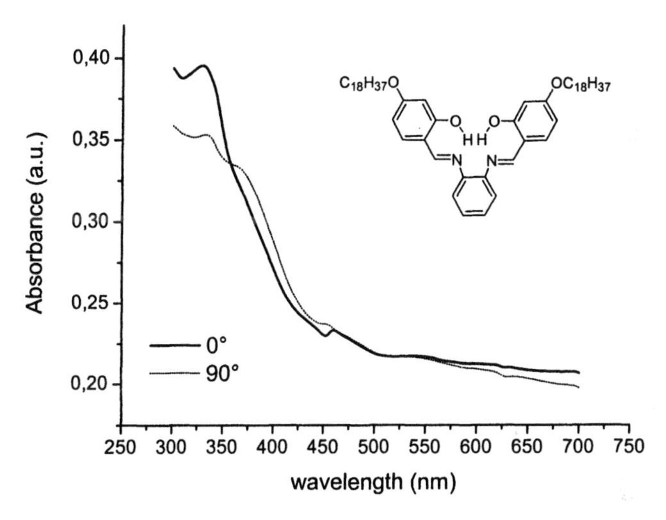

Figure 5. UV-Vis spectra in polarized light of a UHMWPE oriented film (draw ratio = 20) containing the 3 wt.% of the ligand and its chemical structure.

Comparing both spectra, this absorption resulted only slightly dichroic suggesting the poor anisotropy of the ligand electronic transitions probably due to its centrosymmetric structure (18). In addition, the scattering of the radiation denoted the presence of microsized ligand aggregates (10) into the polymer matrix due to a partial phase separation induced probably by the presence of polar -OH groups. These last might also affect the orientation of the molecules along the macromolecular chains thus limiting their anisotropic behaviour.

In the UHMWPE oriented film (draw ratio = 30) containing the 3 wt.% of $NiC_{18}OSalophen$ the bands attributed to the intraligand electron transitions resulted red-shifted of about 40-50 nm due to the interaction with the metal (Figure 6) (19). In addition, a new broad absorption at higher wavelengths, in the range from 400 to 500 nm, is observed (Figure 6). This absorption is attributed to the charge-transfer (CT) transitions from the excited filled d orbital electrons of the metal to the empty π^* orbitals of the imine chromophore of the ligand (27) and is demonstrated to be responsible for the non-linear optical response of Bis(salicylaldiminate) metal complexes (23).

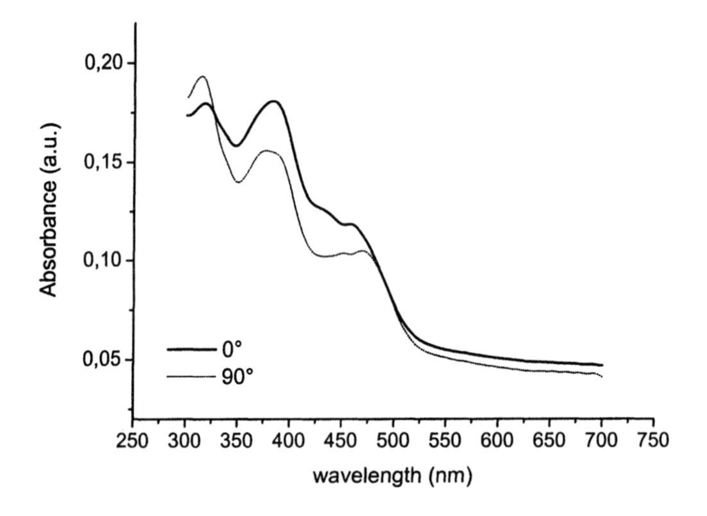

Figure 6. UV-Vis spectra in polarized light of a 3 wt.% $NiC_{18}OSalophen$/UHMWPE oriented film (draw ratio = 30)

The π π^* intraligand transitions appeared no influenced by the light polarization direction even after complex formation, whereas a dichroic, even if just pronounced, behaviour of the CT band can be detected. This indicates that the total dipole moment of the d π^* transition, oriented along the symmetry axis of the molecule (19, 28), is partially aligned along the drawing direction of the polymer.

In Figure 7, the UV-Vis spectra in polarized light of oriented UHMWPE films with draw ratio respectively 20 (up) and 30 (down) containing the 3 wt.% of $CuC_{18}OSalophen$ are reported.

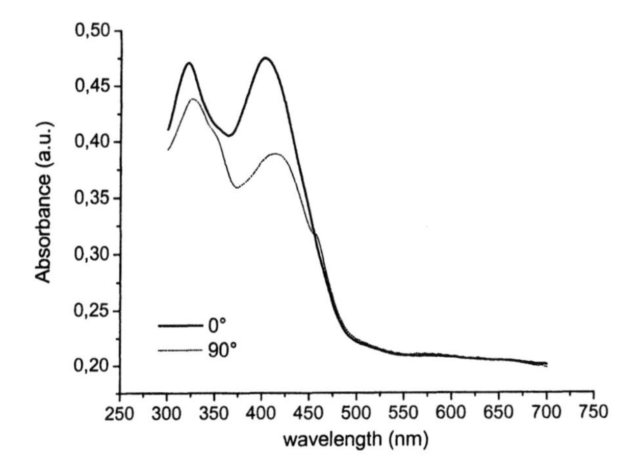

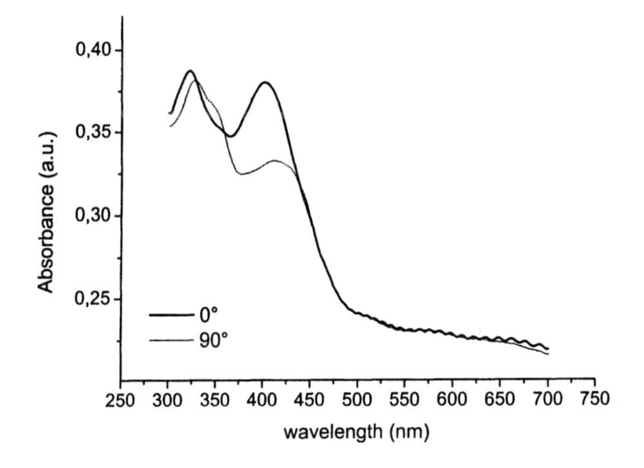

Figure 7. UV-Vis spectra in polarized light of 3 wt.% $CuC_{18}OSalophen$/UHMWPE (oriented films with draw ratio = 20, up and 30, down)

With respect to the corresponding Nickel complex the π π* electronic transitions attributed to the ligand coalesce into a single broad band pointed at about 320 nm, as observed in solution (22). In addition, the charge-transfer transition results blue-shifted of about 50 nm and exhibits an appreciably increased intensity and resolution. Moreover this band showed increased dichroism and resulted scarcely influenced by the drawing extent. On the other hand, as for Nickel complex composites, the intraligand absorption band appeared poorly influenced by the orientation process.

On passing to Cobalt complexes dispersed into the UHMWPE oriented matrix, the UV-Vis absorption spectra change again (Figure 8).

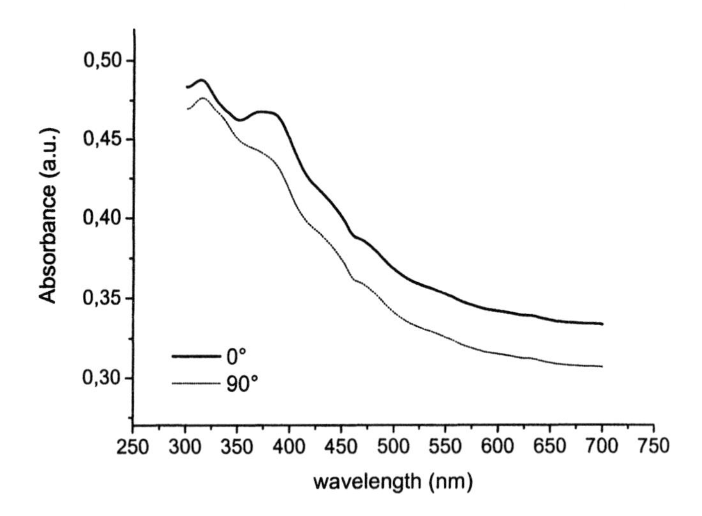

Figure 8. UV-Vis spectra in polarized light of a 3 wt.% $CoC_{18}OSalophen$/UHMWPE oriented film (draw ratio = 20)

The π π^* intraligand electronic transitions result further blue-shifted to about 310 nm, scarcely resolved and characterized by a complete isotropic behaviour. On the other hand, the anisotropy of the CT absorption band, centred at about 375 nm, appeared more distinct than those of Nickel and Copper complexes dispersed in oriented polyethylene. Moreover, additional evident shoulders came out at higher wavelengths, in the range between 450-600 nm, as in solution (29).

The particular enhanced anisotropy of the d π^* transition band of Cobalt and Copper Bis(salicyladiminate) complexes with respect to Nickel system may be derived from a different contribute to the electronic transition by the metal centre. In fact, Bis(salicyladiminate) Cobalt and Copper complexes are known to exhibit larger non-linear optical response associated to the CT transition than Nickel derivatives due to the different charge distribution on the excited electronic states (19, 22).

This behaviour leads to d π* electronic transitions characterized by increased dipole moment for Cobalt and Copper with respect to Nickel complexes (19, 22). The higher hyperpolarizability and transition dipole moment of the d π* electronic transition may represent the reason why $CoC_{18}OSalophen$ and $CuC_{18}OSalophen$ dispersed in oriented UHMWPE show appreciably higher dichroic behaviour for the CT absorption band than $NiC_{18}OSalophen$ when excited by a linearly polarized radiation.

Conclusions

Bis(salicyladiminate) Nickel(II), Copper(II) and Cobalt(II) complexes (Figure 1b) were prepared by functionalizing the ligand moiety with long and linear alkyl chains able to confer the metal complex a good compatibility with the polyethylene matrix as evidenced by previous dispersion studies performed on Schiff-Base metal(II) complexes (Figure 1a).

The strategy based on increasing the optical properties by extending the conjugation path of the molecule, adopted in organic highly anisotropic terthiophene based chromophores, was here applied to modify the ligand structure with the introduction of a phenyl ring. The insertion of this aromatic unit did not change the phase behaviour of the complexes in polyethylene with respect to Schiff-Base Nickel and Copper complexes/UHMWPE binary films but, on the other hand, strongly contributed to the increase of the organic moiety conjugation leading to materials characterized by a special anisotropy of the charge-transfer transition band.

Indeed, the Nickel, Copper and Cobalt complexes with the $C_{18}OSalophen$ ligand in polyethylene films showed after polymer matrix orientation an anisotropic behaviour attributed to the above charge-transfer d π* electronic transition, whose effect was particularly influenced by the electronic nature of the metal centre.

These results reported here provide an excellent starting basis for the preparation of polyethylene composite films with molecularly dispersed metal complexes where the supramolecular interactions between the dispersed phase and the continuous matrix is confirmed by the optical response.

The presence of these metal species with a controlled morphology can be considered as a way to prepare polyethylene films showing new optical and catalytical properties.

Acknowledgments

Dr. Paolo Elvati and Mr. Piero Narducci are kindly acknowledged for the helpful discussions and SEM measurements.

References

1. Land, E. H. US 2,289,714, 1942.
2. Land, E. H. US 2,289,713, 1948.
3. Land, E. H.; Grabau, M. US 2,440,105, 1948.
4. Land, E. H.; West, C. D. *Colloid Chemistry* **1946**, *6*, 160-190.
5. Land, E. H. US 2,454,515, 1942.
6. Broer, D. J.; Van Haaren, J. A. M. M.; Van de Witte, P.; Bastiaansen, C. *Macromol. Symp.* **2000**, *154*, 1-13.
7. Montali, A.; Bastiaansen, C.; Smith, P.; Weder, C. *Nature* **1998**, *392*, 261-264.
8. Dirix, Y.; Tervoort, T. A.; Bastiaansen, C. *Macromolecules* **1997**, *30*, 2175-2177.
9. Dirix, Y.; Tervoort, T. A.; Bastiaansen, C. *Macromolecules* **1995**, *28*, 486-491.
10. Tirelli, N.; Amabile, S.; Cellai, C.; Pucci, A.; Regoli, L.; Ruggeri, G.; Ciardelli, F. *Macromolecules* **2001**, *34*, 2129-2137.
11. Ciardelli, F.; Cellai, C.; Pucci, A.; Regoli, L.; Ruggeri, G.; Tirelli, N.; Cardelli, C. *Polym. Adv. Techol.* **2001**, *12*, 223-230.
12. Pucci, A.; Ruggeri, G.; Moretto, L.; Bronco, S. *Polym. Adv. Technol.* **2002**, *13*, 737-743.
13. Pucci, A.; Moretto, L.; Ruggeri, G.; Ciardelli, F. *e-Polymers* **2002**, Paper No 15, Paper No 15.
14. Pucci, A.; Ruggeri, G.; Cardelli, C.; Conti, G. *Macromol. Symp.* **2003**, *202*, 85-95.
15. Palmans, A. R. A.; Eglin, M.; Montali, A.; Weder, C.; Smith, P. *Chem. Mater.* **2000**, *12*, 472-480.
16. Montali, A.; Palmans, A. R. A.; Eglin, M.; Weder, C.; Smith, P.; Trabesinger, W.; Renn, A.; Hecht, B.; Wild, U. P. *Macromol. Symp.* **2000**, *154*, 105-116.
17. Pucci, A.; Elvati, P.; Ruggeri, G.; Liuzzo, V.; Tirelli, N.; Isola, M.; Ciardelli, F. *Macromol. Symp.* **2003**, *204*, 59-70.
18. Di Bella, S.; Fragala, I.; Ledoux, I.; Diaz-Garcia, M. A.; Lacroix, P. G.; Marks, T. J. *Chem. Mater.* **1994**, *6*, 881-883.
19. Di Bella, S.; Fragala, I.; Ledoux, I.; Diaz-Garcia, M. A.; Marks, T. J. *J. Amer. Chem. Soc.* **1997**, *119*, 9550-9557.

20. Aiello, I.; Ghedini, M.; La Deda, M.; Pucci, D.; Francescangeli, O. *Eur. J. Inorg. Chem.* **1999**, 1367-1372.
21. Chen, D.; Martell, A. E. *Inorg Chem.* **1977**, *26*, 1026-1030.
22. Di Bella, S.; Fragala, I.; Ledoux, I.; Marks, T. J. *J. Amer. Chem. Soc.* **1995**, *117*, 9481-9485.
23. Lacroix, P. G. *Eur. J. Inorg. Chem.* **2001**, 339-348.
24. Sano, T.; Nishio, Y.; Hamada, Y.; Takahashi, H.; Usuki, T.; Shibata, K. *J. Mater. Chem.* **2000**, *10*, 157-161.
25. Leung, A. C. W.; Chong, J. H.; Patrick, B. O.; MacLachlan, M. J. *Macromolecules* **2003**, *36*, 5051-5054.
26. Goodhew, P. J.; Humphreys, J.; Beanland, R. In *Electron Microscopy and Analysis*; Taylor & Francis: London, 2001.
27. Bosnich, B. *J. Amer. Chem. Soc.* **1968**, *90*, 627-632.
28. Di Bella, S.; Fragala, I. *Synth. Met.* **2000**, *115*, 191-196.
29. Hitchman, M. A. *Inorg. Chem.* **1977**, *16*, 1985-1993.

Chapter 4

Functional Materials Based on Self-Assembled Comb-Shaped Supramolecules

Gerrit ten Brinke[1] and Olli Ikkala[2]

[1]Laboratory of Polymer Chemistry, Materials Science Centre, University of Groningen, Nijenborgh 4, 9747AG Groningen, The Netherlands
[2]Department of Engineering Physics and Mathematics, Center for New Materials, Helsinki University of Technology, P.O. Box 2200, FIN–02015 HUT, Espoo, Finland

In this paper we will review the main features of our approach to create functional materials using self-assembly of hydrogen-bonded comb-shaped supramolecules. Typically, the supramolecules consist of homopolymers or diblock copolymers, where short chain amphiphiles are hydrogen-bonded to the homopolymers or to one of the blocks of the copolymer. The side chains serve several important goals, such as improved processing, additional ordering leading to hierarchically ordered materials, functionalization and swelling. Moreover, if required, they can be easily dissolved away after having served their role in the self-assembling process. Applications include proton conductivity (anisotropic, switching), electronically conducting nanowires, polarized luminance, dielectric stacks, membranes, and nano objects.

Polymer molecules involving chemically different chain moieties within a single molecule, either by covalent linking or by physical bonds, tend to self-assemble in periodic structures with a characteristic length scale in the nanometer range. A large number of different architectures is available, such as diblock copolymers, triblock copolymers, graft copolymers, their supramolecular versions, etc. Together these create a very flexible tool for preparing materials for nanotechnology applications (1-9). This flexibility includes a variety of different periodic structures with the possibility to tune the length scale(s) from 1-1000nm. Another level of complexity is introduced by combining block copolymers with supramolecular concepts (10). In particular, comb-shaped supramolecules obtained by attaching side groups by physical interactions (ionic, hydrogen bonding, metal coordination, etc) to homopolymers or block copolymers have been thoroughly investigated over the last decade (11-20). In this review we will restrict ourselves to comb-shaped supramolecules involving hydrogen bonded short chain amphiphiles. In many cases the presence of such short side chains will result in microphase separation with a characteristic length scale in the order of 3-5nm. If diblock copolymers are used in combination with amphiphiles that form hydrogen bonds to one of the blocks, hierarchically ordered materials are formed with a large length scale corresponding to the microphase separation between both blocks and another short length scale due to the microphase separation within the domains formed by the comb-shaped blocks (19-20). An important additional feature, implied by the supramolecular nature of the side chain bonding, is the possibility to remove the side chains in a straightforward manner after they have served their role in the self-assembling process. The concept is illustrated in Figure 1.

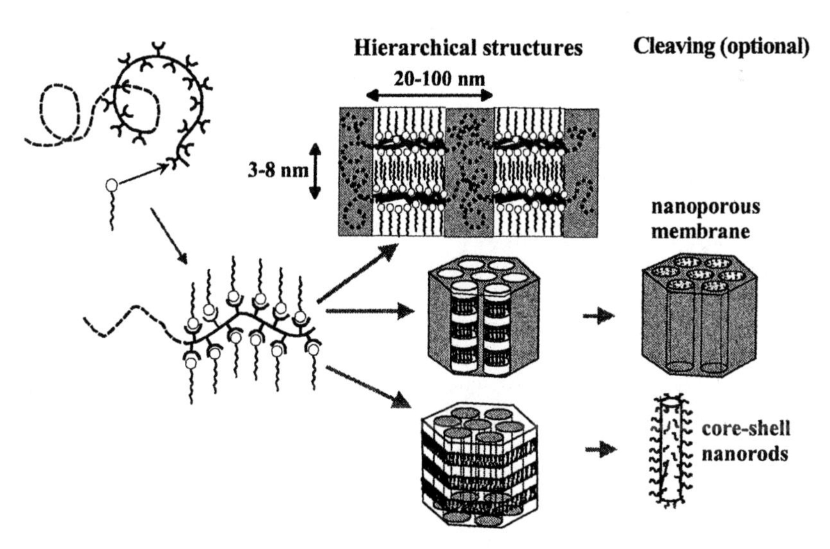

Figure 1. Concept to construct hierarchically self-assembled supramolecules structures with selected applications based on removal of side chains (20).

Comb-Shaped Supramolecules in Action

During the last decade the hydrogen based comb-shaped supramolecules formed the basis for the development of a series of possible applications (20). These make use of specific properties that can be attributed to the presence of the side chains, such as self-assembly, swelling, processability, cleavage, etc.

Homopolymer-based comb-shaped supramolecules

Homopolymer-based examples of functionality include (anisotropic) proton conductivity, polarized luminance and electronically conducting nanowires.

In the first case poly(4-vinylpyridine) (P4VP) is protonated by toluene sulfonic acid (TSA) to obtain an acid-base complex that exhibits temperature activated protonic conductivity. To this stoichiometric salt pentadecylphenol (PDP) is hydrogen bonded and a comb-shaped supramolecule is formed (see Figure 2 without PS block).

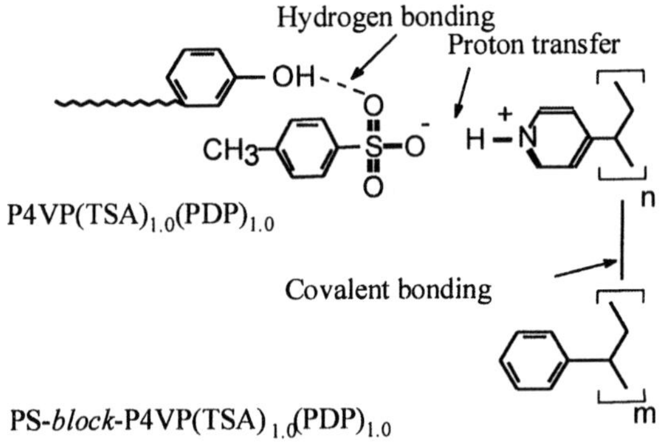

Figure 2. Scheme of PS-b-P4VP(TSA)(PDP) complex

This $P4VP(TSA)_{1.0}(PDP)_{1.0}$ complex self-assembles into a lamellar morphology below ca. 100°C, which can then be exploited to prepare anisotropically proton conducting materials (Figure 3). The side chain induced self assembly together with the acid doping are the essential ingredients underlying this application.

Polarized luminance has been demonstrated using poly(2,5-pyridinediyl) (PPY) in combination with camphersulphonic acid (CSA) and octylgallate (OG). PPY is first protonated with CSA, which leads to an efficient photoluminescence quantum yield. Further hydrogen bonding with octylgallate (OG), allowing multiple hydrogen bonds, leads to comb-shaped supramolecules that can easily be processed and that self-assemble in the form of a layered structure. Due to the fluid nature of the thin films, the rodlike entities can be oriented by simple shear. SAXS demonstrates that indeed highly ordered structures are formed. After evaporation of OG a corresponding well-ordered solid film is obtained exhibiting polarized luminescence. Here processability is combined with structure formation and cleavability (22-25).

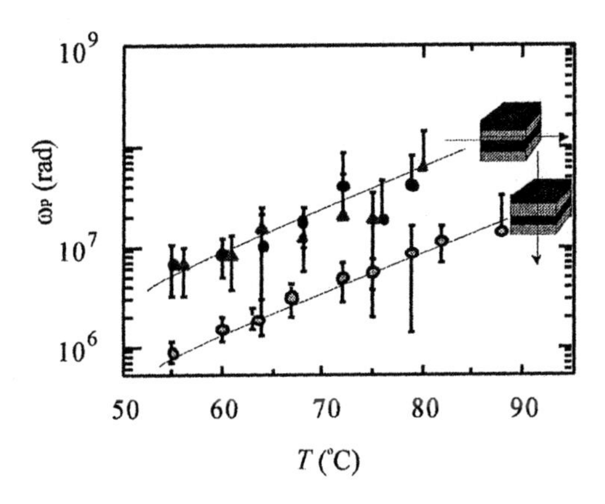

Figure 3. Anisotropic proton conductivity in macroscopically aligned $P4VP(TSA)_{1.0}(PDP)_{1.0}$. ω_p denotes charge hopping frequency (21).

Electronically conducting nanowires have been constructed on the basis of polyaniline (PANI) (26). Polyaniline (PANI) is one of the most interesting polymers in the area of electronically conducting polymers. Protonation of PANI by strong acids is known to render conducting polymer salts. Although rather stiff, PANI is by no means a rigid rod polymer. Therefore, it is to be expected that electrical conductivity of PANI-based systems may greatly benefit from confinement of PANI chains within narrow cylinders. To create cylindrical self-assembled structures hydrogen bonded polyaniline supramolecules were made combining ionic- and hydrogen bonding. This combination serves a dual purpose: the protonation introduces the electrical conductivity and the hydrogen bonding allows self-assembly. The iminic nitrogens of PANI were first nominally fully protonated using camphorsulfonic acid (CSA) to yield $PANI(CSA)_{0.5}$ and then hydrogen bonded to 4-hexylresorcinol (Hres). Both hydroxyl groups seem to be required to prevent macrophase separation. As demonstrated by the SAXS data, the resulting rodlike supramolecules indeed self-organize into hexagonally ordered cylindrical structures for a number y of Hres units per aniline repeat unit in the range y=0.5-2.0. Upon formation of the cylinders that contain 3-4 PANI molecules per cylinder, the electrical conductivity increases two orders of magnitude as shown in Figure 4. As in the case of proton conductivity, this application uses the side chain induced self-assembly and doping.

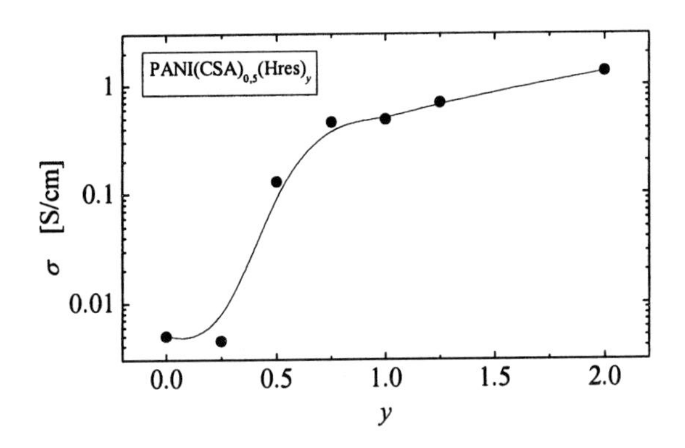

Figure 4. The strong increase in conductivity σ as a function of the amount of Hres coincides with the formation of PANI containing cylinders (26).

Diblock copolymer-based comb-shaped supramolecules

If, instead of homopolymers, suitable block copolymers are used as starting materials, characteristic two length scale hierarchical structures may be formed upon complexation with amphiphiles (19-20). These systems form the basis for a variety of applications. Besides the hierarchical structure formation, characteristic features used advantageously include improved processing, doping, swelling and cleaving. The structure formation and subsequent cleaving of side chains is schematically illustrated in Figure 1. The best-studied example involves a diblock copolymer of polystyrene (PS) and poly(4-vinyl pyridine) where pentadecylphenol is hydrogen bonded to the P4VP block: PS-b-P4VP(PDP) (see scheme of Figure 2 without the TSA). Usually, a microphase separated morphology, consisting of PS and P4VP(PDP) domains, is present throughout the experimental temperature range with a characteristic length scale in the order of 10-50 nm. Under stoichiometric conditions, i.e. one PDP molecule per pyridine group, below ca. 65°C a short length scale layered structure, due to the comb-shaped nature of the P4VP(PDP)$_{1.0}$ blocks, is formed inside the

P4VP(PDP) domains (27). Lamellar-*within*-lamellar, lamellar-*within*-cylinders and lamellar-*within*-spheres, as well as the complimentary structures have all been demonstrated and imaged by TEM (19, 28, 29). A characteristic example is presented in Figure 5.

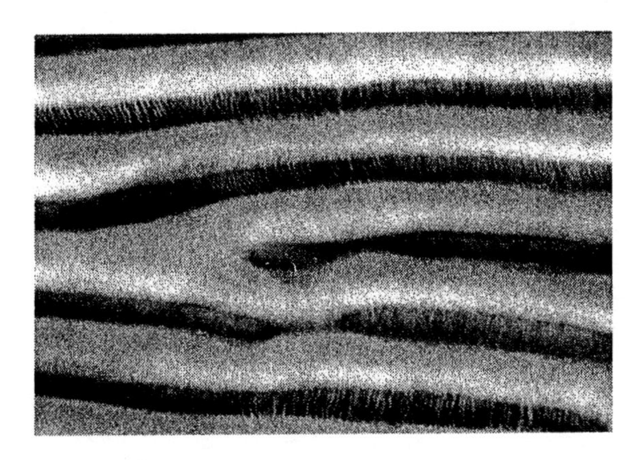

Figure 5. Lamellar-within-lamellar self-assembled morphology of PS-b-P4VP(NDP), where NDP is nonadecylphenol (29).

An interesting example, where the presence of a *hierarchical* structure is explicitly used, involves proton conductivity. As already discussed before, complexation of P4VP with a strong acid, such as methane or toluene sulfonic acid (TSA), creates a proton conducting material. If a stoichiometric complexation with TSA is taken and subsequently combined with the hydrogen bonding PDP amphiphiles, self-assembly leads to a layered structure below ca. 140°C, where the polar layers are proton conducting. If instead of homopolymer P4VP, a suitable P4VP-*b*-PS block copolymer is used, a hierarchically ordered *lamellar*-within-*lamellar* material, similar to the one illustrated in Figure 5, is obtained. After applying an appropriate oscillatory shear protocol to such a PS-b-P4VP(TSA)(PDP) sample (see scheme in Figure 2), the macroscopic ordering is improved, although far from perfect (30). As a result a material is obtained exhibiting different proton conductivity in the three different directions (21). Figure 6 presents the conductivity data.

If in stead of TSA, MSA is used, a much more complex phase behavior results involving different structural transitions and concurrent temperature switching in proton conductivity (19). Below ca. 100°C, again a lamellar-*within*-lamellar structure is present. Above 100°C, the short length scale lamellar structure disappears at the corresponding order-disorder transition temperature. Higher temperatures improve the miscibility of PDP with PS until they become completely miscible at ca 135°C. Since, the miscibility of PDP with P4VP(MSA) decreases, PDP gradually diffuses into the PS phase thereby increasing its relative volume fraction. As a consequence, an order-order transition to a hexagonally ordered cylindrical morphology occurs with P4VP(MSA) forming the cylinders. The sequence of transitions and the corresponding conductivity data are presented in Figure 7.

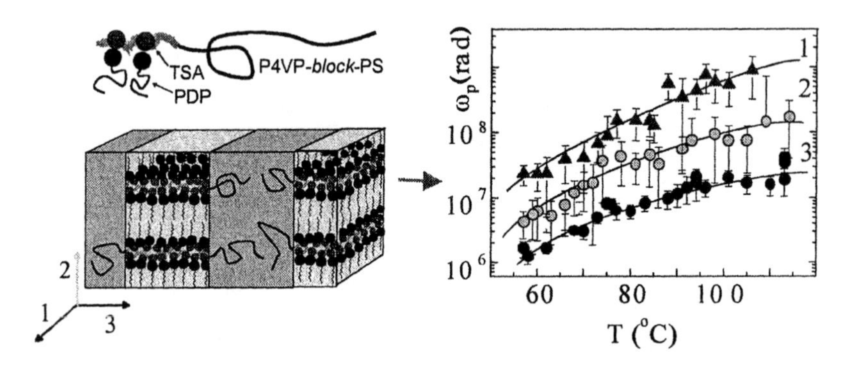

Figure 6. Anisotropic proton conductivity of hierarchically structured lamellar-within-lamellar PS-bP4VP(TSA)$_{1.0}$(PDP)$_{1.0}$ (21).

Self-assembly in block copolymer systems leads to well-ordered structures which are potentially interesting for photonic crystals applications (31-32). The transport of electromagnetic radiation can be manipulated using photonic band gap materials, which contain periodic structures with sufficiently high dielectric contrast. There are, however, several important problems to be solved before this potential can be fully realized. One of this is the large periodicity required. A large periodicity requires high molecular weight block copolymers, which are notoriously difficult to prepare in a single crystal-like state. In connection with these two requirements, a large periodicity and a well-ordered structure, the comb-shaped supramolecules concept is expected to have some advantages as well. The presence of supramolecular side chains gives rise to strong stretching of the polymer backbone in the microphase separated state and thus to enhanced spacing of the ordered structure (33). Moreover, they do so while increasing the molecular mobility at the same time. Recently we demonstrated that in the case of high molar mass PS-*b*-P4VP diblock copolymers, PS-*b*-P4VP(DBSA)$_y$, where DBSA denotes dodecyl benzene sulfonic acid, forms self-assembled one-dimensional optical reflectors for an excess of DBSA, i.e. $y > 1.0$ (34). This is demonstrated in Figure 8 showing a series of transmission data. The obvious next step consists in exploiting the complex phase behavior of PS-*b*-P4VP(MSA)(PDP), discussed above in connection with switching proton conductivity, to prepare materials with reversible switching band gaps (35). This application forms a perfect example of the very effective swelling induced by the side chains to obtain the large periodicities required.

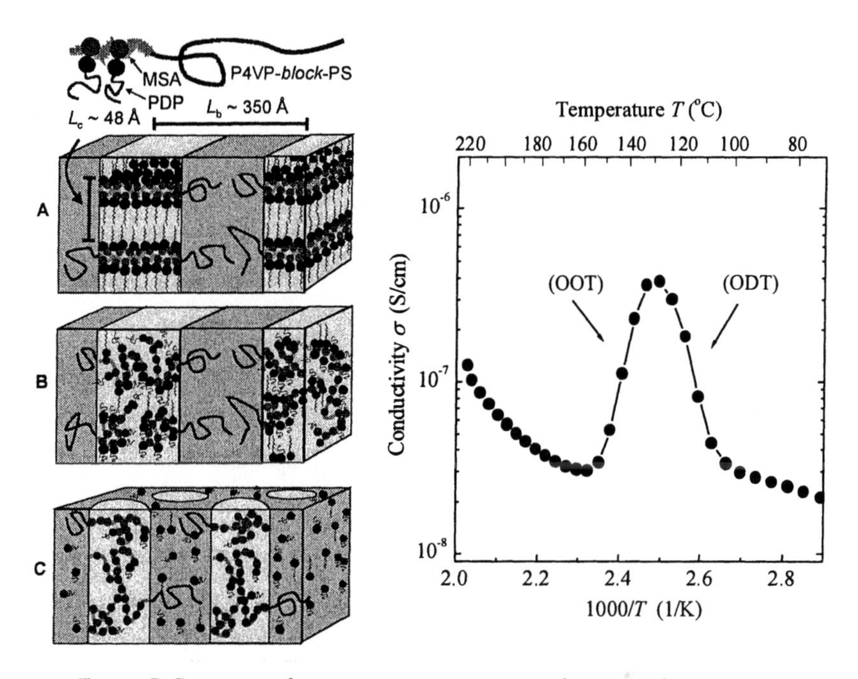

Figure 7. Sequence of transitions occurring as a function of temperature in PS-b-P4VP(MSA)$_{1.0}$(PDP)$_{1.0}$. A: T<100°C, B: 100<T<140°C, C: T>140°C. Right: Corresponding proton conductivity (19).

Self-assembled periodic structures of block copolymers offer unique possibilities to prepare nanoporous membranes. Block lengths may be selected in such a way that hexagonally ordered cylindrical structures are formed. Taking the minority block to be degradable allows for a straightforward way to produce such membranes. Liu and co-workers (36) prepared thin films with nanochannels from poly(*tert*-butyl acrylate)-*block*-poly(2-cinnamoylethyl methacrylate), where the *tert*-butyl groups are cleavable by hydrolysis and poly(2-cinnamoylethyl methacrylate) is photo-crosslinkable. The latter property is used to crosslink the matrix.

Our concept to prepare hierachically structured materials consisting of supramolecules, obtained by physically bonding amphiphilic molecules to one block of a block-copolymer, allows for an alternative route towards nanoporous films. In fact, the cleavability of the supramolecular side chains make these systems ideally suited. A most simple example concerns the preparation of membranes containing hollow self-assembled cylinders with polymer brushes at the wall. To allow crosslinking of the membrane afterwards, diblock copolymers of polyisoprene and poly(2-vinylpyridine) (PI-*b*-P2VP) were selected as starting material.

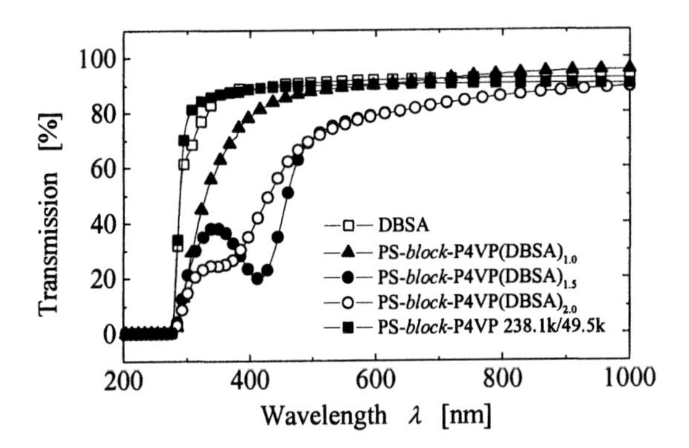

Figure 8. UV-Vis transmission graphs for PS-b-P4VP(DBSA)$_y$ for a high molar mass block copolymer, which leads to very long periodicities in excess of 100nm enabling the formation of a bandgap (34).

To increase the tendency to microphase separation between the PI block and the P2VP-based comb-shaped block octylgallate (OG) was selected as the hydrogen bonding amphiphile (see scheme Figure 9). For appropriate block molar masses, a cylindrical morphology was obtained where the cylinders are formed by the comb-shaped P2VP(OG) blocks. The order of the structure may extend over macroscopic distances by the application of suitable external fields, e.g. large amplitude oscillatory shear. Figure 10 shows a SAXS picture demonstrating the macroscopic alignment of the cylinders achieved by shear. It was taken during in-situ synchrotron X-ray measurements (ESRF, Dubble), with a specially designed rheometer placed in the beamline (37-39). This so-called "tooth" rheometer provides a rather well-defined shear profile and allows the SAXS measurements to be taken tangentially and radially. After the alignment, the supramolecular side chains, accounting for no less than 75% of the material inside the cylinders, can easily be dissolved afterwards, leaving a porous membrane.

Figure 9. Scheme of octylgallate hydrogen bonded to the P2VP block of a PI-b-P2VP diblock copolymer.

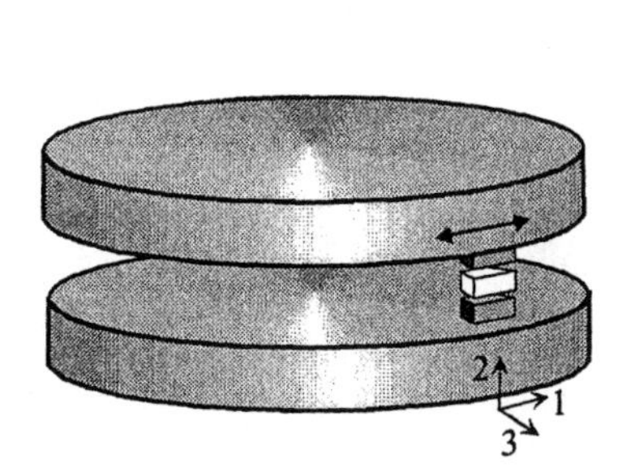

Figure 10. Cartoon of tooth rheometer design and tangential X-ray picture (incoming beam along 1-direction) of PI-b-P2VP(OG)$_{0.75}$, demonstrating the macroscopic alignment of the cylinders parallel to the shear direction (37-39).

Self-assembly of block copolymers has been used extensively to prepare individual polymeric "nano-objects" (40-44). The comb-shaped supramolecules discussed at length in this review allow a novel and general concept to prepare core-corona type aggregates. The procedure to such nano-objects is illustrated in Figure 1 for nano-rods. In the case of PS-*b*-P4VP(PDP) supramolecules self-assembling into a hexagonally ordered cylindrical morphology with the comb-shaped P4VP(PDP) forming the majority phase and PS the cylinders, the dissolution of the side chains will lead to "hairy rods". Figure 11 presents an explicit example of nanorods consisting of a PS core and a P4VP corona (43-44).

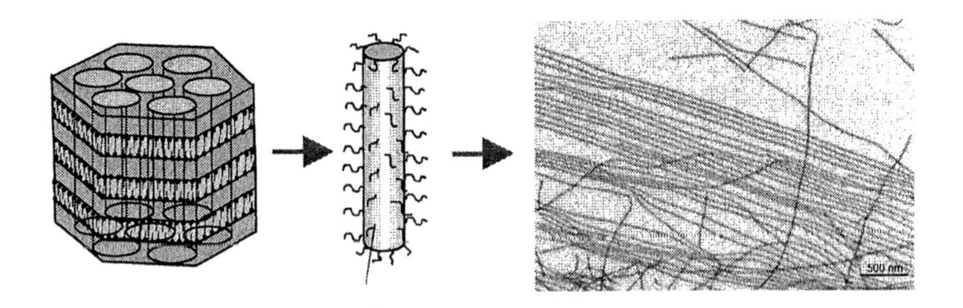

Figure 11. Preparation of hairy rod objects from PS-b-P4VP(PDP)$_{1.0}$ (44).

In summary, we have shown that the combination of self-assembly at different length scales leads to structural hierarchies that may be exploited to create functional materials. In the present review this was accomplished by combining block copolymer self-assembly with supramolecular concepts. The physically bonded short chain amphiphiles lead to self-assembly at a length scale of a few nm. An order of magnitude larger length scale is provided by the block copolymer itself. Very recently this has been extended to include an even larger length scale upon using colloidal particles (45).

Acknowledgment

It has been a pleasure to collaborate on these subjects with many distinguished scientists and skillful graduate students whose names appear in the cited references.

References

1. Nanosciences; Special Issue of Scientific American, **2001**, *285*(3).
2. *The Physics of Block Copolymers*; Hamley, I.W.; Oxford University Press: New York, 1998.

3. Bates, F.S.; Fredrickson, G.H. *Annu. Rev. Phys. Chem.* **1990**, *41*, 525.
4. Bates, F.S.; Fredrickson, G.H. *Physics Today* **1999**, *52*, 32.
5. Muthukumar, M.; Ober, C.K.; Thomas, E.L. *Science* **1999**, *277*, 1225.
6. Fasolka, M.J.; Mayes, A.M. *Annu. Rev. Mater. Res.* **2001**, *31*, 323.
7. Park, C.; Yoon, J.; Thomas, E.L. *Polymer* **2003**, *44*, 6725.
8. Antonietti, M. *Nature Materials* **2003**, *2*, 9.
9. Hamley, I. W. *Angew. Chem. Int. Ed.* **2003**, *42*, 1692.
10. *Supramolecular Chemistry: Concepts and Perspectives*; Lehn, J.M.; VCH, Weinheim, 1995.
11. Kato, T.; Frechet, J.M.J. *Macromolecules* **1989**, *22*, 3818.
12. Navarro-Rodriquez, D.; Guillon, D.; Skoulios, A. *Makromol. Chem.* **1992**, *193*, 3117.
13. Kato, T.; Nakano, M.; Moteki, T.; Uryu, T.; Ujiie, S. *Macromolecules* **1995**, *28*, 8875.
14. Brandys, F.A.; Bazuin, C.G. *Chem. Mater.* **1996**, *8*, 83.
15. Kato, T.; Mizoshita, N.; Kanie, K. *Macromol. Rapid Commun.* **2001**, *22*, 797.
16. Kato, T. *Science* **2002**, *295*, 2414.
17. Faul, C.F.J.; Antonietti, M. *Adv. Mater.* **2003**, *15*, 673.
18. Thünemann, A.F. *Prog. Polym. Sci.* **2002**, *27*, 1473.
19. Ruokolainen, J.; Mäkinen, R.; Torkkeli, M.; Mäkelä, T.; Serimaa, R.; Ten Brinke, G.; Ikkala, O. *Science* **1998**, *280*, 557.
20. Ikkala, O.; Ten Brinke, G. *Science* **2002**, *295*, 2407.
21. Mäki-Onnto, R.; De Moel, K.; Polushkin, E.; Alberda van Ekenstein, G.; Ten Brinke, G.; Ikkala, O. *Adv. Mater.*, **2002**, *14*, 357.
22. Ikkala, O.; Kaapila, M.; Ruokolainen, J.; Torkkeli, M.; Serimaa, R.; Jokela, K.; Horsburgh, L.; Monkman, A.; Ten Brinke, G. *Adv. Mater.* **1999**, *11*, 1206.
23. Knaapila, M.; Ikkala, O.; Torkkeli, M.; Jokela, K.; Serimaa, R.; Dolbnya, I.P.; Bras, W.; Ten Brinke, G.; Horsburgh, L.E.; Palson, L-O., Monkman, A.P. *Appl. Phys. Lett.* **2002**, *81*, 1489.
24. Knaapila, M.; Torkkeli, M.; Jokela, K.; Kisko, K.; Horsburgh, L.E.; Palsson, L-O.; Seeck, O.H.; Dolbnya, I.P.; Bras, W.; Ten Brinke, G.; Monkman, A.P.; Ikkala, O.; Serimaa, R. *J. Appl. Cryst.* **2003**, *36*, 702.
25. Knaapila, M.; Stepanyan, R.; Horsburgh, L.E.; Monkman, A.P.; Serimaa, R.; Ikkala, O.; Subbotin, A.; Torkkeli, M.; Ten Brinke, G. J. *Phys. Chem. B.* **2003**, *107*, 14199.
26. Koskonen, H.; Ruokolainen, J.; Knaapila, M.; Torkkeli, M.; Serimaa, R.; Ten Brinke, G.; Bras, W.; Monkman, A.; Ikkala, O. *Macromolecules*, **2000**, *33*, 8671.
27. Ruokolainen, J.; Torkkeli, M.; Serimaa, R.; Komanschek, B.E.; Ikkala, O.; Ten Brinke, G. *Phys. Rev. E.*, **1996**, *54*, 6646.

28. Ruokolainen, J.; Tanner, J.; Ikkala, O.; Ten Brinke, G.; Thomas, E.L. *Macromolecules* **1998**, *31*, 3532.
29. Ruokolainen, J.; Ten Brinke, G.; Ikkala, O. *Adv. Mater.* **1999**, *11*, 777.
30. Riikka Mäkinen, Janne Ruokolainen, Olli Ikkala, Karin de Moel, Gerrit ten Brinke, Walter De Odorico, Manfred Stamm, *Macromolecules*, **2000**, *33*, 3441.
31. *Photonic Crystals: Molding the Flow of Light*; Joannopoulos, J.D.; Meade, R.D.; Winn, J.N, Princeton University Press, Princeton, 1995.
32. Urbas, A.; Fink, Y.; Thomas, E.L. *Macromolecules* **1999**, *32*, 4748.
33. Vasilevskaya, V.V.; Gusev, L.A.; Khokhlov, A.R.; Ikkala, O.; ten Brinke, G. *Macromolecules* **2001**, *34*, 5019.
34. Kosonen, H.; Valkama, S.; Ruokolainen, J.; Torkkeli, M.; Serimaa, R.; Ten Brinke, G.; Ikkala, O. *Eur. Phys. J. E.* **2003**, *10*, 69.
35. Valkama, S.; Kosonen, H.; Ruokolainen, J.; Torkkeli, M.; Serimaa, R.; Ten Brinke, G.; Ikkala, to be published.
36. Liu, G. *Adv. Mater.* **1997**, *9*, 437.
37. Polushkin, E.; Alberda van Ekenstein, G.; Dolbnya, I.P.; Bras, W.; Ikkala O.; Ten Brinke, G. *Macromolecules* **2003**, *36*, 1421.
38. Polushkin, E.; Alberda van Ekenstein, G.O.R.; Ikkala, O.; G. ten Brinke *Rheologica Acta* **2004**, *43(4)*.
39. Bondzic, S.; de Wit, J.; Polushkin, EW.; Schouten, A.J.; ten Brinke, G.; Ruokolainen, J.; Ikkala, O.; Dolbnya, I.; Bras, W. *Macromolecules* **2004**, accepted.
40. Yu, Y.; Eisenberg, A. *J. Am. Chem. Soc.* **1997**, *119*, 8383.
41. Desbaumes, L.; Eisenberg, A. *Langmuir* **1999**, *15*, 36.
42. Liu, G.; Ding, J.; Qiao, L.; Guo, A.; Dymov, B.P.; Gleeson, J.T.; Hashimoto, T.; Saijo, K. *Chem. Eur. J.* **1999**, *5*, 2740.
43. De Moel, K.; Alberda van Ekenstein, G.O.R.; Nijland, H.; Polushkin, E.; Ten Brinke, G.; Mäki-Ontto, R.; Ikkala, O. *Chem. Mater.* **2001**, *13*, 4580.
44. Alberda van Ekenstein, G.O.R.; Polushkin, E.; Nijland, H.; Ikkala, O.; Ten Brinke, G. *Macromolecules* **2003**, *36*, 3684.
45. Mezzenga, R.; Ruokolainen, J.; Fredrickson, G.H.; Kramer, E.; Moses, D.; Heeger, A.J.; Ikkala, O. *Science* **2003**, *299*, 1872.

Chapter 5

Helical Conformations of Conjugating Polymers and Their Control of Cholesteric Order

Akio Teramoto[1,2], Kazuto Yoshiba[1], Shigeru Matsushima[1], and Naotake Nakamura[1]

[1]Research Organization of Science and Engineering and Department of Applied Chemistry, Ritsumeikan University, Nojihigashi 1–1–1, Kusatsu, Siga 525–8577, Japan
[2]Current address: Department of Biological and Chemical Engineering, Faculty of Engineering, Gumma University, 1–5–1, Tenjin-cho, Kiryu, Gumma 355–8515, Japan

The optical activity of chiral helix-forming polymers, which derives from uneven populations of the left-handed and right-handed helical conformations, depends on solvent condition as well as molecular weight. This molecular weight dependence has been analyzed by a statistical mechanical theory of the linear Ising model, taking into account the restriction of the conformation of the terminal units. These chiral helix-forming polymers form cholesteric liquid crystals at high concentrations, which are characterized by the cholesteric pitch P. P changes with the conformational transition, which is well explained by a statistical mechanical theory of P considering chiral repulsive and attractive interactions among helical molecules.

In solution most linear polymers do not take ordered conformations and are usually randomly coiled, whereas some polymers are rigid and even rod-like, with α-helical polypeptides as typical examples. However there are a number of polymers situated between these two extremes, which are not very flexible nor rod-like. They are referred to as stiff polymers or semi-flexible polymers depending on their situation or their stiffness. It is established that the conformation of such a *non-flexible* polymer is well described by the wormlike chain model(*1-6*), which is characterized by the persistence length *q*. This parameter tells how close is the average shape of the polymer to a straight rod. The finite stiffness of the polymer arises from its local rod-like structure, which is in many cases formed of a helical conformation. Typical helices studied extensively are single-stranded helices such as α-helical polypeptides(*7-11*),polyisocyanates(*12-14*), and polysilylenes(*15,16*), double-stranded helices xanthan(*17-19*) and succinoglycan(20), and the triple helix of schizophyllan (*21-23*).

Another point to note here is that the conformation of a linear polymer, or its microscopic structure, determines the macroscopic structure of its solutions in two ways. In one way helical polymers tend to be rod-like, and such rod-like polymers, which have large *q*, form lyotropic liquid crystals at high concentrations. This means that the microscopic property such as *q* controls the liquid crystal formation(*24-27*). In the other way if the polymer has a helical conformation of one screw sense, its solution shows large optical activity and becomes cholesteric when it forms a liquid crystal. Indeed this occurs for polymers with chiral groups and/or nematic liquid crystals doped with chiral solvents(*12-14*), and the cholesteric pitch *P* of the solution changes when its conformation changes with temperature and/or solvent. Theories based on the wormlike chain model have been proposed to explain the liquid crystal formation(*25-27*) and cholesteric order(*28,29*) in stiff polymer solutions.

We have been interested in dilute solution properties of linear polymers, particularly where helical conformations are of importance. These polymers, if chiral, show remarkable optical activity and undergo conformational transitions of their helical conformations with the change in solvent condition.

$$\left[\mathrm{NH{-}\underset{R}{\overset{H}{C}}{-}\overset{O}{\overset{\|}{C}}}\right]_n \qquad \left[\mathrm{\underset{R}{N}{-}\overset{O}{\overset{\|}{C}}}\right]_n \qquad \left[\mathrm{\underset{R2}{\overset{R1}{Si}}}\right]_n$$

1 **2** **3**

Scheme 1. α-Polypeptides (**1**), polyisocyanates (**2**), polysilylenes (**3**)

Such polymers include polypeptides **1**, polyisocyanates **2**, and polysilylenes **3** in addition to triple-helical polysaccharide schizophyllan, where the side chain for **2** is partially or totally chiral, and those for **3**, R1, and R2, one or either of them is chiral.

In addition to the dilute solution properties, helical polymers are usually rod-like and solutions of such polymers at high concentrations form lyotropic liquid crystals, which are cholesteric for polymers with some asymmetric center. Beside α-helical polypeptides(*30-34*), we find liquid crystalline behavior of such polymers as schizophyllan(*35*), polyisocyanates(*12-14,36-39*), and polysilylenes (*40,41*). The present paper reviews these dilute solution and concentrated solution properties of such polymers from experimental and theoretical points of view, referring mainly to our previous and present results. In the following, typical experimental results for polysilylenes(*40-45*), schizophyllan(*28,29,46,47*) and polyisocyanates(*48-52*), along with theoretical analysis are presented. Consult ref. 9-11 for helix-coil transitions in polypeptides.

Theory of Conformational Transitions in Dilute Solution

We consider a polymer chain of N repeat-units, and each unit can be in either of the P (left handed or right handed) and M(right handed or left handed) conformations: P and M can be helical conformations of different senses, or one can be ordered and the other is disordered. Thus the polymer chain consists of an alternating sequence of M-chains and P-chains. The partition function Z_N of this chain is written(*53,54*)

$$Z_N = \mathbf{A}\mathbf{M}^{N-1}\mathbf{B} \qquad (1)$$

Here **M** is the statistical weight matrix defined by

$$\mathbf{M} = \begin{pmatrix} u_{\mathrm{M}} & vu_P \\ vu_{\mathrm{M}} & u_P \end{pmatrix} \qquad (2)$$

where u_{M} and u_{P} are the statistical weights of the units in the right-handed helical conformation and left-handed helical conformation, respectively, and v is the transition probability between the two helices. They are related to the respective free energies dependent on temperature T by

$$u_{\mathrm{M}}(T) = \exp\left[-G_{\mathrm{M}}(T)/RT\right] \qquad u_{\mathrm{P}}(T) = \exp\left[-G_{\mathrm{P}}(T)/RT\right]$$

$$v(T) = \exp(-\Delta G_r / RT) \tag{3}$$

where $G_M(T)$, and $G_P(T)$ are the free energies for the M- and P-units, respectively, and $\Delta G_r(T)$ for the reversal unit. It can be shown that only the difference $2\Delta G_h(T) = G_P(T) - G_M(T)$ is of physical significance, and $\Delta G_r(T)$ is measured from their average. The vectors **A** and **B** specify the situations of two terminal units and are defined by

$$\mathbf{A} = \begin{pmatrix} au_M & u_P \end{pmatrix} \qquad \mathbf{B} = \begin{pmatrix} b \\ 1 \end{pmatrix} \tag{4}$$

Here $a=0$ or $b=0$ specifies that the initial or terminal unit does not take the M-conformation(*53,54*). For example, the terminal units of an α-helical polypeptide are not involved in the helical conformation, which is shown by setting $a=b=0$. The fraction f_M of the M-units in the chain and the number n_r of the alterations of the sequences are obtained by the standard procedure in statistical mechanics as

$$f_M = (1/N)\frac{\partial(\ln Z_N)}{\partial(\ln u_M)} \quad \text{and} \quad n_r = \frac{\partial(\ln Z_N)}{\partial(\ln v)} \tag{5}$$

As shown previously(*53,54*), the transition behavior, namely the f_M-N relation changes with the restriction on the terminal units conformations. Figure 1 shows some results of theoretical calculations for f_M, where the left panel is for the case of $a=b=1$ and the right panel is for the case of $a=b=0$; f_M changes from

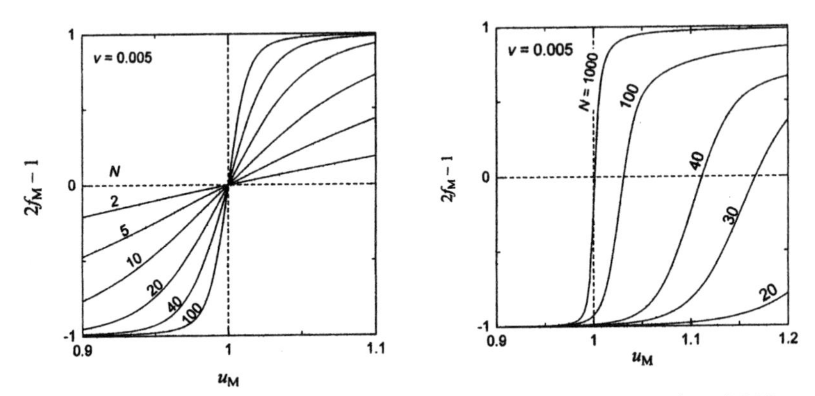

Figure 1. Theoretical dependence of $2f_M$-1 on u_M and N with v=0.005. Left: Case of a=b=1 Right: Case of a=b=0

0 to 1 as $2f_M$-1changes from –1 to +1. They show entirely different u_M dependence with varying N. In the former case, f_M becomes zero at a critical point u_M=u_P irrespective of N, or this behavior is a definite proof for the former situation. In the latter case the critical point (or critical temperature) moves toward larger u_M as N decreases. Lifson et al.(*49*) formulated the helix reversal in chiral polyisocyanates as the former case. However the situation depends on polymer and solvent and the theory does not tell which would be the case for individual polymers. Indeed the latter is the case with helix-coil transitions in polypeptide and order-disorder transition in aqueous schizophyllan, a triple-helical polysaccharide. This choice may be justified because in these polymers the ordered conformations are stabilized by hydrogen bonds formed along the helix axis between neighboring residues, and terminal residues cannot be ordered because they have no such partner on one side. We will show how to solve this problem from experimental data combined with the theory. Experimentally the optical activity is expressed by specific rotation [α] or molar ellipticity [θ], or Kuhn's dissymmetry ratio g_{abs}(=[θ] /ε; ε, the molar absorption coefficient), where they change parallel with each other. When the polymer chain consists of left-handed (M) and right-handed (P) helices only, with the fractions f_M and f_P(=1-f_M), then these parameters measure 2 f_M –1. This formulation is essentially the same as that of Lifson, et al.(*49*) except for the explicit consideration of the terminal conformation. Assuming the left-handed and right-handed helices are symmetric, g_{abs} (or [θ]) can be converted to f_M by g_{abs} (or [θ])/g_M(or $[\theta]_M$) =$2f_M$-1, where g_M(or $[\theta]_M$) is the corresponding value for a perfect M helix.

Experimental Results and Discussion

Polysilylenes are composed of Si-Si main chains with two pendent groups on each Si atom. These main chains tend to be planar because of their σ-conjugation, but to be skewed off planar due to the interactions of pendent groups with themselves and with the main chain, resulting in helical conformations. When these pendent groups are chiral or there exists chiral perturbations, the helical conformations are biased on one sense, and the solution shows large optical activity. This is actually the case with stereo-specifically deuterated poly(hexyl isocyanate)s discovered by Green et al.(*12-14,48-52*). On the other hand Fujiki and collaborators(*15*) have synthesized a number of such polysilylenes showing large optical activity. Among them poly{n-hexyl-[(S)-3-methylpentyl]silylene}(PH3MPS) (semi-flexible, q=6.1 nm in isooctane at 25°C) and poly{[(R)-3,7-dimethyloctyl]-[(S)-3-methylpentyl]silylene}(PRS) (rod-like, q=103 nm in isooctane at 25°C) have been studied in great detail(*41-45*). It was found that their absorption spectra virtually show no transition in spite of a large change of g_{abs} in isooctane. Thus we have considered as in the case of chiral polyisocyanates that these polymers are essentially helical and the change in g_{abs} is due to the change in the proportion of left-handed and right-handed helices. The fraction f_M of M units is calculated from g_{abs} or from [θ] by

$$f_M(T) = (1/2)(g_{abs}/g_M + 1) = (1/2)([\theta]/[\theta]_M + 1) \qquad (7)$$

Here the subscript signifies the corresponding value for the complete M helix. Using an appropriate value for g_M, or $[\theta]_M$, the experiment can be compared with the theory.

Figures 2 and 3 plot Kuhn's dissymmetry ratio g_{abs} for the two polymers PH3MPS and PRS as functions of temperature and chain length N. It is seen that $2f_M$-1 changes with both temperature (left panel) and chain length (right panel). Particularly this remarkable chain length dependence indicates these polymers to be linear cooperative systems. In addition PRS is unique in that it undergoes a

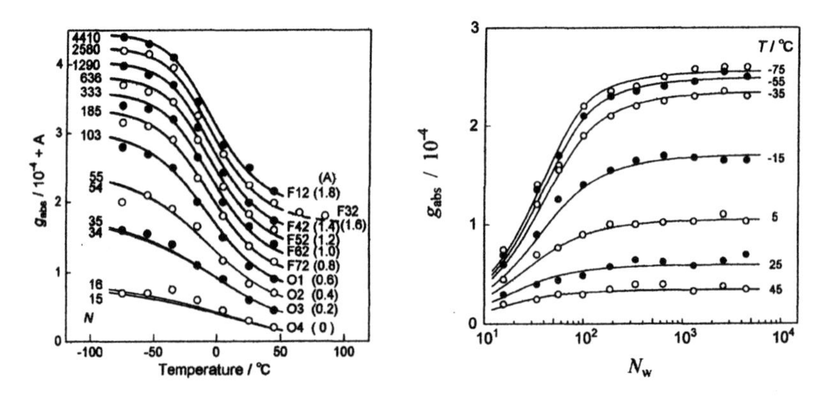

Figure 2. Changes of g_{abs} with temperature and molecular weight for PH3MPS in isooctane)(44). Symbols: experimental data; curves, theoretical values. Synbols and curves(left), shifted upward by the number in the parentheses.

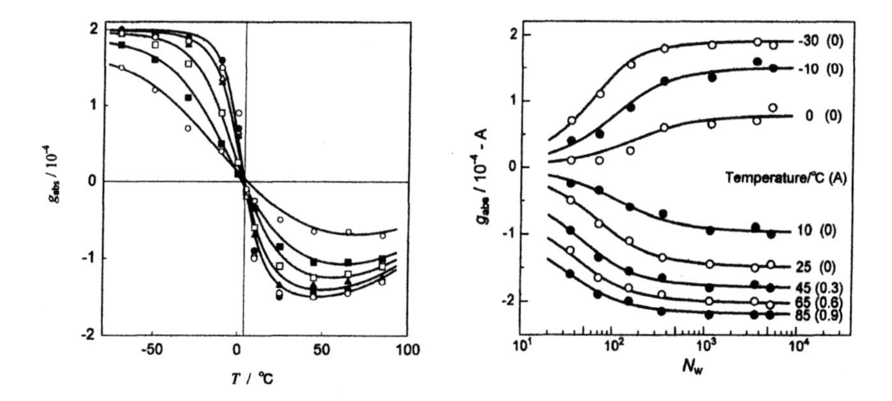

Figure 3. Changes of g_{abs} with temperature and molecular weight for PRS in isooctane)(45). Symbols: experimental data; curves, theoretical values. Curves and symbols, shifted downward by the number in the parentheses (right).

helix sense inversion such that one sense is favored over the other below a critical temperature T_c, but the other is favored above T_c. This helix sense inversion is a general phenomenon shown to occur in other chiral polysilylenes(*15*), copolyisocyanates(*14,37-39*) and polyisocyanates(*55-57*).

Indeed Hino et al.(*57*) have found that poly{3[(S)-2-methylbutoxy]phenyl isocyanate} (P3MBuOPI) has a T_c around –30°C. Recently we confirmed their finding using fractionated samples of different molecular weights(*58*). Figure 4 shows results of circular dichroism $[\theta]_{270}$ for five P3MBuOPI samples with different N_w in tetrahydrofuran (THF), exhibiting T_c at –32°C irrespective of molecular weight (*58*). This behavior is the same as that found with PRS. It must be noted in the intrinsic viscosity data of Figure 5 that despite this helix sense inversion, [η] changes smoothly even around T_c, indicating that their global conformations show no sign of transition. Examined in detail, PRS is stiffest(45), but PH3MPS is relatively flexible(*42,43*), and as T is lowered, the former becomes slightly stiffer whereas the latter appreciably stiffer. On the other hand, P3MBuOPI is much flexible (q=3.$_0$nm) compared with these poly(silylene)s. This flexibility may be ascribed to reduced conjugation in the main chain amide bonds due to conjugation to side chain phenyl groups. Its [η] increases gradually with increasing temperature, and, taken literally, it tends to shrink at lower temperature. Such a trend has never been observed in other systems. The values of q for polysilylenes have been analyzed in terms of the broken helix model(*59,60*).

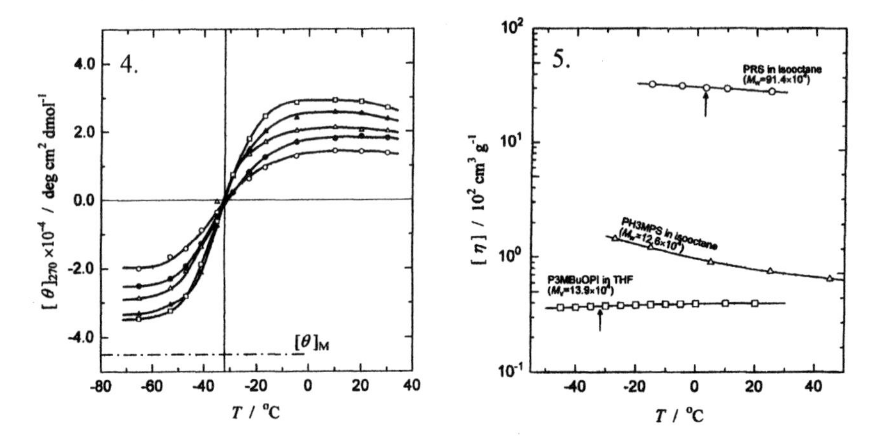

Figure 4. Temperature dependence of $[\theta]_{270}$ for poly{3[(S)-2-methylbutoxy]phenyl isocyanate} (P3MBuOPI) in terahydrofuran. (left)(58). N_w = weight-average degree of polymerization.

Figure 5. Temperature dependence of intrinsic viscosity. (right) PRS in isooctane(45); PH3MPS in isooctane(44); P3MBuOPI in tetrahydrofuran)(58). Allows, locations of the critical temperature T_c.

Theoretical Analysis

Basic considerations. Lifson et al.*(49)* derived a theory of conformational transitions to describe the transition behavior of stereo-specifically deuterated poly(hexyl isocyanate)s [αPdHIC: poly((R)-1-deuterio-n-hexyl isocyanate)*(51)* and βPdHIC : poly((R)-2-deuterio-n-hexyl isocyanate)*(50)*, where it is assumed that terminal units of the polymer chain are allowed to take both left-handed and right-handed helical conformations. This corresponds to the case with a=b=1 mentioned above. Although this assumption has been used also in other analyses*(44,45,50,51)*, it's validity is not obvious experimentally nor theoretically. In the following we show it is indeed valid for the polyisocyanates and PH3MPS.

Data Analysis. The theory describes $2f_M$-1 as a function of N with u_M and v as solvent dependent parameters, whereas the experiment gives $2f_M$-1 as a function of (M=NM_o, M_o=the residue molecular weight) at a specified solvent condition. Thus comparison of these two sets of information will give the values of u_M and v at the specified condition if they are consistent with one another.

Therefore focusing on the dependence of g_{abs} or [θ] on molecular weight, these data are analyzed in terms of the helix reversal model theory, which is specified by setting a=b=1 and u_M=1/u_P. Actually only the ratio u_M/u_P is of physical significant. Thus f_M at a fixed T is a function of N, with u_M/u_P =exp(-2$\Delta G_h/RT$) and v as adjustable parameters dependent on the solvent condition employed. The helix sense reversal of PRS is explained in terms of a double-well potential for the rotation about Si-Si bonds, which give rise to the difference in energy and entropy contributions between the left-handed and right-handed helices, such that the sum of these contributions (ΔG_h) vanishes at T_c(*45*). Solid curves for g_{abs} vs N_w in Figures 2 and 3 represent the theoretical values with the optimal values for u_M and v(*50*), which follow the data points precisely. This validates the theory and provides the parameters without additional assumption. Copolymers can be treated by an approximate theory more conveniently,(53-55) although detailed simulations can be made based on the matrix method (*53,54,64*).

Figure 6 shows $[\alpha]_{300}$ for βPdHIC in 1-chlorobutane at 0°C(*59*). Here the solid curve represents the theoretical values obtained by fitting the data to the theory with a=b=1 and follows the data points (symbols) closely. The same data are compared with the theory with a=b=0 for various values of v (dashed lines). However 2f_M-1 becomes negative at low N for any value of v, indicating this condition to be invalid. The same is true for PH3MPS, and indeed no combination of u_M and v could fit the data. Thus this is also a type of polymer with a=b=1. The importance of the conformation of the terminal residue has been exemplified theoretically(*53,54*) with data for achiral polyisocynanates obtained with chiral initiators (*61,62*).

The main chains of these polymers consist of alternating sequences of the left-handed and right-handed helices, and the average sequence length $< l_v >$ is determined by the combination of the parameters u_M, v and N. It has been shown that $< l_v >$ for infinite N is about 37nm (184 Si atoms) for PRS at the critical temperature of 3°C and about 7nm (36 Si atoms) for PH3MPS. On the other hand, it is about 77nm (444 units=888 atoms) for βPdHIC and 52 nm (300 units) for (P3MBuOPI) in terahydrofuran. This large difference in $< l_v >$ between the two types of polymer is due to the difficulty in helix reversal, i.e, the difference in ΔG_r. In polyisocyanates a helix reversal consists of a few units, which contain a cis-trans conformation change of an amide bond, costing a high energy(*63*). It should be remarked that all the three polymers, which undergo the helix sense reversal, show no sign of global conformation transition at T_c. It is seen in Figure 4 that $[\theta]_{270}$ shows a remarkable N dependence at a fixed temperature. This similar behavior has been found on chiral polyisocyanates and polysilylenes. The behavior of P3MBuOPI is similar to that of βPdHIC shown in Figure 6 and is more remarkable than the polysilylenes, which is ascribed to large ΔG_r for polyisocyanates. Although it does now show helix sense reversal below 20°C,

$2\Delta G_h$ in dichloromethane suggest T_c would be around 50°C. Molecular characterizations such as UV absoption spectra, $2\Delta G_h$ and ΔG_r, and their correlations to the global conformation (q) of helical polysilylenes are theoretically analyzed elsewhere(*16*,60) in connection with those of poly(alkyl isocyanate)s

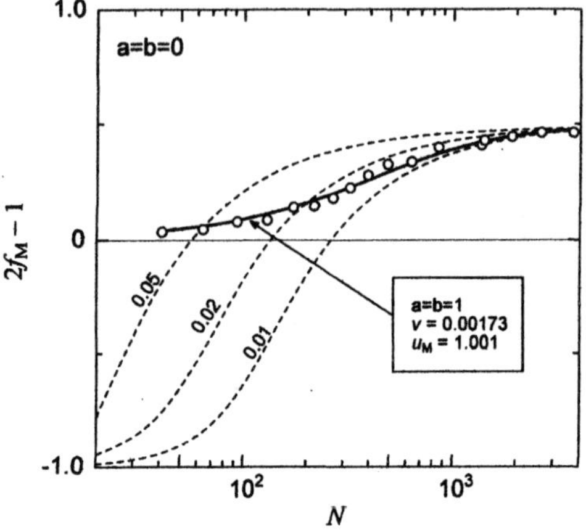

Figure 6. $2f_M$*-1 obtained from optical rotation of poly{(R)-2-deuterio-n-hexyl isocyanate} in 1-chlorobutane.*

Symbols, experimental data (50), solid curve, theoretical values with u_M*=1.01 and v= 0.00174 and a=b=*1, *(50) dashed curves, theoretical values with v=0.05 and* u_M*=1.0282, 0.02 and 1.0112, and 0.01 and 10058.*

Microscopic-Macroscopic Structure Correlation

Schizophyllan is a β-1,3-D-glucan with one glucose unit as the side chain per three main-chain glucose units directed outward from the helix core(*21-23*). At low temperature an ordered structure is formed of the side chains with water molecules in the vicinity, which is disordered as the temperature is raised. (*46,47,66-73*) This is a typical linear cooperative transition as found by heat capacity in the left panel of Figure 7.(*71,73*) The triple-helix is rigid and intact in water and forms a cholesteric liquid crystal, whose pitch P or cholesteric wave number q_c ($q_c=2\pi/P$) changes with the transition as seen in the right panel of the same figure (*47*). This change in q_c with T and concentration(*28,29,46,47*) has been explained by the theory(*28,29,47*) with chiral repulsive and attractive interactions among the helices. It has been shown that q_c changes with temperature and concentration without conformational transition and even changes its sign for polypeptide solutuions(*32-34*). Such changes are also well explained by the theory as the result of the cooperation of the repulsive and attractive interactions amplified by the concentration factors(*28,29,47*).

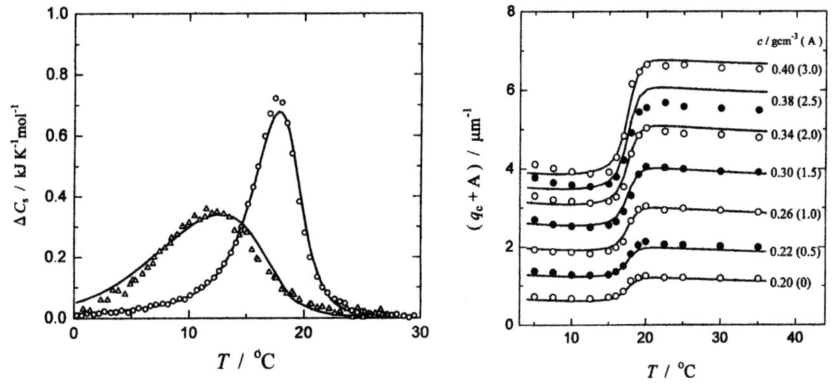

Figure 7. *Order-disorder transition in aqueous solutions of schizophyllan. Left: Heat capacity data for two samples of different molecular weights(M_w =97,800, 184,000 (47,71,73). Right: Temperature dependence of cholesteric wavelength for* D_2O *solutions of schizophyllan KR-1A(M_w=237,000) at the indicated concentrations. Solid curves, theoretical values. In the right panel, the data points and theoretical curves are shifted vertically by the values in the parentheses for viewing clarity(47).*

Solutions of such stiff polysilylenes as poly[n-hexyl-(S)-(2-metylbutyl)silylene] and poly[n-decyl-(2-methylpropyl)silylene](*15,40,41*) have been found to become liquid crystalline at high concentrations. It was therefore expected that a solution of such a polymer with a chiral dope would become cholesteric as found with polyisocyanate solutions and if the dope undergoes a helix sense reversal, the pitch of the doped solution would change its sign. To check this possibility, the cholesteric pitch P of a nematic solution of □poly[n-decyl-(2-methylpropyl)silylene] □ doped with PRS(RS-6) was measured at various temperatures(*74*). Figure 8 shows the temperature dependence q_c of the solution. It is seen that q_c decreases rapidly with lowering temperature and tends to vanish, or P tends to diverge, at lower temperature. This change in P parallels that of g_{abs} of a PRS sample at infinite dilution shown in the insert, although the reversal of sense of P could not be observed. Thus we conclude that the microscopic structure change as revealed with g_{abs} induces the change in cholesteric structure. Similar observations have been made by Green's group(*14,37-39*) on co-polyisocyanates covering a wide range of T_c.

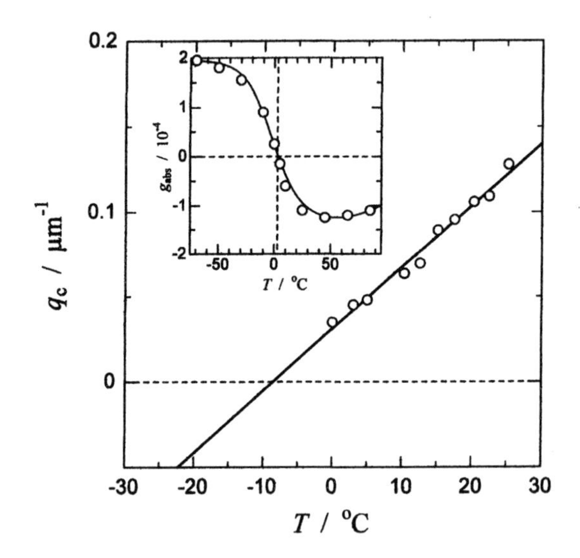

Figure 8. *Temperature dependence of cholesteric wavelength q_c of an isooctane solution of poly[n-decyl-(2-methylpropyl)silylene] (ZID2XA, M_w=49,000) doped with PRS(RS-6, M_w =39, 000(74). The insert shows the curve of g_{abs} for the doped PRS in dilute solution(45).* c=*0.3395 g/cm^3;* *ZID2XA: RS -6=0.1937:0.0168(74).*

So far helix-forming polymers are shown to be generally semiflexible and locally of helical conformations. The imbalance in the populations of left-handed and right-handed helical conformations gives rise to remarkable optical activity and induces cholesteric order in nematic liquid crystals. All these results are explained consistently by the general framework of theories based on linear Ising model of chain conformations(*53,54*) for a variety of polymers including a-helical polypeptides, a triple-helical polysaccharide schizophyllan, chiral polysilylenes, and chiral polyisocyanates(*10,11,49-54,58,59,66,67,70-73*) and the theory of cholesteric liquid formation considering chiral repulsive and attractive interaction among polymer chains(*28,47,53,54*). With these types of theoretical analyses, further prediction may be facilitated with a fewer data.

Acknowledgement. A.T. thanks Yamashita Sekkei Inc. for the Chair-Professorship at Ritsumeikan University and the Chancellor Dr. Hachiro Kawamoto of Ritsumeikan University for his encouragement.

References and Notes

1. Kratky, O.; Porod, G. *Recl. Trav. Chim. Pays-Bas* **1949,** *68,* 1106.
2. Murakami, H.; Norisuye, T.: Fujita, H. *Macromolecules* **1980,** *13,* 345-352.
3. Kuwata, M; Norisuye, T.: Fujita, H. *Macromolecules* **1984,** *17,* 2731-2734.
4. Norisuye, T.: Fujita, H. *Polym. J.* **1982,** *14,* 143.
5. Norisuye, T. *Prog. Polym. Sci.* **1993,** *18,* 543-584.
6. Yamakawa, H. *Helical wormlike chains in polymer solutions.* Springer Berlin, **1997.**
7. Doty, P.; Holtzer, A.M.; Bradbury, J.H.; Blout, E.R. *J. Am. Chem. Soc.* **1954,** *76,* 4493.
8. Bamford, C.H.; Elliott, A.; Hanby, W.E. *Synthetic Polypeptides,* Academic Press, NY, **1956.**
9. Urnes, P.; Doty, P. *Adv. Protein Chem.* **1961,** *16,* 408.
10. Teramoto, A.; Fujita, H. *Adv. Polym. Sci.* **1975,** *18,* 65-149.
11. Teramoto, A.; Fujita, H. *J. Macromol. Sci.-Rev. Macromol. Chem.* **1976,** *C15,* 165-278.
12. Green, M. M.; Peterson, N. C.; Sato, T.; Teramoto, A.; Cook, R.; Lifson, S. *Science* **1995,** *268,* 1860-1866.
13. Green, M. M.; Park, Ji-Wong; Sato, T.; Teramoto, A.; Lifson, S.; Selinger, R. L. B.; Selinger, J. V. *Angew. Chem. Int. Ed.* **1999,** *38,* 3138-3154.
14. Cheon, K.S.; Selinger, J.V.; Green, M.M. *Angew. Chem, Int. Ed.* **2000,** *39,* 1482-1485.
15. Fujiki, M.; Koe, J.R.; Terao, K.; Sato, T.; Teramoto, A.; Watanabe, J. *Polym. J.* **2003,** *35,* 297-344; Fujiki, M. *J. Am. Chem. Soc.* **1996,** *118,* 7424-7425.
16. Sato, T.; Terao, K.; Teramoto, A. Fujiki, M. *Polymer* **2003,** *44,* 5477-5495.
17. Sato, T.; Norisuye, T.; Fujita, H. *Polym. J.* **1984,** *16,* 341-350.
18. Sato, T.; Kojima, S.; Norisuye, T.; Fujita, H. *Polym. J.* **1984,** *16,* 423-429.
19. Sato, T.; Norisuye, T.; Fujita, H. *Macromolecules* **1984,** *17,* 2696-2700.
20. Kido, S.; Nakanishi, T.; Norisuye, T.; Kaneda, I.; Yanaki, T. *Biomacromolecules* **2001,** *2,* 952-957.
21. Norisuye, T.; Yanaki, T.; Fujita, H. *J. Polym. Sci., Polym. Phys. Ed.* **1980,** *18,* 547-558.
22. Yanaki, T.; Norisuye, T.; Fujita, H. *Macromolecules* **1980,** *13,* 1462-1466.

23. Norisuye, T. *Macromol. Symp.* **1995**, *99*, 31-42.
24. Onsager, L. *Ann. N.Y. Acad. Sci.* **1949**, *51*, 627-630.
25. Odijk, T. *Macromolecules* **1986**, *19*, 2313.
26. Sato, T.; Teramoto, A. In *Ordering in Macromolecular Systems*; A. Teramoto, M. Kobayashi and T, Norisuye, Eds., Springer-Verlag, Berlin Heidelberg, **1994**, pp. 155-169.
27. Sato, T.; Teramoto, A. *Adv. Polym. Sci.* **1996**, *126*, 85-216.
28. Sato, T.; Nakamura, J.; Teramoto, A.; Green, M.M. *Macromolecules* **1998**, *31*, 1398-1405.
29. Green, M.M.; Zanella, A.; Gu, H,; Sato, T.; Gottarelli, G.; Jha, S. K.; Spada, G. P.; Schoevaars, A. M.; Feringa, B.; Teramoto, A. *J. Am. Chem. Soc.* **1998**, *120*, 9810-9817.
30. Robinson, C.; Ward, J. C; Beevers, R.B. *Discuss. Faraday Soc.* **1958**, *25*, 29.
31. Robinson, C. *Tetrahadron* **1961**, *13*, 219.
32. Toriumi, H.; Yahagi, K. Uematsu, I. *J. Polym. Sci., Polym. Phys. Ed.* **1980**, *19*, 1167-1169.
33. Toriumi, H.; Minakuchi, S.; Uematsu, I. *Mol. Cryst. Liq. Cryst.* **1983**, *94*, 267-284.
34. Uematsu, I; Uematsu, Y. *Adv. Polym. Sci.* **1985**, *41*, 35.
35. Van, K.; Norisuye, T. Teramoto, A. *Mol. Cryst. Liq. Cryst.* **1981**, *78*, 123-134.
36. Gu, Hong, Ph.D. thesis, Osaka University, Osaka, Japan, **1997**.
37. Li, J.; Schuster, G.B.; Cheon, K. S.; Green, M. M.; Selinger, J. V. *J. Am. Chem. Soc.* **2000**, *122*, 2603-2612.
38. Green, M.M.; Cheon, K.S.; Yang, S.Y.; Park, J.W.; Liu, W.H.; Swansburg, S. *Account Chem. Res.* **2001**, *34*, 672.
39. Tang, K.; Green, M.M.; Cheon, K.S.; Selinger, J.V.; Garetz, A. *J. Am. Chem.* Soc. **2003**, 125, 7313.
40. Watanabe, J.; Kamee, H.; Fujiki, M. *Polym. J.* **2001**, *33*, 495-.
41. Natsume, T.; Wu, L.; Sato, T.; Terao, K.; Teramoto, A.; Fujiki, M. *Macromolecules* **2001**, *34*, 7899-7904.
42. Terao, K.; Terao, Y.; Teramoto, A.; Nakamura, N.; Terakawa, I.; Sato, T.; Fujiki, M. *Macromolecules* **2001**, *34*, 2682-2685.
43. Terao, K.; Terao, Y.; Teramoto, A.; Nakamura, N.; Fujiki, M.; Sato, T. *Macromolecules* **2001**, *34*, 4519-4525.
44. Terao, K.; Terao, Y.; Teramoto, A.; Nakamura, N.; Fujiki, M.; Sato, T. *Macromolecules* **2001**, *34*, 6519-6525.
45. Teramoto, A.; Terao, K.; Terao, Y.; Nakamura, N.; Sato, T.; Fujiki, M. *J. Am. Chem. Soc.* **2001**, *123*, 12303-12310.
46. Teramoto, A.; Yoshiba, K.; Nakamura, N.; Nakamura, J.; Sato, T. *Mol. Cryst. Liq. Cryst.* **2001**, *368*, 373-380.

47. Yoshiba, K.; Teramoto, A.; Nakamura, N.; Sato, T. *Macromolecules* **2003**, *36*, 2108-2113.
48. Green, M.M.; Andreola, C.; Munos, B.; Reidy, M. P. *J. Am. Chem. Soc.* **1988**, *110*, 4063-4064.
49. Lifson, S.; Andreola, C.; Peterson, N. C.; Green, M. M. *J. Am. Chem. Soc.* **1989**, *111*, 8850-8858.
50. Gu, H.; Nakamura, Y.; Sato, T.; Teramoto, A.; Green, M. M.; Andreola, C.; Peterson, N. C.; Lifson, S. *Macromolecules* **1995**, *28*, 1016-1024.
51. Okamoto, N.; Mukaida, F.; Gu, H.; Nakamura, Y.; Sato, T.; Teramoto, A.; Green, M. M.; Andreola, C.; Peterson, N. C.; Lifson, S. *Macromolecules* **1996**, *29*, 2878-2884.
52. Green, M. M. in *Circular Dichroism-Principles and Applications*, N. Berova, K. Nakanishi and R. W. Woody, eds. **2000**, Wiley-VCH, New York, Chapter 17.
53. Gu, H.; Sato, T.; Teramoto, A.; Varichon, L.; Green, M.M. *Polym. J.* **1997**, *29*, 77-84.
54. Teramoto, A. *Prog. Polym. Sci.* **2001**, *26*, 667-720. Here eq (2.10) should be read: $Z_N = \sum_{j=1}^{2} \mathbf{AU}(\lambda_j)\lambda_j^{(N-1)} \mathbf{V}(\lambda_j)\mathbf{B},$
55. Maeda, K.; Okamoto, Y. *Macromolecules* **1998**, *31*, 5164-5166.
56. Maeda, K.; Okamoto, Y. *Macromolecules* **1999**, *32*, 974-980.
57. Hino, K.; Maeda, K.; Okamoto, Y. *J. Phys. Org. Chem.* **2000**, *13*, 361-367.
58. Yoshiba, K.; Teramoto, A.; Sato, T.; Maeda, K.; Okamoto, Y., in preparation for *Macromolecules*.
59. Gu, H.; Nakamura, Y.; Sato, T.; Teramoto, A.; Green, M.M.; Jha, S. K.; Reidy, M.P. *Macromolecules* **1998**, *31*, 6362-6368.
60. Sato, T.; Terao, K.; Teramoto, A. Fujiki, M. *Macromolecules* **2002**, *35*, 5355-5357.
61. Okamoto, Y.; Matsuda, M.; Nakano, T.; Yashima, E. *Polym. J.* **1993**, *25*, 391.
62. Maeda, K.; Matsuda, M.; Nakano, T.; Okamoto, Y. *Polym. J.* **1995**, *27*, 141.
63. Lifson, S.; Felder, C.E.; Green, M.M. *Macromolecules* **1992**, *25*, 4142.
64. Sato, T.; Terao, K.; Teramoto, A. Fujiki, M. *Macromolecules* **2002**, *35*, 2141-2148.
65. Mansfield, M.L. *Macromolecules* **1986**, *19*, 854-859.
66. Itou, T.; Teramoto, A.; Matsuo, T.; Suga, H. *Macromolecules* **1986**, *19*, 1234-1240.
67. Itou, T.; Teramoto, A.; Matsuo, T.; Suga, H. *Carbohydr. Res.* **1987**, *160*, 243-257.
68. Kitamura, S.; Kuge, T. *Biopolymers* **1989**, *28*, 639-654.

69. Kitamura, S.; Ozawa, M. Tokioka, H.; Hara, C.; Ukai, S.; Kuge, T. *Thermochim. Acta*, **1990**, *163*, 89-96.
70. Hayashi, Y.; Shinyashiki, N.; Yagihara, S.; Yoshiba, K.; Teramoto, A.; Nakamura, N.; Miyazaki, Y.; Sorai, M.; Wang, Qi *Biopolymrs* **2002**, *63*, 21-31.
71. Yoshiba, K.; Ishino, T.; Teramoto, A.; Nakamura, N.; Miyazaki, Y.; Sorai, M.; Wang, Qi; Hayashi, Y.; Shinyashiki, N.; Yagihara, S. *Biopolymrs* **2002**, *63*, 370-381.
72. Yoshiba, K.; Teramoto, A.; Nakamura, N.; Kikuchi, K.; Miyazaki, Y.; Sorai, M. *Biomacromolecules* **2003**, *4*, 1348-1356.
73. Yoshiba, Kazuto, Ph.D. thesis, Ritsumeikan University, Kusatsu, Japan, **2003**.
74. Yoshiba, K.; Matsushima, S.; Teramoto, A.; Sato, T., to be submitted to *Macromolecules*.

Chapter 6

Molecular Orbital Theory Predictions for Photophysical Properties of Polymers: Toward Computer-Aided Design of New Luminescent Materials

Zoltán A. Fekete

Department of Physical Chemistry, University of Szeged, H–6701 Szeged, P.O. Box 105, Hungary (email: Capri3@fekete.mailshell.com, fax: (36–62) 544652, telephone/voicemail (1–781) 6235997

Molecular orbital (MO) calculations at the semiempirical quantum chemical level have been utilized to study a number of chromophores and polymer building block model molecules. Various conjugated systems – including poly(phenylene-vinylene) derivatives - are considered for potentially luminescent materials targeted across the visible spectral range. After mapping out the relevant geometries at the ground and corresponding excited states, optical transition (absorption as well as emission) energies and oscillator strengths are estimated. Systematic correlation of these calculated properties with measured ones in known materials may guide designing new advanced polymers.

[1] Dedicated to Professor Frank E. Karasz for the occasion of his 70th birthday

Introduction

Polymers for optical-electronic applications *(1-3)* — such as light-emitting diode (PLED) substrates *(4-7)* or polymer lasers *(8, 9)*— are among the most intensively studied new advanced materials. In the past, discovering of promising new candidate structures (like that of F.E. Karasz' classic segmented blue emitting polymers *(10, 11)* whose backbone is shown on Scheme **1a**), and then establishing structure-activity relations as they are systematically varied, required a lot of ingenuity as well as difficult trial-and-error search. Modern theoretical chemical calculation methods have now reached the stage where they can be used on systems of size

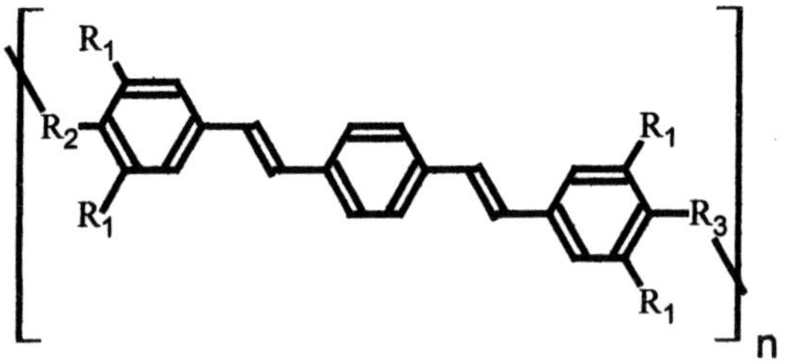

1a: $R_1 = OCH_3$, $R_2 = O—$, $R_3 = O(CH_2)_8—$

1b: $R_1 = H$, $R_2 = —$, $R_3 = CH{=}CH—$

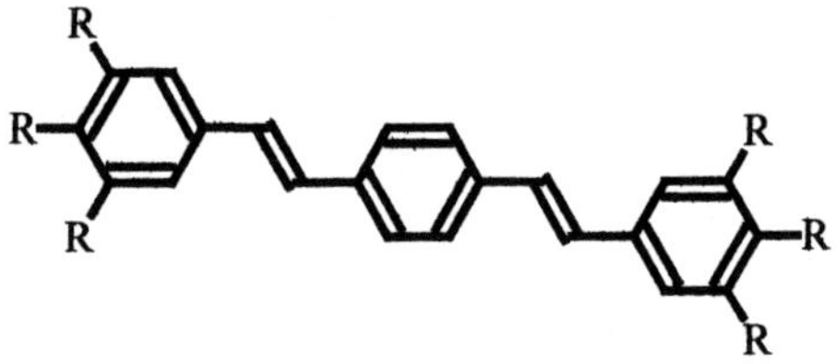

1c: R = H

1d: $R = OCH_3$

Scheme 1. Structures of PPV-related polymers and PPV3 model compounds.

relevant to actual applications (i.e. hundreds to thousands of atoms). In this contribution I demonstrate how to utilize these computational techniques in describing photophysical behavior of polymers, in order to aid the design of materials with custom-tailored electronic-optical properties. A few select example molecules are presented, chosen mostly to relate to Karasz's seminal works in the field of poly(phenylene-vinylene) (PPV, Scheme **1b**) based PLED materials *(3)*. Results on these small models (the trimer PPV3 **1c**, and its bis(trimethoxy) substituted analog **1d**) reveal the molecular orbital basis of photophysical behavior in these systems, as will be shown below. In the future we plan larger scale computations that include larger segments containing several of these chromophores, to model interactions in the solid polymer phase between them as well as with the embedding matrix. Calculations of this type may clarify the nature of the photoexcited states and the efficiency of their generation in PLEDs. The discussion here is limited to semiempirical methods (similarly to most reports that have appeared to date *(12)*) which require much less computational effort than do *ab initio* methods, although with the increase of available computing power the later are also beginning to be applicable to medium-sized systems *(13)*.

Computational methods

A computationally efficient combination of various semiempirical quantum chemical methods was used *(12, 14)*, as they are implemented in the MOPAC *(15)* and Arguslab *(16-18)* program packages. All the software runs on commodity personal computers (with AMD Athlon or Intel Pentium class processors). The polymer molecules were modeled with oligomers of sufficient length. For this investigation typically 3-6 repeat units are included in most cases. It has been demonstrated experimentally (e.g. by distyrylbenzenes in relation to PPV-based polymers *(6, 19)*) that in these systems a trimer often exhibits some of the essential features of the extended conjugated polymer already.

The initial geometries for the ground state (S_0) and then for the excited states (singlet S_1 and triplet T_1) were optimized by the semiempirical AM1 method *(14-16, 20)*. All degrees of freedom were fully relaxed with no symmetry or other geometrical constraints applied. The electron spin was restricted according to the state multiplicity (RHF singlet and triplet calculations *(21)*). In the excited states electron correlation was taken into account by using single-excitation configuration interaction for two electrons in two orbitals (the MECI option in the MOPAC program) in these runs. The potential energy minima located were verified by calculating the Hessian force matrix and checking that no imaginary eigenvalues occurred.

At the appropriate points on the potential energy hypersurfaces obtained, the bandgap is estimated as the HOMO-LUMO energy difference, while optical absorption and emission energies are approximated from the distance between the curves forming the Frank-Condon envelope, as illustrated schematically on Figure 1. Spectral intensity is characterized by transition moments and oscillator strengths for the radiative transitions (${}^{0}S_0 \rightarrow {}^{FC}S_1$, ${}^{0}S_1 \rightarrow {}^{FC}S_0$), calculated from a configuration interaction computation (using Slater-type Gaussian orbital base). The INDO1/S-CI Hamiltonian was used in the calculations on electronic transitions *(16-18, 22-25)*. The Einstein-coefficient B_{21} for stimulated emission was also determined, which is useful for judging feasibility of application as a laser substrate. Radiative lifetimes were calculated as well.

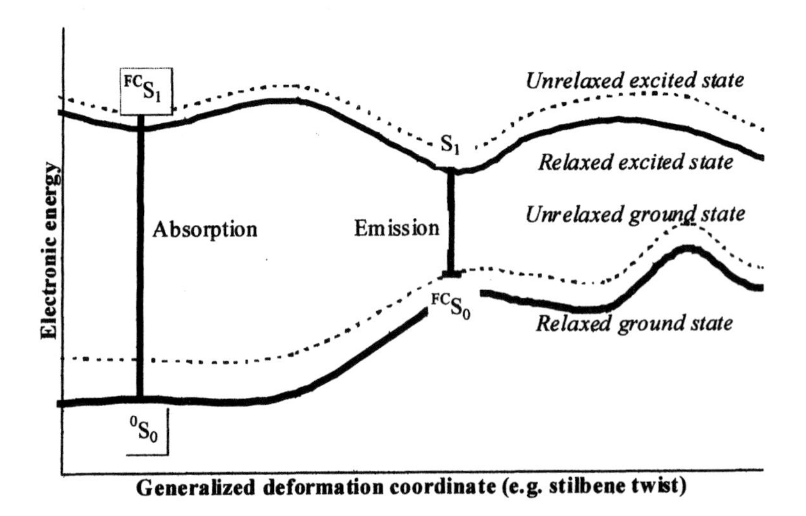

Figure 1. Schematic plot of a cut through potential energy surfaces.

In addition to the above-mentioned static quantities, the dynamics of the evolution of states was also investigated. The dynamic reaction coordinate (DRC) capability of MOPAC, in conjunction with configuration interaction, was used to trace the evolving excited state as it starts from the geometry corresponding to the energy minimum in the ground state. Relaxation following the Frank-Condon vertical transition is simulated in this semiempirical treatment.

Results and discussion

Excited-state molecular dynamics

To illustrate the DRC trajectories calculated, Figures 2a-c show potential energy plots of the S_1 state of 1,4-divinyl-benzene (a model minimal core unit of phenylene-vinylene polymers), evolving from the ground state upon photoexcitation (${}^{0}S_0 \rightarrow {}^{FC}S_1$.transition). The zero of the energy scale applied on this figure is set to the initial point reached by the vertical Frank-Condon transition. Different timescale events can be observed over the course of this 4 picosecond simulation. The fastest changes are demonstrated on Figure 2a, with the first 40 fs shown with 0.1 fs calculation timesteps; similar fast modes superimposed on the slower ones are seen at later times as well. The slower changes can be discerned from Figures 2b and 2c, where averages of 20 points (binned over 2 fs) and of 400 points (binned over 40 fs) respectively are plotted to smooth out the rapid oscillations. Since the calculation yields not only the energy, but the coordinates, MO contributions and partial charges of all atoms as well, much theoretical information can be gathered about the relaxing excited state on the femtosecond-picosecond timescale relevant for photophysical processes. For example, the behavior of "breathers" *(26)* (multi-quanta vibronic states) can be analyzed in extended π–conjugated systems such as PPV-based polymers. Figure 3 illustrates tracing the skeletal motions by displaying the temporal evolution of the C_1-C_2 and C_2-C_3 distances of the phenylene ring in 1,4-divinyl-benzene. The vibronic relaxation of the ground state after $S_1 \rightarrow S_0$ luminescent transition can be studied in the same way.

Conformational mapping of PPV-based polymers

Although their continuous conjugation along the polymer backbone makes these systems stiff, the rather low barrier to rotation around the vinyl-phenyl linkage lends them considerable freedom of movement. The actual extent of this freedom depends on a delicate balance between the tendency of the π–bonding atoms to align co–planarly and the steric hindrance of groups bound to them. It is of interest to map out this conformational behavior theoretically. Semiempirical quantum chemical methods provide a computationally feasible way of calculating a large number of points on the potential energy hypersurface. As a simple example, data on phenylene ring rotation in a methoxy-substituted distyrylbenzene chromophore (Scheme **1d**) is presented. This investigation concerned semi-rigid rotation: the terminal phenylene units were held fixed (as if embedded in a very viscous medium), aligned co–planarly with each other; the middle ring was set at certain angles with respect to them, while geometry of the connecting atoms was optimized with these constraints. The ground and excited state potential energy curves so obtained are shown on Figure 4.

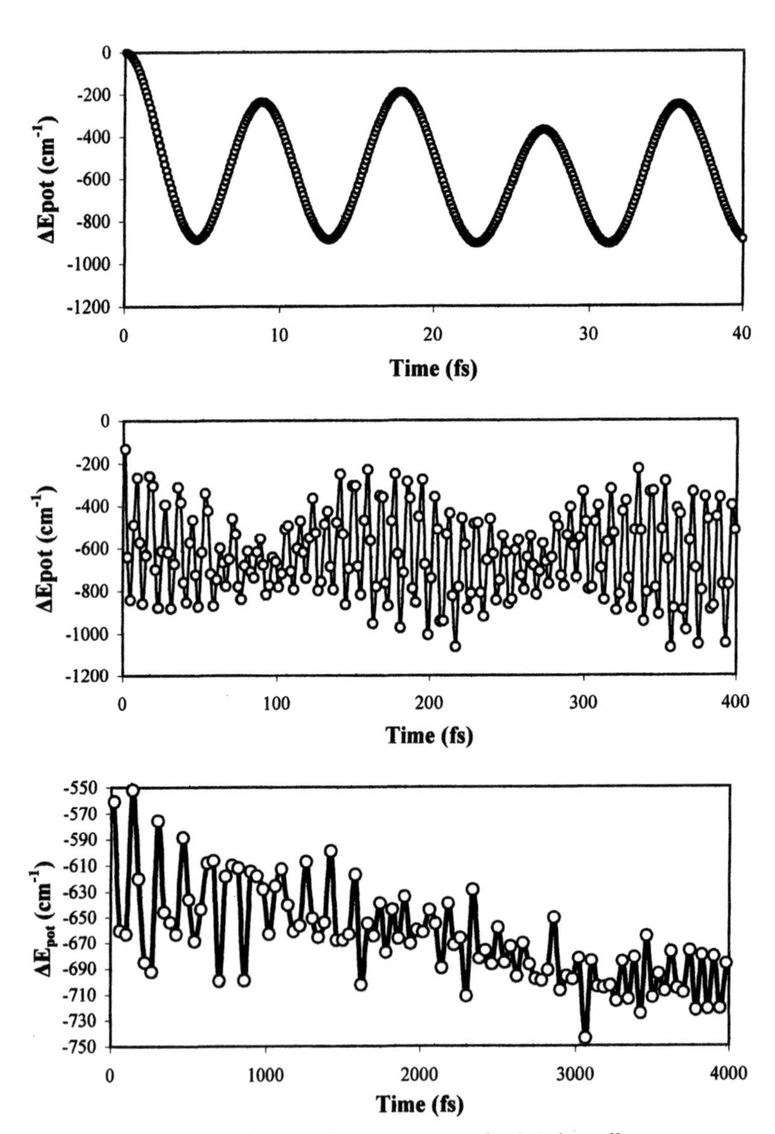

Figure 2a-c. DRC potential energy curves for 1,4-divinylbenzene.

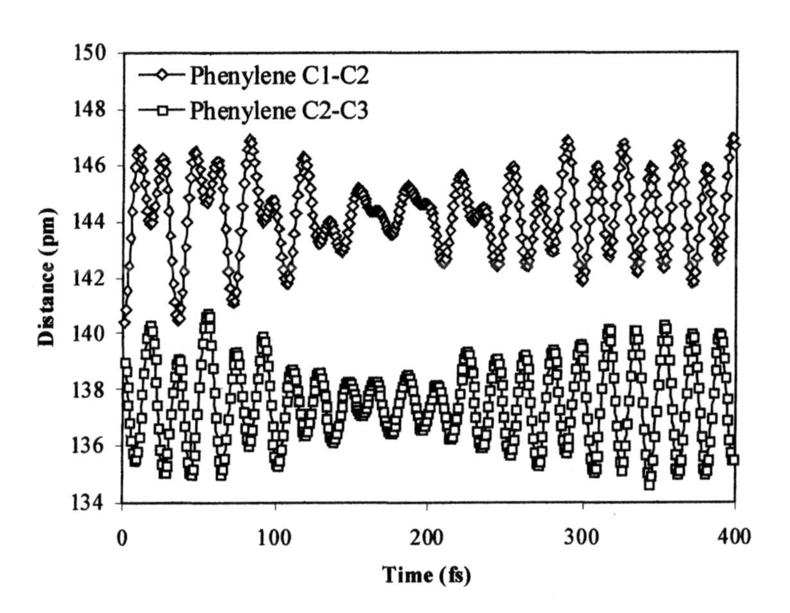

Figure 3. DRC potential energy curves for 1,4-divinylbenzene.

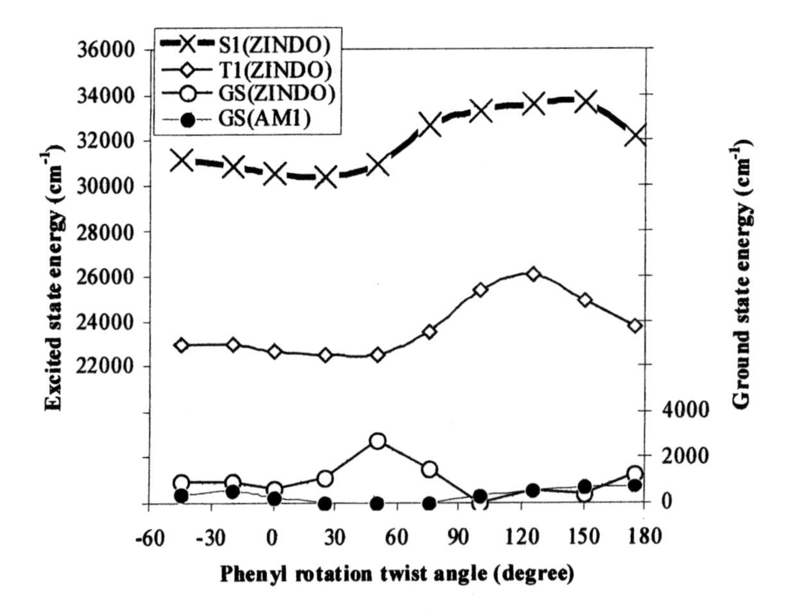

Figure 4. Potential energy curves for phenylene ring rotation in ***1d***.

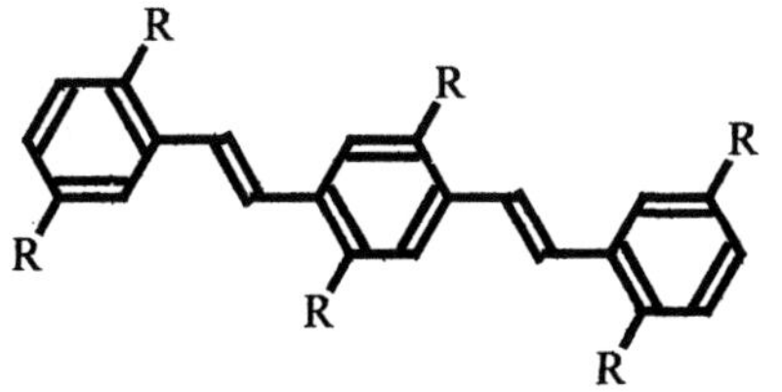

2a: R = H
2b: R = CH_3
2c: R = NH_2
2d: R = OH
2e: R = F
2f: R = Cl
2g: R = CN
2h: R = OCH_3

Scheme 2. Structures of substituted PPV3 model compounds.

Table I. Calculated spectral characteristics of substituted PPV3 molecules

Molecule	λ_a *(nm)*	f_a	λ_e *(nm)*	f_e	τ_r *(ns)*	B_{12} $(10^{21}\ m^3 J^{-1})$
2a	334	1.97	379	2.09	1.03	3.18
2b	-27	0.70	3	2.04	1.07	3.12
2c	70	1.64	25	1.64	1.50	2.65
2d	61	1.84	16	1.84	1.27	2.92
2e	-2	1.82	-1	1.95	1.10	2.96
2f	52	1.99	6	1.99	1.12	3.07
2g	77	1.54	31	1.54	1.64	2.53
2h	59	1.82	13	1.82	1.27	2.86

NOTE: transition wavelengths for **2b-h** are shown relative to those of **2a**

Substituent effects on PPV3-based blue-emitting chromophores

A series of 2,5-disubstituted phenylene-vinylene oligomers (Scheme 2) were investigated to reveal systematic trends with varying the substituents. Calculated properties are listed in Table I.

Systematic changes of the absorption and emission wavelengths (λ_a and λ_e) can be seen with varying the substituents, even though the absolute values calculated (shown in the first line of Table I) appear too low.

Something old, something new – something borrowed, something blue-emitting?

An obvious advantage of theoretical calculations is that they can be carried out for compounds not yet in existence. Predicted properties can guide selection

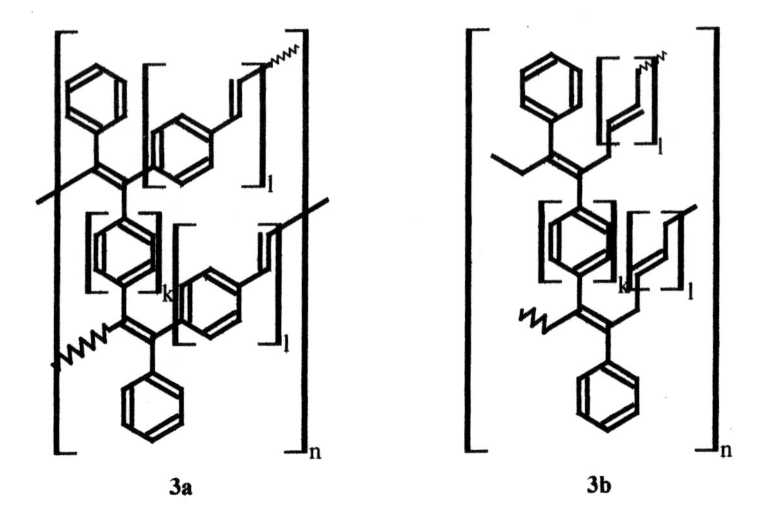

Scheme 3. Structures of crosslinked PV-related polymers and model compounds.

of desirable target molecules to be synthesized. Based on these calculations I propose one interesting new type of polymer[2] (Scheme **3a**), which borrows building blocks from traditional PPV variants, but connects them together in a novel way that may potentially provide for useful luminescent behavior. The chromophore would be the familiar PPV3 trimer segment, but built perpendicularly to the double-stringed backbone. This orientation inhibits energy transfer (since the dipole-dipole overlap is near zero), so that the short segment may be luminescent emitter even in the presence of a longer conjugated segment since it is orthogonal to it. At the same time the uninterrupted conjugation along the backbone provides better carrier mobility than that possible with molecules designed with flexible saturated alkyl spacers. Also the rigid backbone prevents aggregation of the chromophores either with each other or with the backbone. In this respect it resembles ladder-type polymers, but it is more flexible. There is also similarity with the idea of using pendant group chromophores *(7, 27-29)*, but in this case they are held apart at well defined orientations. A variant on this theme is **3b,** which has the simpler polyacetylene backbone instead of a PPV-like one. Calculations are currently in progress for model compounds **4a** and **4b**. As illustrated on Figure 4, displaying the angle between the transition dipole of the intended chromophore and the one polarized along the backbone, there is indeed significant oscillator strength at near-perpendicular direction.

---

[2] Actually producing this kind of polymer would likely be difficult, but what would Frank's next seventy year without challenges?

4a

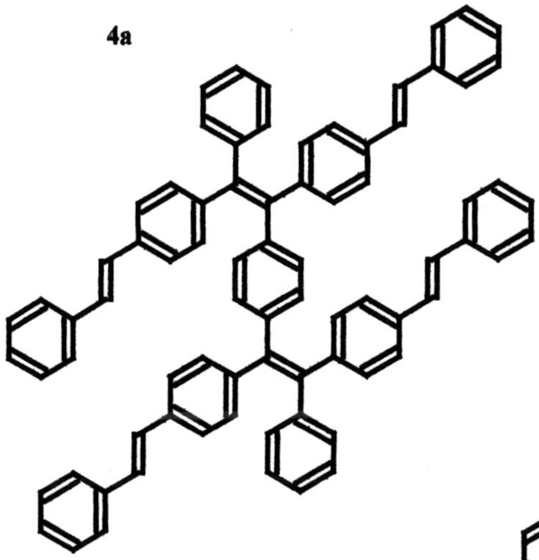

4b

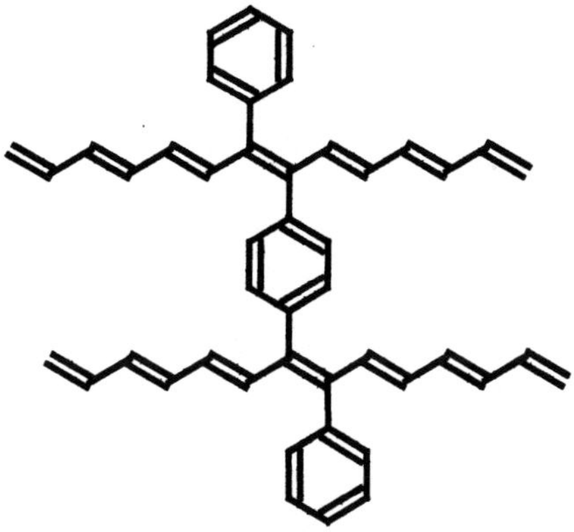

Scheme 4. Structures of model compounds for crosslinked PV-related polymers.

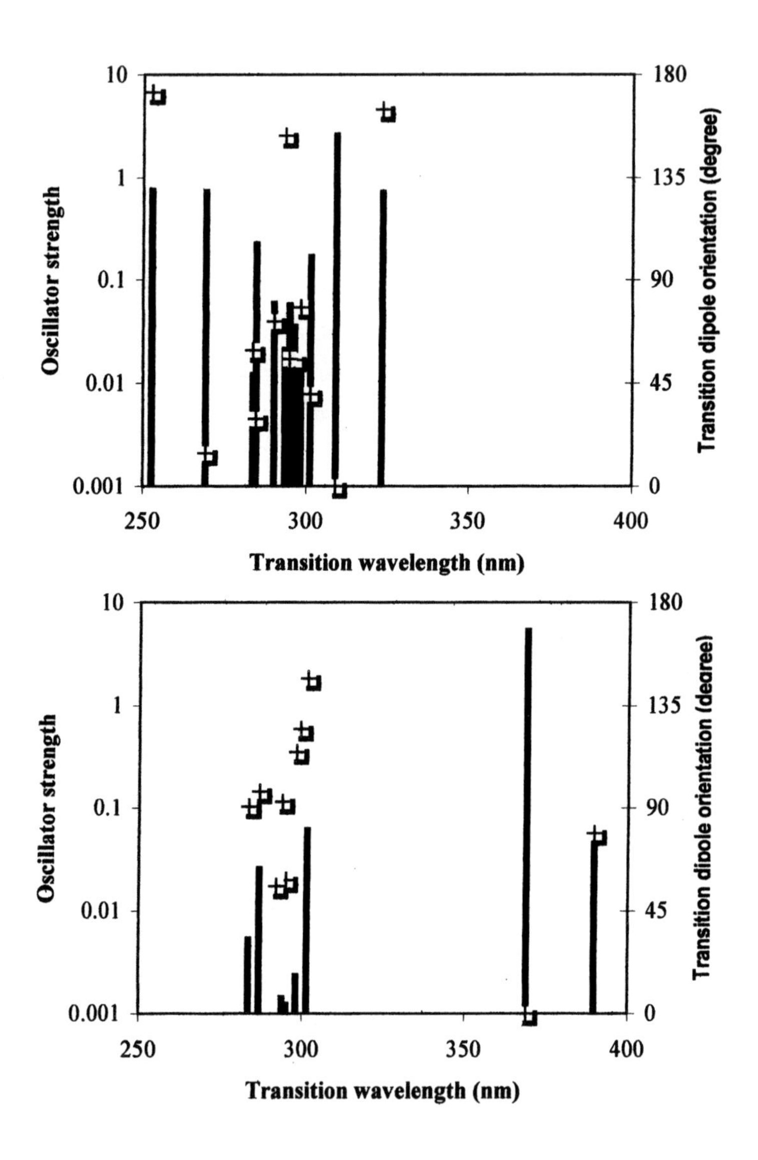

*Figure 4. Calculated transitions in crosslinked PV-related polymers **4a** and **4b**.*

References

1 Heeger, A. J. *Angew. Chem.-Int. Edit.* **2001,** 40, (14), 2591.
2 MacDiarmid, A. G. *Angew. Chem.-Int. Edit.* **2001,** 40, (14), 2581.
3 Akcelrud, L. *Prog. Polym. Sci.* **2003,** 28, (6), 875.
4 Segura, J. L. *Acta Polym.* **1998,** 49, (7), 319.
5 Hu, B.; Yang, Z.; Karasz, F. E. *J. Appl. Phys.* **1994,** 76, (4), 2419.
6 Sarker, A. M.; Kaneko, Y.; Lahti, P. M.; Karasz, F. E. *J. Phys. Chem. A* **2003,** 107, (34), 6533.
7 Aguiar, M.; Akcelrud, L.; Karasz, F. E. *Synth. Met.* **1995,** 71, (1-3), 2187.
8 Hide, F.; DiazGarcia, M. A.; Schwartz, B. J.; Andersson, M. R.; Pei, Q. B.; Heeger, A. J. *Science* **1996,** 273, (5283), 1833.
9 McGehee, M. D.; Heeger, A. J. *Adv Mater* **2000,** 12, (22), 1655.
10 Sokolik, I.; Yang, Z.; Karasz, F. E.; Morton, D. C. *J. Appl. Phys.* **1993,** 74, (5), 3584.
11 Yang, Z.; Sokolik, I.; Karasz, F. E. *Macromolecules* **1993,** 26, (5), 1188.
12 Wang, B. C.; Chang, J. C.; Pan, J. H.; Xue, C. H.; Luo, F. T. *Theochem-J. Mol. Struct.* **2003,** 636, 81.
13 Yu, J. W.; Fann, W. S.; Lin, S. H. *Theor. Chem. Acc.* **2000,** 103, (5), 374.
14 Lahti, P. M. *Int. J. Quantum Chem.* **1992,** 44, (5), 785.
15 Dewar, M. J. S.; Zoebisch, E. G.; Healy, E. F.; Stewart, J. J. P. *J. Am. Chem. Soc.* **1985,** 107, (13), 3902.
16 Thompson, M. A.; Glendening, E. D.; Feller, D. *J. Phys. Chem.* **1994,** 98, (41), 10465.
17 Thompson, M. A.; Schenter, G. K. *J. Phys. Chem.* **1995,** 99, (17), 6374.
18 Thompson, M. A. *J. Phys. Chem.* **1996,** 100, (34), 14492.
19 Liao, L.; Pang, Y.; Ding, L. M.; Karasz, F. E. *Macromolecules* **2001,** 34, (21), 7300.
20 Zhang, J. P.; Lahti, P. M.; Wang, R. S. *J. Phys. Org. Chem.* **1999,** 12, (1), 53.
21 Roothaan, C. C. J. *Rev. Mod. Phys.* **1951,** 23, 69.
22 Ridley, J. E.; Zerner, M. C. *Theor. Chim. Acta* **1976,** 42, (3), 223.
23 Bunce, N. J.; Ridley, J. E.; Zerner, M. C. *Theor. Chim. Acta* **1977,** 45, (4), 283.
24 Zerner, M. C.; Loew, G. H.; Kirchner, R. F.; Muellerwesterhoff, U. T. *J. Am. Chem. Soc.* **1980,** 102, (2), 589.
25 Thompson, M. A.; Zerner, M. C. *J. Am. Chem. Soc.* **1991,** 113, (22), 8210.
26 Tretiak, S.; Saxena, A.; Martin, R. L.; Bishop, A. R. *P Natl Acad Sci USA* **2003,** 100, (5), 2185.
27 Aguiar, M.; Karasz, F. E.; Akcelrud, L. *Macromolecules* **1995,** 28, (13), 4598.
28 Aguiar, M.; Akcelrud, L.; Karasz, F. E. *Synth. Met.* **1995,** 71, (1-3), 2189.
29 Aguiar, M.; Hu, B.; Karasz, F. E.; Akcelrud, L. *Macromolecules* **1996,** 29, (9), 3161.

Chapter 7

Impact of Cyano-Functional Group on Luminescence of Poly(*m*-phenylenevinylene) Derivatives: Its Dependence on Conjugation Length

Liang Liao[1], Yi Pang[1,*], Liming Ding[2], and Frank E. Karasz[2]

[1]Department of Chemistry and Center for High Performance Polymers and Composites, Clark Atlanta University, Atlanta, GA 30314
[2]Department of Polymer Science and Engineering, University of Massachusetts, Amherst, MA 01003

Soluble cyano-substituted poly(1,3-phenylene vinylene) (**8**) and poly[(1,3-phenylene vinylene)-*alt*-tris(1,4-phenylene vinylene)] (**10**) derivatives have been synthesized and characterized, in comparison with a green–emitting poly[(1,3-phenylene vinylene)-*alt*-(1,4-phenylene vinylene)] derivative (**9**). Chromophores in these polymers are well defined as a result of π-conjugation interruption at adjacent *m*-phenylene units, leading to blue-emission for film **8** (λ_{max}≈477 nm) and red-emission for film **10** (λ_{max}≈640 nm). Optical color tuning in these polymers is achieved through controlled insertion of different oligo(*p*-phenylene vinylene) length. Although the chromophore of **10** contains only 4.5 phenylene-vinylene units with cyano-substitution on ~50% of the vinylene units, emission λ_{max} of **10** is comparable to that of cyano-substituted PPV derivative **1**. An LED based on **10** emits red-light (644 nm) with an external quantum efficiency of 1.2%.

Since the discovery of polymeric light-emitting diodes (LEDs),[1] π-conjugated polymers have attracted significant attention over the past decade because of their potential applications in display technologies.[2] In a typical LED device, the emitting polymer layer is sandwiched between two electrodes. Electroluminescence (EL) is achieved by injecting electrons from a cathode into the conduction band and holes from an indium tin oxide (ITO) anode into the valence band of the polymer emissive layer. Combination of the injected holes and electrons leads to formation of excitons, which emit photons upon returning to ground states via radiative relaxation. To achieve highly efficient LEDs, charge carrier injection (including both electrons and holes) as well as charge transport must be balanced, and the energy barriers at the electrode-polymer interface be minimized.[3]

For poly(*p*-phenylenevinylene) (PPV), injection and transport of holes is easier than that of electrons, which make this polymer useful as a hole-transport material[4] in LEDs. The relative poor electron affinity of PPV, which leads to a low rate of electron injection, however, has been a major barrier in the development of PPV-based bright LEDs. An increase in the electron affinity of PPV would lead to improved charge injection and transport. This concept has been successfully demonstrated by comparing poly[(2-methoxy-5-ethylhexyloxy-1,4-phenylene)vinylene] (MEH-PPV, **1a**)[4] with the cyano-substituted MEH-PPV **1b** (Scheme 1); the latter significantly enhanced EL efficiency than **1a**, reaching an external quantum efficiency as high as 4%. In addition, attachment of cyano group shifts the emission color of MEH-PPV (yellow) to longer wavelength, although the unspecified chromophore length in CN-MEH-PPV somewhat hampers the estimation of each cyano group's contribution to the optical properties. Furthermore, the effect of the cyano group on the charge injection may vary with the chromophore conjugation length, which remains to be a poorly understood issue.

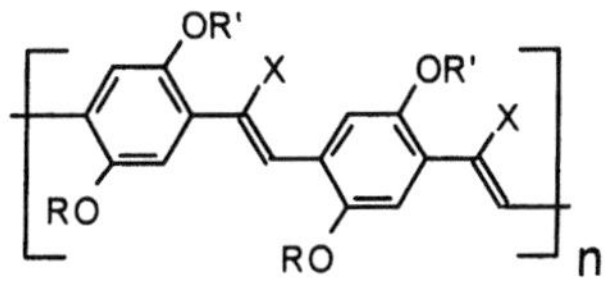

1

a: R= 2-ethylhexyl, R'=methyl, X=H;
b: R=2-ethylhexyl, R'=methyl, X=CN;
c: R=R'=hexyl, X=CN

Scheme 1. Chemical structures of MEH-PPV (**1a**) and CN-MEH-PPV (**1b**).

Recent studies have shown that poly[(1,4-phenylenevinylene)-*alt*-(1,3-phenylenevinylene)] (P*p*PV*m*PV) derivatives **3** are bright green-emitting materials.[5,6,7,8] While replacement of the *p*-phenylene in **3** with an *m*-phenylene leads to the blue-emitting poly(1,3-phenylenevinylene) (P*m*PV) **2**,[9] extension of the *para*-phenylenevinylene block length along the chain provides a yellow-emitting material **4**.[10] To improve the electroluminescence of these polymers, a logical approach is to attach a cyano-functional group on the vinylene bond, which increases the electron affinity of the polymer backbone.[4] On the basis of its relative position to the substituted phenyl rings, there are two possible positions (α and β on the vinylenes of **2-4**) to locate the cyano-group. Since steric interaction between the cyano-substituent at the α-position and the alkoxy group on the phenyl ring (shown in **7**) can cause significant twisting of a π-conjugated polymer backbone,[11,12] a preferred choice is to attach the cyano-substituent at the β-position (Scheme 2). The alkoxy substituent on the phenyl ring of **5** is located at the *ortho*-position relative to the vinyl bond, which permits resonance between the alkoxy and cyano groups to lead to the resonance structure **6**.

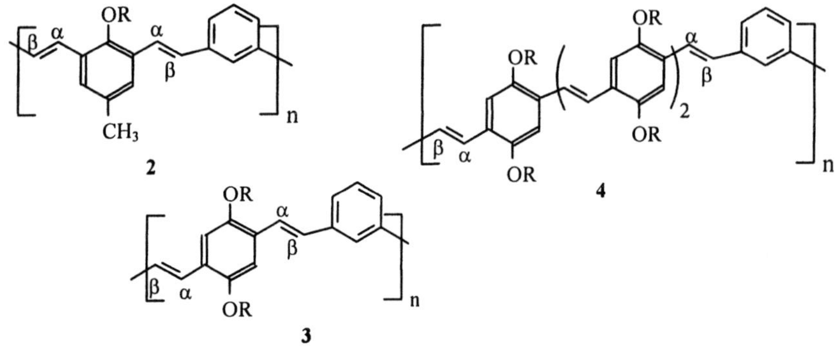

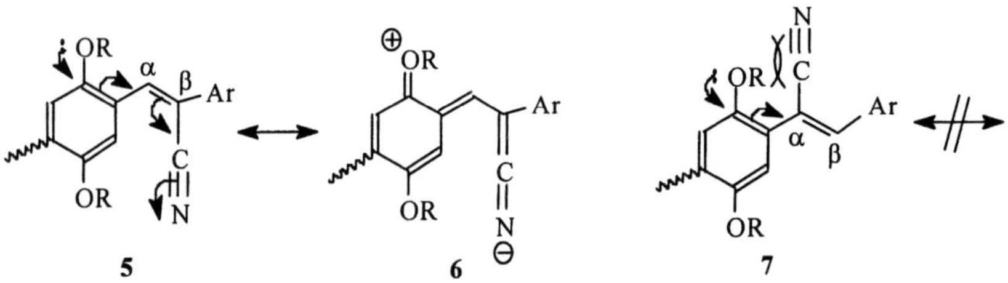

Scheme 2. Ground-state resonance forms for β- and α-cyano-substituted compounds. Charge transfer resonance is forbidden between an alkoxy and α-cyano group.

It should be noted that the cyano-substituent in PPV **1b** is at the α-position relative to the methoxy on one phenyl, but at the β-position relative to the alkoxy on the other phenyl ring. Steric interaction of cyano-substituent with the adjacent methoxy (or alkoxy) in **1b**, which prevents the co-planarity, reduces its electronic interaction with the PPV backbone. To achieve the maximum/optimum effect of a cyano-group on the optical properties of PPV, it is thus desirable to place the cyano-group at the β-position as shown in **8-10**. Our recent study[13] on **9** has shown that the polymer emits yellow light (λ_{max} = 578 nm), which is red-shifted by ~45 nm from its parent polymer **3**. To further evaluate the optical impact of cyano-substituent, we now report the synthesis and optical properties of two specific examples of **8** and **10**.

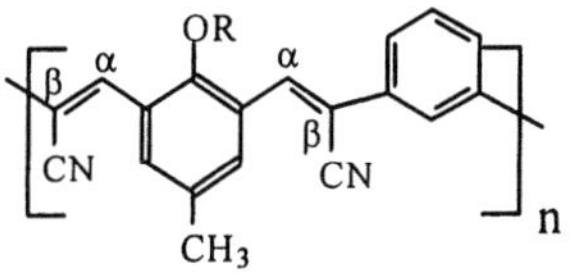

8 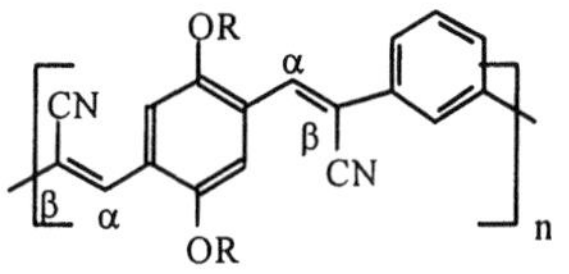 **9**

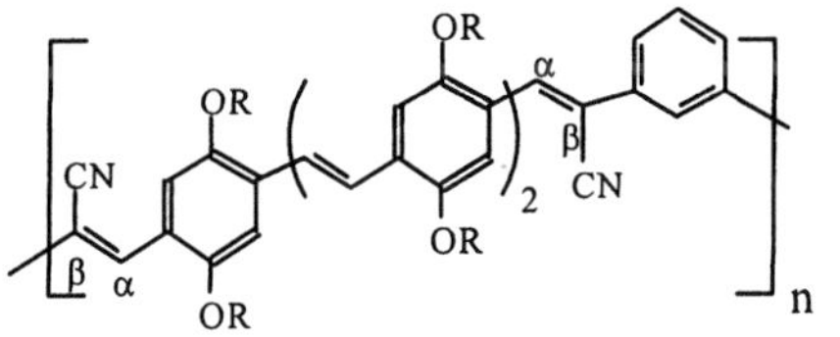

10

(R = n-hexyl)

Polymer synthesis and characterization

The monomer, 2-hexyloxy-5-methylbenzene-1,3-dicarbaldehyde **12**, was prepared from dibromide **11**.[9] Knoevenagel condensation of monomer **12** with 1,3-phenylenediacetonitrile proceeded smoothly in the presence of potassium *t*-butoxide to afford a cyano-substituted poly(1,3-phenylenevinylene) derivative **8**. The polymer **10** was synthesized similarly by condensation of dialdehyde **17**[10] with 1,3-phenylenediacetonitrile. Solubility of **10** in THF was lower than that of **8**, partially attributing to the lower *m*-phenylene linkage in the former. The infrared spectra of polymer films of **8** and **10** showed the characteristic cyano (–C≡N) stretch band at ~2207 cm^{-1} in a medium intensity.[14] Absence of carbonyl absorption at ~1700 cm^{-1} in the polymer films indicated that the reaction was complete. Degree of polymerization (DP) of **8-10** was estimated to be 23, 54, and 4, respectively (Table 1). Lower Mw of **10** was partially attributed to its low solubility in solvents. Elemental analysis results[15] of **8** and **10** were satisfactory.

a) KOAc, Bu_4NBr, $CH_3CN/CHCl_3$; b) $LiAlH_4/THF$; c) PCC/CH_2Cl_2, r.t. 2 h

$BrCH_2$ CH_2Br CH_3 OC_6H_{13} **11** → CH_3 OC_6H_{13} **12**

NC CN; t-BuOK, $(n\text{-}Bu)_4NOH$, t-BuOH/THF → **8**

13 (C_6H_{13}, Br) → $AcOK$, K_2CO_3, $MeCN/CHCl_3$, reflux 2 d → **14** (OAc, C_6H_{13})

13 → 1) PPh_3, toluene, Ar, reflux 3 h; 2) **16**, CH_2Cl_2, LiOEt/EtOH, r.t. 100 min → **17** (OC_6H_{13}, $C_6H_{13}O$)

14 → $LiAlH_4$, THF, r.t. 15 h → **15** (OH, C_6H_{13})

15 → PCC, CH_2Cl_2, r.t. 2 h → **16**

17 → NC CN; t-BuOK, $(n\text{-}Bu)_4NOH$, t-BuOH/THF, 50 °C, 20 min → **10**

Absorption and photoluminescence (PL) of 8. UV-vis absorption and fluorescence of **8** in tetrahydrofuran (THF) are shown in Figure 1, and the results are summarized in Table 1. The absorption spectrum of **8** at 25 °C exhibited a broad band with $\lambda_{max} \approx 319$ nm, which was slightly red-shifted from that of its parent polymer **2** (R=n-hexyl, $\lambda_{max} \approx 308$ nm).[9] Lowering the temperature to −108 °C caused a slight bathochromic shift (by ~3 nm) without resolving the hidden vibronic band structure.

Fluorescence spectrum of **8** in THF solution exhibited a broad emission band at 25 °C (Figure 1). As the polymer solution was cooled, the emission λ_{max} (451 nm at 25 °C) was slightly red-shifted to 455 nm at −108 °C, but notably blue-shifted to 427 nm at −198 °C. The initial small bathochromic shift was attributed to the molecular conformational response to the low temperature. As the temperature was further lowered to −198 °C, the polymer chromophore was frozen into a rigid solid matrix, in which the molecules no longer moved and rotated freely. This rigid environment reduced or eliminated the solvent effect on the excited state, which require reorientation of solvent molecules, thereby causing the emission spectra to shift to the lower wavelength.[16]

Absorption and photoluminescence (PL) of 10. UV-vis absorption of **10** in THF at 25 °C (Figure 2) exhibited a broad absorption band with λ_{max} = 464 nm, which was about 14 nm red-shifted from that of its parent polymer **4**. As the temperature was lowered to −108 °C, the absorption band was further red-shifted by about 18 nm (λ_{max} = 482 nm), attributing to the adoption of more planar conformation at the low temperature. The temperature-induced bathochromic shift from the solution of **10** was notably larger than that from **8**, due to the longer conjugation length in the chromophore of the former. It should be noted that the true chromophores for polymers **8**, **9**, and **10** can be represented by the molecular fragments **18**, **19**, and **20**, respectively, as a result of the effective π-conjugation interruption at *m*-phenylene units. While the fragment **18** contains one cyano substitution, the fragments **19** and **20** include two cyano-substituents. The bathochromic shift in the solution absorption λ_{max} for each cyano substitution can be estimated to be, therefore, ~11 nm for **18**, ~20 nm for **19**,[13] and 7 nm for **20**. In other words, the effect of cyano-substitution on the optical absorption appeared to be weakly dependent on the chromophore conjugation length.

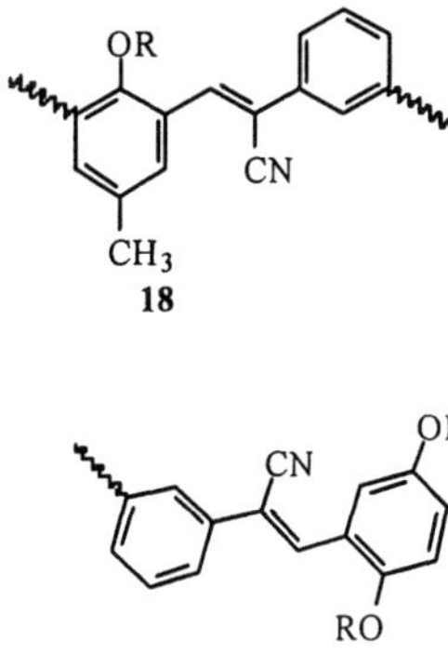

OR CN CH3 **18**

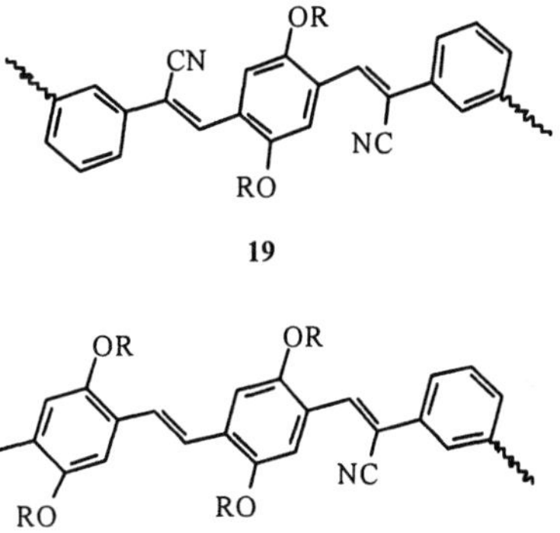

OR CN NC RO **19**

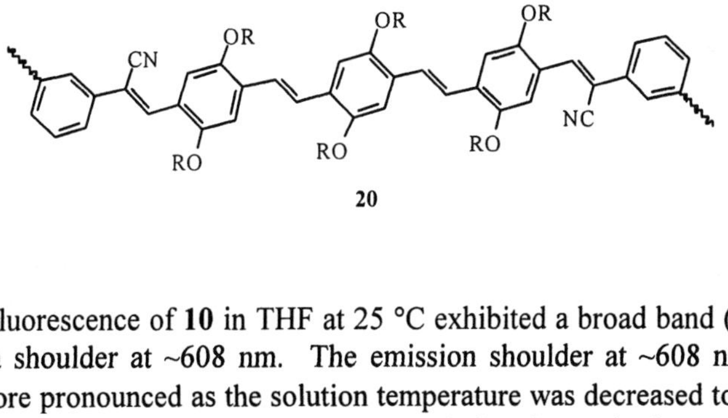

OR OR OR CN NC RO RO RO **20**

Fluorescence of **10** in THF at 25 °C exhibited a broad band (λ_{max} =566 nm) and a shoulder at ~608 nm. The emission shoulder at ~608 nm became slightly more pronounced as the solution temperature was decreased to −108 °C, attributing to the reduced molecular motion and vibration at the low temperature. The emission λ_{max} at −108 °C was red-shifted by ~13 nm to 579 and 624 nm, as

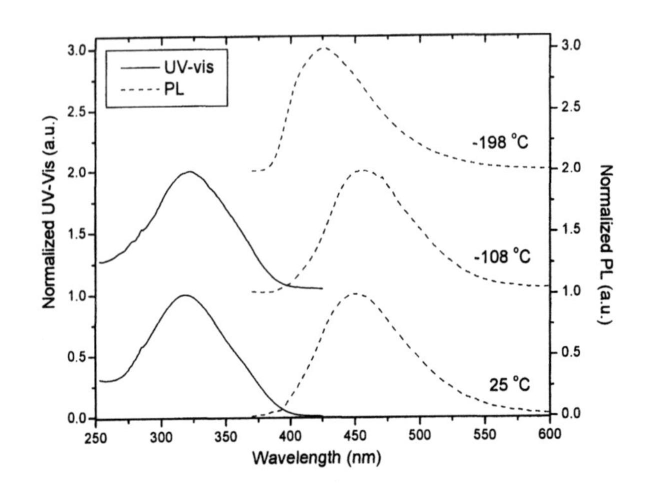

*Figure 1. UV-vis absorption (solid line) and fluorescence (broken line) spectra of polymer **8** in THF at 25 °C, –108 °C, and –198 °C. The spectra at different temperatures are offset for clarity.*

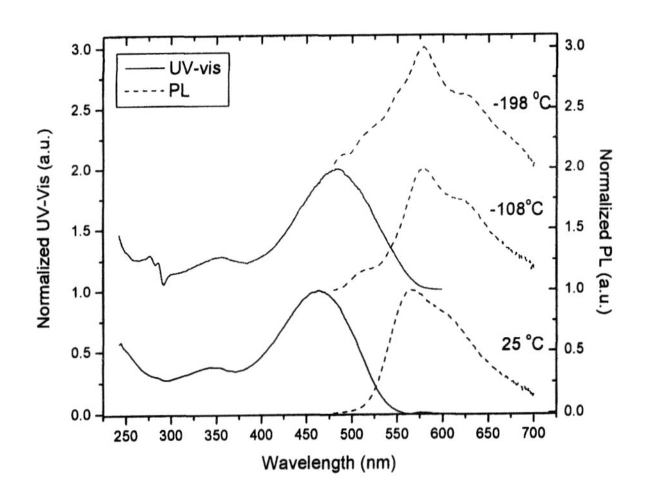

*Figure 2. UV-vis absorption (solid line) and fluorescence (broken line) spectra of **10** in THF at 25 °C, –108 °C, and –198 °C. The spectra at different temperatures are offset for clarity.*

the chromophore adopted a more planar conformation at the low temperature. Further cooling the sample to −198 °C did not reveal additional vibronic structures.

Thin Film Properties. The solid-state spectra were acquired from films spin-cast on quartz plates. The absorption λ_{max} values were at about 325 and 502 nm for the films **8** and **10** (Figure 3), respectively. The film absorption λ_{max} values were red-shifted from their respective solution λ_{max} by about 6 and 40 nm (Table 1), respectively, indicating the strong chromophore-chromophore interaction present even in the ground states. It should be noticed that the bathochromic shift from the solution to film states was about 0 nm and 18 nm in the absorption spectra of their parent polymers **2**[9] and **4**,[10] respectively. Presence of the polar cyano functional group, therefore, significantly contributed to the increased chromophore-chromophore interaction in the films **8** and **10**, which tunes the absorption λ_{max} to the longer wavelength.

Fluorescence spectra of both films **8** and **10** exhibited the similar emission profiles as that of their respective solutions. The emission peaks were centered near 477 and 640 nm for **8** and **10**, respectively, providing blue and red emission. It should be noticed that the emission λ_{max} of **10** (640 nm) was comparable to that of CN-PPV **1c** (660 nm),[17] although the chromophore of the former polymer contained only 4.5 phenylene-vinylene units with ~50% of the vinylene bonds bearing cyano-substituent. In addition, UV-vis absorption λ_{max} of film **10** (λ_{max} = 502 nm) occurred at a slightly longer wavelength than that of film **1c** (λ_{max} = 490 nm), further indicating that both polymers had very similar bandgaps. It is clear that placement of cyano-substitution at β-position of the vinylene bonds offers some advantages in tuning the optical properties of PPV materials.

EL Properties. LED devices of ITO/PEDOT/polymer/Ca/Al configuration were fabricated to examine the EL properties. The double layer devices yielded the blue (λ_{max} = 476 nm), yellow (λ_{max} ≈558 & 587 nm), and red (λ_{max} = 644 nm) emissions for **8**, **9**, and **10**, respectively (Figure 4). The color tuning was achieved through increasing the effective conjugation length in the polymer. The emission peak full widths at half maximum (fwhm) were about 84, 122, and 111 nm for **8**, **9**, and **10**, respectively. Although the chromophores in each polymer are well defined with equal conjugation length, their emission bands were not significantly narrower than that from regular PPV derivatives **1**. Among the possible reasons are molecular aggregation and excimer formation.[18]

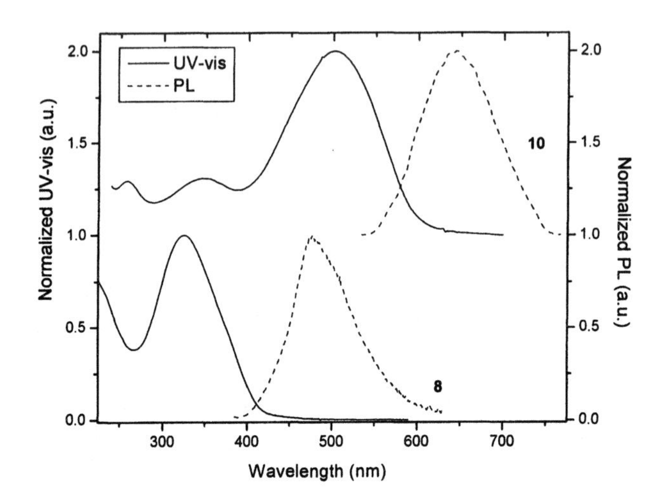

*Figure 3. UV-vis (solid line) and fluorescence (broken line)spectra of thin films **8** (bottom) and **10** (top)*

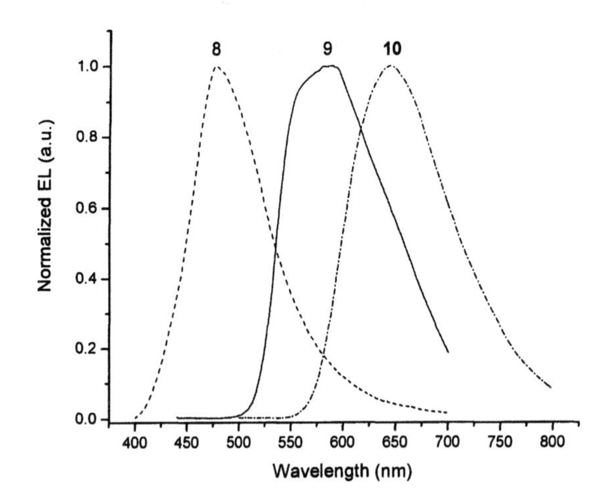

Figure 4. Electroluminescence spectra for ITO/PEDOT/Polymer/Ca/Al devices.

Table 1. Molecular Weight and Optical Properties of Polymers

polymer	*UV-vis Absorption* (λ_{max}, *nm*)		*Fluorescence* (λ_{max}, *nm*)		*EL* (η_{ext})	$\phi_{fl}(\%)^{b}$	$M_w(PDI)$	DP
	THF solution	*film*	*THF*	*film*				
2	308	308	421	438, **461**	**445**, 462 (0.01%)[9]	0.55	8450 (2.1)	13
3	328, **406**[a]	336, **422**	**447**, 475	507, **540**	513, 523 (0.56%)[18]	0.6	45000(2.0)	55
4	450	468	**516**, 545	**570**, 610	**566**, 604(sh) (0.25%)	0.51	63500 (1.7)	37
8	319	325	451	477	476 (0.0085%)	0.75	65000 (7.6)	23
9	360, **447**	381, **485**	521	573	558, **587** (0.03%)	0.41	70700 (2.9)	54
10	343, **464**	351, **502**	566	640	644 (1.2%)	0.56	5430 (1.3)	4

[a] The **bold** number indicates the most intense peak; [b]fluorescence quantum efficiencies were measured in THF solution while excited at 366 nm or 350 nm, in reference to quinine sulfate standard.

It was noted that the emission color of **8**, **9**, and **10** was red-shifted from that of their respective parent polymers, whose EL emission (Table 1) was centered near 445 and 462 nm for **2**,[9] 513 and 523 nm for **3**,[6] and 560 nm for **4**.[10] The bathochromic shift induced by the cyano substitution was estimated to be ~30 nm (from **2** to **8**), 45-64 nm (from **3** to **9**), and 80 nm (from **4** to **10**). The larger effect of cyano substitution in **10** than that in **9** could be rationalized by considering the alkoxy-cyano interaction. As seen from the molecular fragment **19**, the effective chromophore in **9**, both cyano groups are interacting with the alkoxy groups on the same benzene ring, which may dilute the resonance effect and result in a smaller dipole moment. In the chromophore **20**, however, two cyano groups are distant away from each other, and its resonance interaction with an adjacent alkoxy substituent would be independent from each other, thereby generating a larger molecular dipole. In other words, the chromophore-chromophore interaction is anticipated to be stronger in film **10** than that in films **8** and **9**. The larger effect of cyano substitution in **10** is attributed, at least partially, to the stronger resonance interaction between the alkoxy and cyano groups in the chromophore.

Figure 5 shows the dependence of current density and luminance of devices on the applied voltage. The turn-on voltages for devices of **8**, **9**, and **10** are 8, 4.5, and 4 V, respectively. The higher turn-on voltage for the device of **8** appears to be related to the short conjugation length (or larger band gap) in the chromophore. The external quantum efficiency was determined to be 0.0085%, 0.03%, and ~1.2% for **8**, **9**, and **10**, respectively. Although the current densities were at a comparable level in both devices of **9** and **10**, the brightness of the latter was improved by about two orders of magnitude.

Conclusion

We have synthesized and examined cyano-substituted poly(*p*-phenylenevinylene) derivatives **8-10**, in which *m*-phenylene units are regularly inserted along the main chain to provide uniform chromophores of different conjugation length. In these polymers, the cyano-substituents are attached at a specified location (i.e., β-position on vinylene) to achieve the maximum resonance effect between an electron-donating alkoxy and an electron-withdrawing cyano group. Such effect appears to be larger when the two adjacent cyano-groups in the chromophore are separated from each other, as seen in the molecular fragment **20**. This improved resonance interaction has been demonstrated to be useful in color tuning, leading to red-emitting material **10** (λ_{max}≈644 nm), whose chromophore contains only 4.5 PV units. The

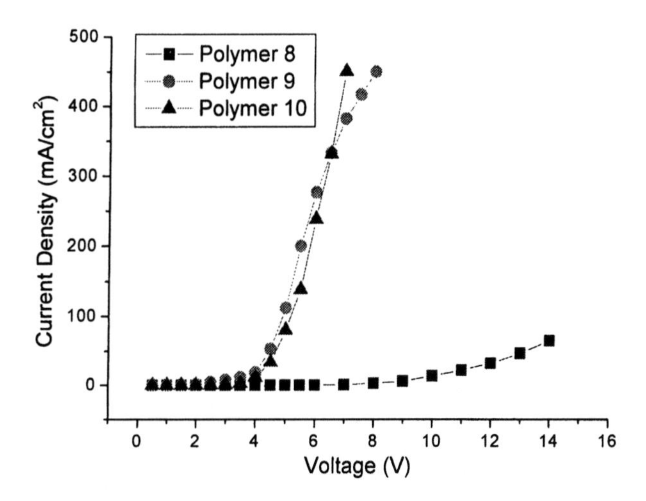

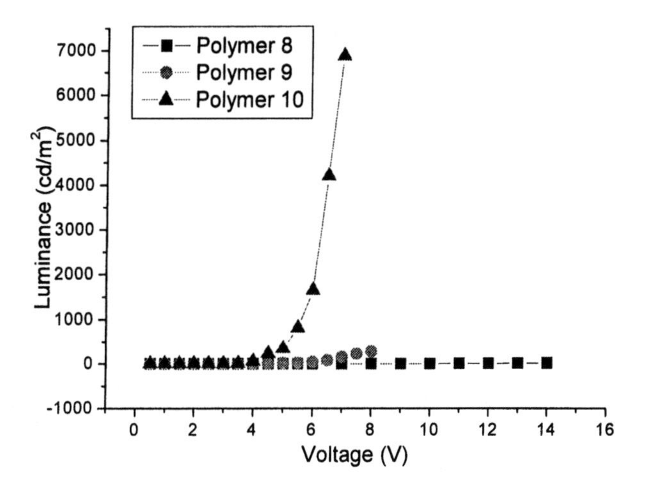

Figure 5. Voltage dependence of current density and luminance for ITO/PEDOT/Polymer/Ca/Al devices.

external quantum efficiency of the device using **10** reaches ~1.2%. Since the charge injection in **10** is not balanced with the device configuration of ITO/PEDOT/**10**/Ca/Al, the EL efficiency is expected to be further improved under optimum conditions. Comparable EL efficiency between **10** and CN-MEH-PPV **1b** suggests that 50% of cyano substitution is sufficient to reach the desirable level of charge injection properties.

Acknowledgement

Support of this work has been provided by AFOSR (Grant No. F49620-00-1-0090), NIH/NIGMS/MBRS/SCORE (Grant No. S06GM08247), and NASA (Grant No. NCC3-911).

References and Notes

1. Burroughes, J. H.; Bradley, D. D. C.; Brown, A. R.; Marks, R. N.; MacKay, K.; Friend, R. H.; Burn, P. L.; Holmes, A. B. *Nature* **1990**, *347*, 539-541.
2. Kraft, A.; Grimsdale, A. C.; Holmes, A. B. *Angew. Chem. Int. Ed.* **1998**, *37*, 402-428.
3. Parker, I. D.; Pei, Q.; Marrocco, M. *Appl. Phys. Lett.* **1994**, *65*, 1272-1274.
4. Greenham, N. C.; Moratti, S. C.; Bradley, D. D. C.; Friend, R. H.; Holmes, A. B. *Nature* **1993**, *365*, 628-630.
5. Pang, Y.; Li, J.; Hu, B.; Karasz, F. E. *Macromolecules* **1999**, *32*, 3946-3950.
6. Liao, L.; Pang, Y.; Ding, L.; Karasz, F. E. *Macromolecules* **2002**, *35*, 6055-6059.
7. Ohnishi, T., Doi, S., Tsuchida, Y., and Noguchi, T. Polymer Light-Emitting Diodes Utilizing Arylene-Vinylene Copolymers as Light-Emitting Materials. In *Photonic and Optoelectronic Polymers;* Jenekhe, S. A., Wynne, K. J., Eds.; ACS Symposium Series 672 ; American Chemical Society: Washington, DC, 1997: pp345-357.
8. Drury, A.; Maier, S.; Rüther, M.; Blau, W. J. *J. Mater. Chem.* **2003**, *13*, 485-490.
9. Liao, L.; Pang, Y.; Ding, L.; Karasz, F. E. *Macromolecules* **2001**, *34*, 7300-7305.
10. Liao, L.; Pang, Y.; Ding, L.; Karasz, F. E.; Smith, P. R.; Meador, M. A. *Synthesis of dialdehyde **17** and polymer **4** will be described elsewhere.*
11. Hanack, M.; Behnisch, B.; Häckl, H.; Martinez-Ruiz, P.; Schweikart, K.-H. *Thin Solid Films* **2002**, *417*, 26-31.

12. Hoholoch, M.; Maichle-Mössmer, C.; Hanack, M. *Chem. Mater.* **1998**, *10*, 1327-1332.
13. Liao, L.; Pang, Y.; Ding, L.; Karasz, F. E. *J. Polym. Sci.: Part A: Polym. Chem.* **2003**, *41*, 3149-3158.
14. Lin-Vien, D.; Colthup, N. B.; Fateley, W. G.; Gresselli, J. G. The Handbook of Infrared and Raman Characteristic Frequencies of Organic Molecules; Academic: Boston, 1991; Chapter 8.
15. Anal. Calcd for $C_{19}H_{28}O_5$ (polymer **8**): C, 81.49; H, 6.56; N, 7.60. Found: C, 80.86; H, 6.57; N, 7.71. Anal. Calcd for $C_{70}H_{94}N_2O_6$ (polymer **10**): C, 79.53; H, 8.94; N, 2.64. Found: C, 78.25; H, 8.70; N, 2.55.
16. Lakowicz, J. R. *Principles of Fluorescence Spectroscopy*; Kluwer Academic: New York, 1999; Chapter 6.
17. Rumbles, G.; Samuel, I. D. W.; Collison, C. J.; Miller, P. F.; Moratti, S. C.; Holmes, A. B. *Synthetic Metals* **1999**, *101*, 158-161.
18. Jenekhe, S. A.; Osaheni, J. A. *Science* **1994**, *265*, 765-768.
19. Liao, L.; Pang, Y.; Ding, L.; Karasz, F. E. *J. Polym. Sci: Part A: Polym. Chem.* **2004**, *42*, 1820-1829.

New Polymers in Biomedical Application

Chapter 8

Interaction of Poly(ethylene oxide) and Poly(perfluorohexylethyl methacrylate) Containing Block Copolymers with Biological Systems

H. Hussain[1], K. Busse[1,*], A. Kerth[2], A. Blume[2], N. S. Melik-Nubarov[3], and J. Kressler[1]

Departments of [1]Engineering Science and [2]Chemistry, Martin-Luther-University Halle-Wittenberg, D–06099 Halle, Saale, Germany
[3]School of Chemistry, M.V. Lomonosov Moscow State University, Moscow 119899, Russian Federation

Amphiphilic di- and triblock copolymers containing hydrophilic poly(ethylene oxide) and hydrophobic poly(perfluorohexylethyl methacrylate) are forming micelles and larger aggregates in aqueous solution as demonstrated by various scattering techniques. Measuring the surface pressure (π)-area (A) isotherms in conjunction with infrared reflection absorption spectroscopy (IRRAS) the surface coverage of pure block copolymers and the penetration into lipid monolayer as model membranes (1,2-diphytanoyl-*sn*-glycero-3-phosphocholine) can be followed during compression. Since the block copolymers have potential application for chemosensitizing effects in anti tumor treatment, the nontoxicity was determined using K562 human cells.

Amphiphilic block copolymers form various supramolecular structures such as spherical micelles, vesicles, cylindrical micelles and other complex aggregates in solution (*1*). Though both the diblock and triblock copolymers form micelles in water, yet the association behavior of the respective block copolymers is significantly different. Diblock copolymers prefer to form individual micelles with little tendency for cluster formation, except at very high concentration, while the triblock copolymers with hydrophobic end-blocks have tendency to form intermicellar network structure, caused by inter-micellar bridging (*2*). Triblock copolymers in a solvent selective for the middle block are assumed to form flower-like micelles with the middle block looping in the micelle corona at low concentration (*3*).

Amphiphilic block copolymers of poly(ethylene oxide) (PEO) as hydrophilic block and fluorine containing hydrophobic block might be of great potential interest because of the very peculiar properties of fluorine-containing materials such as low surface energy, high contact angle, reduced coefficient of friction, bio-compatibility and oleo- and hydrophobicity (*4*). A number of investigations have been reported in the literature on the aggregation behavior of the amphiphiles having PEO as the hydrophilic block and fluorine containing hydrophobic segments (*5-8*).

Amphiphilic block copolymers in water readily adsorb at an interface. The polymers are arranged in a way that the hydrophobic block anchors at the surface or interface and the hydrophilic block extends into the solution. In addition, amphiphilic block copolymers have many applications in biophysics, biomedicine, and biotechnology because of their penetration into lipid membranes (*9*, *10*). When exposed to nonionic block copolymer surfactants (poloxamers), the recovery rate of electrical or burn injuries have been found considerably faster (*11*, *12*). Investigations on the penetration of lipid model membranes by amphiphilic molecules have developed a better understanding of some complex physiological issues such as the function of lung surfactants in mammalian lungs, which has led to the development of therapeutic agents for the respiratory distress syndrome, a condition resulting from a deficiency of lung surfactants (*13*, *14*), and might prove helpful in solving other complicated biomedical challenges such as the chemo-sensitizing mechanism of the anticancer activity of some amphiphilic polymers such as pluronic type surfactants, in multidrug resistance tumor cells (*15*).

In this chapter, self-association in water, and interaction with model membranes of PEO and poly(perfluorohexylethyl methacrylate) (PFMA) containing amphiphilic di- and triblock copolymers are discussed. Furthermore, the cytotoxicity of the block copolymers was tested.

Experimental Section

Synthesis of Block Copolymers

The block copolymers were synthesized by atom transfer radical polymerization (ATRP) as reviewed elsewhere (*16*). The generalized chemical structure of the

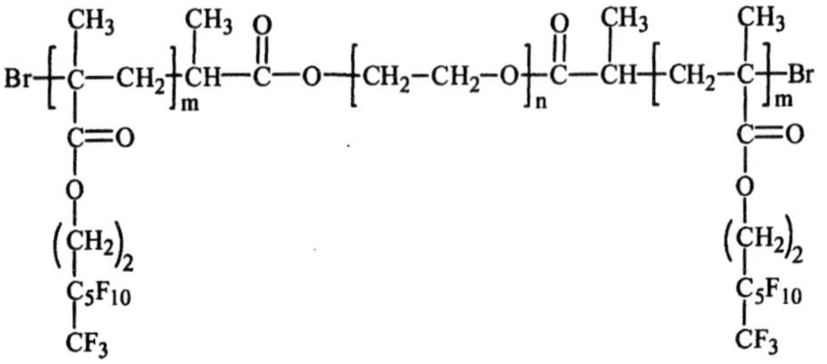

Figure 1. Chemical structure of the PFMA-block-PEO-block-PFMA copolymer.

Table I. Molecular Characteristics of the Copolymers.

Copolymer[+]	M_n(kg/mol) (SEC results)	wt.-% PFMA (^{1}H NMR results)	M_w/M_n
PEO_2F12-D	2.3	12	1.1
PEO_5F15-D	4.9	15	1.03
$PEO_{10}F5$	11.5	5	1.1
$PEO_{10}F9$	15.6	9	2.1
$PEO_{10}F11$	15.9	11	1.99
$PEO_{20}F14$	23.6	14	1.2

[+]In the abbreviation scheme PEO_xFy x represents the PEO molar mass (kg/mol) and y the PFMA wt.% in the block copolymer. To differentiate di- from triblock copolymers, -*D* has been added when a monofunctional macroinitiator was used.

triblock copolymer and the characteristic data of the block copolymers are given in Figure 1, and Table I, respectively.

Dynamic Light, Small Angle Neutron and X-ray Scattering

The aqueous polymer solutions were prepared in double distilled water. Clear solutions were obtained after overnight stirring at room temperature. However, in some cases a few minutes ultrasonic treatment in addition to stirring was given as well to get clear solutions. Dynamic light scattering (DLS) measurements were performed with an ALV-5000 goniometer at $\lambda = 532$ nm at an angle of 90°. The correlation functions were analyzed by the CONTIN method (*17*), giving information on the distribution of decay rate (Γ). Apparent diffusion coefficients were determined and the corresponding apparent hydrodynamic radii ($R_{h,app}$, radius of the hydrodynamically equivalent sphere) calculated via Stokes-Einstein equation. Small angle neutron scattering (SANS) have been performed at D22, ILL Grenoble, and KWSII, FZ Jülich. Small angle X-ray scattering (SAXS) measurements were done at BW4 beamline at HASYLAB, Hamburg. The obtained scattering data are corrected for background scattering and Lorentz corrected.

Surface Pressure and Infrared Reflection Absorption Spectroscopy

All experiments were performed with a Wilhelmy film balance (Riegler & Kirstein, Berlin, Germany). The Teflon trough system of the Infrared Reflection Absorption Spectroscopy (IRRAS) setup consisted of a sample trough and an additional reference compartment connected by three small holes to the sample trough to ensure an equal height of the water surface in both troughs (Riegler & Kirstein). The lipid and the polymers were dissolved in chloroform and known amounts were spread onto ultrapure water. The lipid (1,2-diphytanoyl-*sn*-glycero-3-phosphocholine) (DPhPC) (chemical structure is given in Figure 2) and the polymers were dissolved in chloroform and known amounts were spread onto water surface. The compression speed of the barriers was 0.03 nm^2 per lipid molecule per minute and the experiments were performed at 20°C ± 0.5°C. To keep the air humidity constant, the experimental setup was kept in a

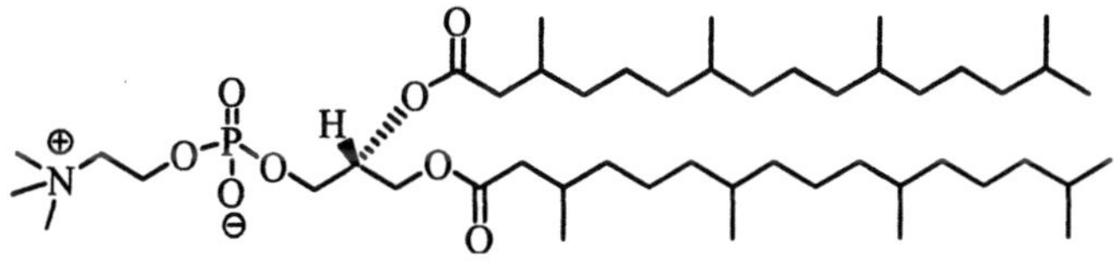

Figure 2. Structure of 1,2-diphytanoyl-sn-glycero-3-phosphocholine (DPhPC).

closed container. To ensure the full evaporation of the solvent, the compression was started 15 min after the spreading of the chloroform solution. For monolayer penetration experiments, the lipid monolayer was obtained as discussed above and then a known amount of the polymer solution in water was injected through the film into the subphase.The surface pressure (π)-area (A) isotherm was measured after waiting for 30 min in each case. The same procedure was followed to study the behavior of the block copolymer chains at the surface in the presence of lipid monolayer by IRRAS.

Infrared spectra were recorded with an Equinox 55 FT-IR spectrometer (Bruker, Karlsruhe, Germany) connected to an XA 511 reflection attachment (Bruker) with the above mentioned trough system and an external narrow band MCT detector. The monolayer films were compressed to a desired area per molecule, the barriers were then stopped and the IRRA spectra were recorded at a spectral resolution of 4 cm^{-1} using Blackman-Harris-4-Term apodization and a zero filling factor of 2. For each spectrum 1000 scans were summed over a total acquisition time of about 4.5 min. The single beam reflectance spectrum of the reference trough surface was ratioed as background to the single beam reflectance spectrum of the monolayer on the sample trough to calculate the reflection absorption spectrum as $-\log(R/R_0)$. The reflection-absorption of the ν(O–H) band was calculated by an integration of the band between 3000 and 3500 cm^{-1}. In the region above 3500 cm^{-1} often a lower signal to noise ratio and/or superimposed rotational vibrational bands of water vapor were observed. This was therefore excluded from the integration procedure (*18*).

Cytotoxicity Measurements

For cytotoxicity experiments the block copolymer samples were dialyzed extensively against distilled water and lyophilized. They were filtered through sterile 0.2 μm pore size Millipore filters for sterilization. The polymer concentration of the sterilized solutions was determined using $BaCl_2/KI_3$ technique that was previously reported for PEO homopolymer and was found to be applicable for measuring concentration of any polymer containing ether bonds (*19, 20*). K562 human erythroleucemia cells were cultured in RPMI 1640 medium (Sigma) supplemented with 10 vol.% fetal calf serum in an atmosphere containing 5 vol.% CO_2 and 96% humidity. The cells density during culturing was maintained in the 200 000–600 000 cells per mL range.

The toxic effect of the polymers on the sensitive cells was determined as reported in literature (*21, 22*) The polymer aqueous solutions were diluted with serum-free medium RPMI1640 (Sigma) up to the required concentration and 100 μL of the preparations were added to 100 μL of K562 cells (~ 10 000 cells per well) cultured in 96-well plates. The cells were incubated with the polymers for one hour and then centrifuged to remove the polymer. 200 μL of the fresh polymer-free medium, supplemented with 10 vol.% of fetal calf serum was added to the cells and they were incubated for three days. Then 50 μL of methyl tetrazolium blue (MTT) dye was added to the cells up to final concentration of 0.2 μg/mL. The samples were incubated for four hours and sedimented on the centrifuge. The dye was reduced by mitochondria to give a colored product generally known as formazan. The medium was removed and 100 μL of DMSO was added to each well to dissolve formazan. Optical densities of the samples were measured on Titertek Multiscan multi-channel photometer (wavelength 550 nm) (Titertek, USA) and the relative amount of cells was calculated as a ratio of the optical density in the sample with polymer to that of control without any additives.

Results and Discussion

Direct evidence for the presence of micelles in solution can be obtained from DLS investigations (*23*). In Figure 3 typical measurements for (a) diblock and (b) triblock copolymer solutions are depicted. Different modes, called fast (I), intermediate (II), and slow (III), corresponding to aggregates with different apparent hydrodynamic radii $R_{h,app}$, can be observed. The aggregates of intermediate size, $R_{h,app}$ between 15 and 31 nm in the solutions of samples investigated, can be regarded as micelles with hydrophobic block making the core and hydrophilic PEO block constitutes the corona of the micelle. The fast mode can be attributed to the single chains in the solution and the slow mode

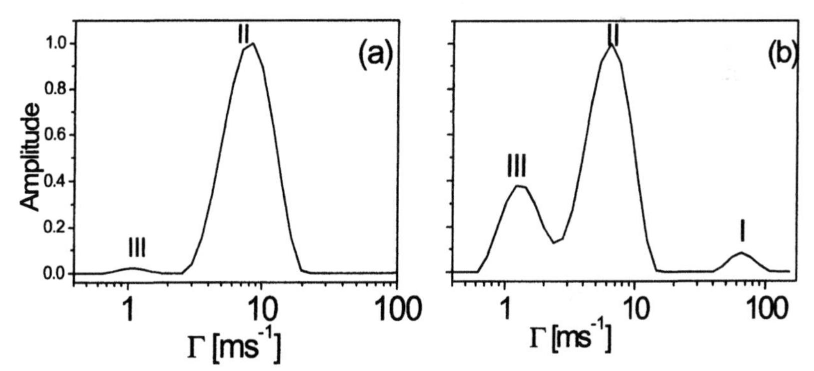

Figure 3. DLS measurement of sample (a) PEO_2F12-D (4 g/L) and (b) $PEO_{10}F11$ (2 g/L). The experiments were carried out at 20 °C. The numbers I, II, and III indicate the fast, intermediate and the slow mode, respectively.

can be assigned to large clusters (typically $R_{h,app}$ larger than 80 nm). Reduced $R_{h,app}$ values of micelles for triblock copolymers compared to those of diblock copolymers with approximately the same PEO block length suggests that the PEO middle block in the triblock copolymer micelles form loop in the micellar corona, resulting in the formation of flower-like micelles as has been suggested for block copolymers with this type of architecture (*3*, *24*). Several groups have reported the presence of large aggregates in addition to regular micelles in aqueous solution of PEO containing block copolymers (*25*). However, in our investigations, the presence of large aggregates or clusters were more evident in triblock copolymer solution, an expected observation for this type of amphiphilic triblock copolymers.

Details of the inner structure of the clusters can be obtained by SANS and SAXS measurements. Independent of the concentration, measured in the range between 0.05 and 1 g/mL, a very similar scattering behavior was found (Figure 4). The best approximation can be done by a cubic lattice structure of spheres, either bcc or sc. This is in agreement to the results of Francois et al. (*26*), who found also a cubic structure in solutions of hydrophobically end-capped PEO in a broad concentration range. The typical lattice constant is in the range of 20-30 nm, i.e. a typical micelle diameter.

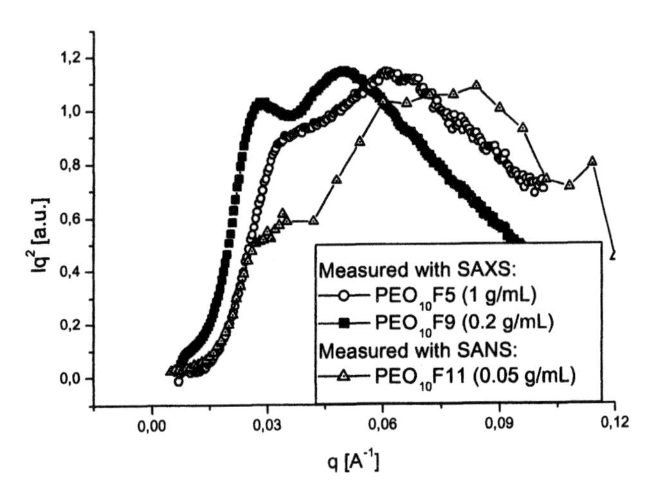

Figure 4. SAXS and SANS traces of Lorentz corrected scattering intensities for three samples of different hydrophobic content and concentration. The approximation can be obtained by assuming a cubic lattice of micelles with a typical lattice constant of 20-30 nm.

Interfacial Properties

Figure 5 shows the experimental π/A isotherms for PEO_5F15-D, $PEO_{10}F11$, and $PEO_{20}F14$ block copolymers (*27*). Due to the limited compression range of our trough, the monolayer had to be deposited at several different surface areas to explore the complete isotherm. The overlapping of the different parts of the obtained isotherms was within the experimental error. The isotherms in Figure 5 have the same appearance irrespective of the block copolymer architecture.

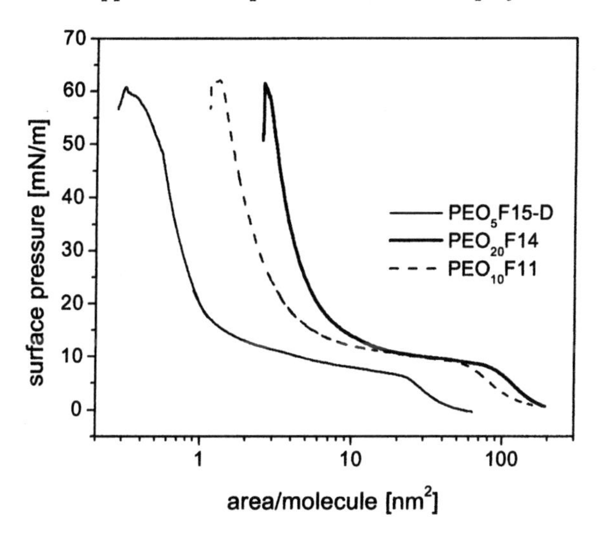

Figure 5. Surface pressure-area isotherms of three different block copolymers at 20°C. The film collapse area is ~ 0.31, 1.31, and 2.65 nm^2/molecule for PEO_5F15-D, $PEO_{10}F11$, and $PEO_{20}F14$, respectively.

The collapse pressure of the monolayer was approximately 60 mN/m for all observed copolymers. For each copolymer isotherm, different regions can be distinguished that correspond to different chain conformations or conformational transitions as described by scaling theories (*28*). A typical liquid expanded phase can be recognized at low surface pressures, where the PFMA hydrophobic segment anchors the polymer chain to the surface while the PEO due to its amphiphilic nature is assumed to adopt a flattened conformation at the interface. This phase can be considered as a self-similar adsorbed layer (SSAL) or due to its appearance as a 'pancake'. This is the characteristic phase for the attractive monomer case (*29*) (i.e. the soluble block has an affinity for the water surface). In contrast, the non-attractive interfaces repel the soluble block from the surface already at low surface coverage, i.e. the mushroom regime. A pseudo-plateau is seen at $\pi \sim 9$ mN/m for triblock which does not exist in the non-attractive case. This pseudo-plateau has also been reported for the PEO homopolymer chains (*30*) and it is assumed to be associated with the dissolution of PEO chains into the water subphase with compression. A first order phase transition has been predicted by theories (*31, 32*) for such systems. However, with a slight but continuous variation of the surface pressure π with A in the present systems, it is difficult to interpret it as a true first order transition. There may be different reasons for this continuous variation of π with A such as the polydispersity of the PEO block (*33*). The plateau region is visible for all our PEO block systems. Faure et al. (*34*) have reported similar behavior for poly(styrene)-*block*-poly(ethylene oxide) copolymers, and interpreted the phase transition corresponding to the plateau region as a first-order transition for block copolymers containing long PEO chains (i.e. PS_{32}-PEO_{700}) with less and less first order character of the transition as the chain length decreases (*35*). The molecular areas A_t at the onset of the respective transition are between 22 nm^2/molecule for the diblock copolymer and 85 nm^2/molecule for $PEO_{20}F14$. With the assumption of a dense copolymer monolayer at the onset of a quasi-SSAL quasi-brush transition (*32*) and an area a_{PEO} = 0.20 nm^2 per PEO monomer, the molecular areas A_{calc} can be obtained ($A_{calc} = Na_{PEO}$ where N is the monomer number) (*36*). The Ligoure quasi-SSAL regime is assigned to a dense pancake structure with a considerable PEO chain overlapping and the respective areas per molecule A_{calc} are in good agreement with our experimental results.

At higher surface coverage the isotherms show a very large increase of the surface pressure. In this regime of the isotherm, a brush conformation is expected. The brush can be viewed as tightly packed PEO chains in the water, anchored at the surface by the PFMA blocks. The area per molecule at the film collapse increases as PEO_5F15-D < $PEO_{10}F11$ < $PEO_{20}F14$ as shown in Figure 5. However, from the data it is difficult to infer the influence of PEO chains at the surface on film collapse. The behavior of $PEO_{10}F11$ and $PEO_{20}F14$ at the interface is quite different from the telechelic poly(ethylene oxide) polymers end-capped with hydrophobic alkane groups. For example, the maximum surface pressure of the brush conformation depended on the PEO chain length, while we observe approximately the same surface pressure at the film collapse for all the three copolymers investigated. This difference may be due to the large difference in hydrophobicity of fluorine-containing PFMA and alkane hydrocarbon chains (*37*). The PFMA block anchors the PEO chains more strongly to the surface as compared with the hydrocarbon chains.

Penetration of Amphiphilic Block Copolymers into Lipid Monolayers

Compression of pure DPhPC film leads to a phase transition to the liquid-expanded phase which is complete at an area of approximately 1.12 nm^2 per lipid molecule. The film collapses at $\pi \sim 40$ mN/m and $A \sim 0.66$ nm^2 per molecule. After injecting the polymer solution into the subphase under a fully expanded DPhPC film, a shift towards a higher area/molecule in the isotherms can be observed. This clearly indicates the penetration of the copolymer chains into the DPhPC monolayer. A considerable increase in surface pressure of the fully expanded DPhPC film is observed with the injection of the copolymer solution into the water subphase. The shift (towards higher area per molecule) in the isotherm increases with the copolymer trough concentration. In addition, at high surface pressures, depending upon the amount of the injected polymer solution, the isotherms of the polymer penetrated DPhPC revert to the pure DPhPC isotherm, indicating the expulsion of the polymer chains from the lipid film. A similar behavior of the isotherms has been reported for poloxamers (triblock copolymer of PEO and PPO) inserted lipid monolayer systems (*9*). At high surface pressures, the isotherms of the poloxamer penetrated lipid monolayer reverted to that of the pure lipid, and this was explained by the elimination of the copolymer chains from the lipid monolayer.

The integrated reflection-absorption intensity (calculation details are given in experimental section) of the corresponding ν(O–H) bands within the IRRA spectra of pure DPhPC monolayer (open symbols) and $PEO_{10}F11$ block copolymer penetrated DPhPC monolayer (filled symbols) are given in Figure 6. The block copolymer concentration was 50 nM in the penetration experiment. The initial increase in intensity of the ν(O–H) band of $PEO_{10}F11$ penetrated DPhPC monolayers with compression can be attributed to increase in surface density and the subsequent stretching of the PEO chains in water subphase due to the steric repulsion between the neighboring chains resulting in a more dense and extended conformation (inset *a* in Figure 6). However, due to the presence of the lipid molecules at the surface, the number density of the polymer chains would not be high enough to form a true brush-like conformation.

At around 0.92 nm^2 area/lipid molecule (inset *b*), the maximum retention surface pressure is reached. The block copolymer is still embedded in the monolayer film, but with further compression, the intensity of ν(O–H) bands decreases. Therefore, the polymer chains do not retain their position in the lipid monolayer. At around 0.88 nm^2 area/lipid molecule, the intensity of the ν(O–H) band from the $PEO_{10}F11$ penetrated DPhPC monolayer approached to pure DPhPC monolayer indicating an almost complete exclusion of the polymer chains from the lipid monolayer (inset *c*). A similar trend can be followed by the ν(C–O) band from the IRRA spectrum of the $PEO_{10}F11$ block copolymer penetrated DPhPC monolayer during compression. This indicates that the mechanism of $PEO_{10}F11$ penetration into the DPhPC monolayer might depend on the interaction of the block copolymer and the hydrophobic part of the lipid monolayer, i.e. the PFMA block might penetrate the hydrophobic acyl chains of the lipid monolayer, and that the block copolymer chains have no preferential interaction with the lipid molecule head group, since otherwise IRRA spectra would be expected to show PEO bands at strong compression.

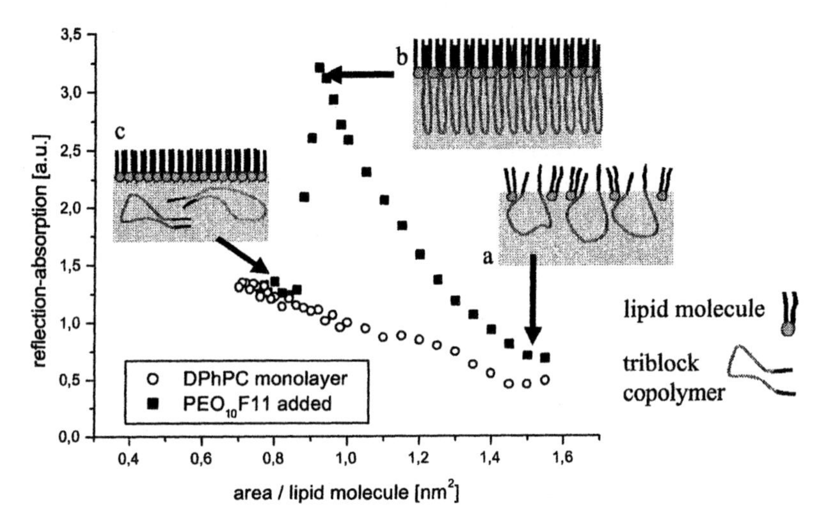

Figure 6. Comparison of the reflection-absorption intensity of the ν(O–H) band from IRRA spectra of pure DPhPC monolayer (o) *and $PEO_{10}F11$ penetrated DPhPC monolayer* (■) *on water surface during compression. The insets schematically represent the behavior of the triblock copolymer chains at the air/water interface in the presence of a lipid monolayer during compression (details in the text).*

Cytotoxicity Results

The way the toxicity measurements (i.e. the toxic effect of the block copolymers on living cells) of the block copolymers have been carried out here, mimics the situation of rapid clearance of the polymer from the blood. In many cases, it is really the case: the concentration of the substance administered into the animal blood decreases 5-10-fold during the first hour (*38*). Hence, the results of our investigations would give information about the "acute" toxicity of the copolymers. Figure 7 shows the block copolymer concentration dependent viability of the K562 human erythroleucemia cells (the percentage of remaining living cells). Each experiment was repeated three times. The results indicate that all the copolymers are practically intoxic under the test conditions.

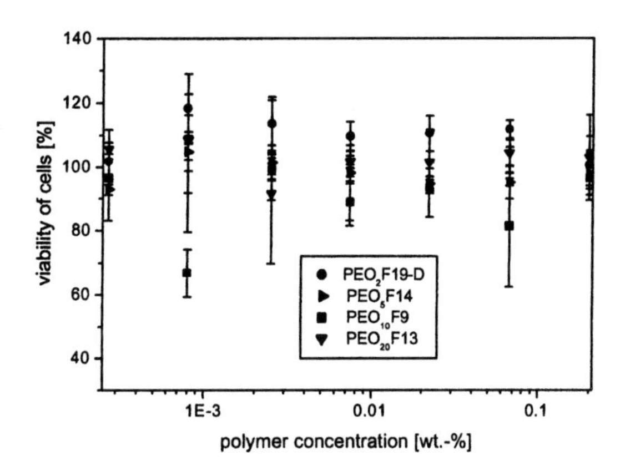

Figure 7. Concentration dependent cytotoxicity of different block copolymers as mentioned in the inset, for K562 human erythroleucemia cells. The y-axis shows the percentage of remaining living cells.

The cell viability exceeding 100 %, as can be seen in Figure 7 for some samples, is within the experimental error. The common inaccuracy in these studies is as large as 15 to 20 %. Therefore, the sample that exhibits above 100 % of living cells simply means that the polymer has no toxicity and nothing else. The data of sample $PEO_{10}F9$ in Figure 7 shows a relatively low cell viability value at lower concentration. However, this cannot be assumed as the real toxicity of the sample. Because, usually toxicity manifests as a regular decrease in the amount of living cells with the increase in the concentration of the toxic agent. For example, exactly under similar test conditions, (i.e. one hour incubation and subsequent culturing for three days), the pluronic copolymers of L61 and L81 showed a real toxicity sigmoidal curve like behavior, having viability of cells value ~ 0 at 0.1 wt.% copolymer concentration (*39*). Therefore, the behavior of the sample $PEO_{10}F9$ with relatively low viability of cell value as compared to other samples at lower copolymer concentration does not represent the real toxicity of the sample.

Acknowledgment

We gratefully acknowledge the Deutsche Forschungsgemeinschaft (DFG, SFB 418), the Bundesministerium für Bildung und Forschung (BMBF, Project 03KRE3HW) and VolkswagenStiftung (AZ: 77742) for financial support.

References

1. Förster S. *Top. Curr. Chem.* **2003**, *226*, 1-28.
2. Thuresson, K.; Nilsson, S.; Kjoniksen, L.; Walderhaug, H.; Lindman, B.; Nystrom, B. *J. Phys. Chem. B* **1999**, *103,* 1425-1436.
3. Maiti, S.; Chatterji, P. R. *J. Phys. Chem. B* **2000**, *104*, 10253-10257.
4. Matsumoto, K.; Kubota, M.; Matsuoka, H.; Yamaoka, H. *Macromolecules* **1999**, *32*, 7122-7127.
5. Zhou, J.; Zhuang, D.; Yuan, X.; Jiang, M.; Y. Zhang, Y. *Langmuir* **2000**, *16,* 9653-9661.
6. Tae, G.; Kornfield, J. A.; Hubbell J. A.; Johannsmann, D.; Esch, T. E. H. *Macromolecules* **2001**, *34*, 6409-6419.
7. Imae, T. *Kobunshi Ronbunshu* **2001**, *58*, 178-188.
8. Busse, K.; Kressler J.; Eck, D.; Höring, S. *Macromolecules* **2002**, *35*, 178-184.
9. Maskarinec, S. A.; Hannig, J.; Lee, R. C.; Lee, K. Y. C. *Biophys. J.* **2002**, *82*, 1453-1459.
10. Bi, X.; Flach, C. R.; Gil, J. P.; Plasencia, I.; Andreu, D.; Oliveira, E.; Mendelsohn, R. *Biochemistry* **2002**, *41*, 8385-8395.
11. Hannig, J.; Zhang, D.; Canaday, D. J.; Beckett, M. A.; Austumian, R. D.; Weichselbaum, R. R.; Lee, R. C. *Radiat. Res.* **2000**, *154*, 171-177.
12. Terry, M. A.; Hannig, J.; Carrillo, C. S.; Beckett, M. A.; Weichselbaum, R. R.; Lee, R. C. *Ann. N. Y. Acad. Sci.* **1999**, *888*, 274-284.
13. Cai, P.; Flach, C. R.; Mendelsohn, R. *Biochemistry* **2003**, *42*, 9446-9452.
14. Brockman, J. M.; Wang, Z. D.; Notter, R. H.; Dluhy, R. A. *Biophys. J.* **2003**, *84*, 326-340.
15. Alakhov, V. Y.; Moskaleva, E. Y.; Batrakova, E. V.; Kabanov, A. V. *Bioconj. Chem.* **1996**, 7, 209-216.
16. Hussain, H.; Budde, H.; Höring, S.; Busse, K.; Kressler, *J. Macromol. Chem. Phys.* **2002**, 203, 2103-2112.
17. Provencher, S. W. *Macromol. Chem.* **1979**, *180*, 201-209.
18. Kerth, A.; Gericke, A.; Blume, A. To be published.

19. Skoog, B. *Vox Sang* **1979**, *37*, 345-349.
20. Cole, S.C.; Christensen, G. A.; Olson, W. P. *Anal. Biochem.* **1983**, *134*, 368-373.
21. Carmichael, J.; DeGraff, W. G.; Gazdar, A. F.; Minna, J. D.; Mitchell J. B. *Cancer Res.* **1987**, *47*, 936-942.
22. Cole, S. P. C. *Cancer Chemother. Pharmacol.* **1986**, *17*, 259-263.
23. Hussain, H.; Busse, K.; Kressler, J. *Macromol. Chem. Phys.* **2003**, *204*, 936-946.
24. Liu, T.; Zhou, Z.; Wu, C.; Nace, V. M.; Chu, B. *J. Phys. Chem. B* **1998**, *102*, 2875-2882.
25. Brown, W.; Schillen, K.; Hvidt, S.; *J. Phys. Chem.* **1992**, *96*, 6038-6044.
26. Francois, J.; Maitre, S.; Rawiso, M.; Sarazin, D.; Beinert, G.; Isel, F., *Colloids and Surfaces, A: Physicochemical and Engineering Aspects* **1996**, *112(2/3)*, 251-265.
27. Hussain H.; Kerth, A. Blume, A.; Kressler, J.; *J. Phys. Chem. B* **2004**, *108*, 9962-9969.
28. de Gennes, P. G. *Macromolecules* **1980**, *13*, 1069-1075.
29. Paeng, K.; Choi, J.; Park, Y.; Sohn, D. *Colloids Surf.* **2003**, *220*, 1-7.
30. Barentin, C.; Muller, P.; Joanny, J. F.; *Macromolecules* **1998**, *31*, 2198-2211.
31. Alexander, S.; *J. Phys. (Paris)* **1977**, *38*, 977-981.
32. Ligoure, C.; *J. Phys. II* **1993**, *3*, 1607-1617.
33. da Silva, A. M. G.; Filipe, E. J. M.; d'Oliveira, J. M. R.; Martinho, J. M. G. *Langmuir* **1996** *12*, 6547-6553.
34. Faure, M. C.; Bassereau, P.; Desbat, B. *Eur. Phys. J. E* **2000**, *2*, 145-151.
35. Faure, M. C.; Bassereau, P.; Lee, L. T.; Menelle, A.; Lheveder, C. *Macromolecules* **1999**, *32*, 8538-8550.
36. O'Connell, A. M.; Koeppe, R. E.; Andersen, O. S.; *Science* **1990**, *250*, 1256-1259.
37. Ravey, J. C.; Stebe, M. J.; *Colloids Surf. A* **1994**, *84*, 11-31.
38. Batrakova, E. V.; Dorodnych, T. Y.; Klinskii, E. Y.; Kliushnenkova, E. N.; Shemchukova, O. B.; Goncharova, O. N.; Arjakov, S. A.; Alakhov, V. Y.; Kabanov, A. V.; *Br. J. Cancer* **1996**, *74*, 1545-1552.
39. Demina, T. (unpublished results).

Chapter 9

Elucidating the Kinetics of β-Amyloid Fibril Formation

Nadia J. Edwin, Grigor B. Bantchev, Paul S. Russo*, Robert P. Hammer, and Robin L. McCarley

Department of Chemistry and Macromolecular Studies Group, Louisiana State University, Baton Rouge, LA 70803

ABSTRACT: The formation of β-Amyloid peptide ($A\beta_{1-40}$) aggregates was monitored by dynamic light scattering. Various sizes of materials may be present throughout the aggregation process, but small scatterers are difficult to detect in the presence of large ones. Fluorescence photobleaching recovery studies on 5-carboxyfluorescein-labeled $A\beta_{1-40}$ peptide solutions readily confirmed the presence of large and small species simultaneously. The effects of dye substitution on the aggregation behavior of $A\beta_{1-40}$ peptide are subtle, but should not prevent further investigations by fluorescence photobleaching recovery or other fluorescence methods.

The brains of individuals with Alzheimer's disease (AD) are characterized by insoluble extracellular amyloid plaques (1) made up of β-amyloid peptide. Collectively called Aβ, these fragments contain 39-43 amino acids cleaved from a larger and harmless membrane protein, the β-amyloid precursor protein (βAPP) whose normal physiological function remains uncertain (2). The $A\beta_{1-40}$ and $A\beta_{1-42}$ fragments are the predominant forms (3, 4). Both generate fibrils *in vitro* by a seeded polymerization mechanism (5). As there is evidence that protofibrils—low oligomeric species that arise early in the aggregation process—may be the toxic form(6), methods to study the mechanism of protein aggregation should be able to monitor continuously the kinetics of fibril growth and identify the early and intermediate stages of assembly. Physiological factors that induce the aggregation of soluble Aβ are of interest in determining the cause of Aβ fibril formation (7), and it is known that the peptide tends to self-assemble under conditions of low and neutral pH. The peptide remains in a monomeric or low oligomeric state at high pH values.

Experimental data relating to the early stages of amyloid aggregation are difficult to obtain (8). Methods based on mass transport must be able to measure both rapidly and slowly diffusing peptide aggregates within the same sample. The most common of these, dynamic light scattering (DLS), measures the diffusion of molecules by tracking fluctuations in the intensity of the scattered light. The data are heavily weighted toward larger particles in the distribution; thus, the method is most sensitive to aggregated Aβ. Even with this limitation, DLS has been used successfully in kinetics studies of Aβ aggregation (9-15).

Fluorescence photobleaching recovery (FPR) is a well-established technique (16-19) for the measurement of diffusion coefficients. It requires the attachment, preferably covalent, of a fluorophore to the molecules of interest. No matter how this is done, the resulting signals are not as heavily weighted towards the larger sizes of a polydisperse sample as they are in DLS. In the case of Aβ, it has been shown that labeling exclusively at the *N*-terminus of the peptide minimizes modifications to the peptide conformation caused by the fluorophore's presence and preserves the original biological activity (20-21). The present study, which is designed to investigate the formation of Aβ aggregation using both DLS and FPR, marks the first application of the highly selective FPR method to amyloid self assembly in solution.

Materials and Methods

Reagents and Chemicals

Peptides -- β-amyloid(1-40) and 5-carboxyfluorescein β-amyloid(1-40) were purchased from Anaspec, Inc. (San Jose, CA). For DLS, β-amyloid(1-40)

(Catalog No. 24236, Lot No. 15716) was used. For FPR, β-amyloid(1-40) and 5-carboxyfluorescein β-Amyloid(1-40) (Catalog No. 20698, Lot No. 12707 and Catalog No. 23513, Lot No. 17235 respectively) were used. Phosphoric acid (99.999%, Catalog No. 34,524-5) and semiconductor-grade potassium hydroxide (99.99%, Catalog No. 30,656-8) were obtained from Aldrich Chemical Co. Sodium chloride (99.999%, Catalog No. 10862) was obtained from Alfa Aesar. The filters used were from Whatman, 0.02 μm (Anotop 10, Catalog No. 6809-1102). The phosphate buffer used for DLS studies was bought as a ready salt from Fisher Scientific (Catalog No. B81).

Sample Preparation

Dynamic Light Scattering

The samples were prepared according to the method of Aucoin (22). Briefly, stock solutions were prepared by dissolving the peptide in filtered 10 mM KOH (pH 11) and vortexing with a Daigger Vortex Genie 2 until completely dissolved. β-amyloid$_{1\text{-}40}$ is known to exist in a monomeric state under sufficiently basic conditions (23). In some cases, the sample was measured at such pH values. Otherwise, the stock was adjusted to the appropriate pH with phosphate buffer and double-filtered through 0.02 μm filters into a disposable, polystyrene cuvette that had been thoroughly rinsed with deionized water. The samples were prepared and kept during the experiment under nitrogen to preserve their original pH.

Fluorescence Photobleaching Recovery

Samples for FPR were prepared in a similar manner to DLS, except in the case of the 5-carboxyfluorescein-labeled peptide, Lot No. 17235, the sample was difficult to solubilize, so it was sonicated in a Branson Model No. 2510 bath sonicator. The sonication was done in five second cycles to prevent heat from affecting the sample. A phosphate-buffered saline solution, pH 7.4 (0.5 M phosphoric acid, 1.5 M sodium chloride, 5 M potassium hydroxide) was filtered through a 0.02 μm filter, after which it was mixed with the desired volume of peptide stock to make a final solution of 100 μM peptide in PBS, pH 7.4 (the final concentrations were 0.05 M phosphoric acid, 0.15 M sodium chloride, 0.5 M potassium hydroxide). Water was first added to the peptide filtrate prior to adding the desired amount of PBS needed for 50 mM in order to prevent rapid aggregation of sample in such high ionic strength buffer. The samples were loaded in 0.2-mm-path-length rectangular microslides (Vitrocom) by capillary action, and the microslides were flame-sealed. Solutions for circular dichroism were prepared in like manner.

Experimental Section

Dynamic Light Scattering. The solution was placed in a disposable polystyrene cuvette, and dynamic light scattering measurements were taken with a Zetasizer Nano ZS, (Malvern Instruments, England). The apparatus is equipped with a HeNe laser (632.8 nm) and fiber optics detector positioned at an angle of 173°. The data were collected with a typical acquisition time of 2 minutes and analyzed with Laplace inversion software supplied with the apparatus (v1.1). The software reports an effective hydrodynamic diameter, which is the diameter of a spherical particle that has the same diffusion coefficient as that computed from the correlograms. Peaks at sizes larger than 2 μm were ignored as artificial or beyond the range of interest of this study. The intensity values, uncorrected for number or mass concentration, were recorded from the software.

FPR Measurement. A brief summary of the FPR apparatus follows; details have appeared elsewhere (16). A fringed pattern is bleached into the sample by intense laser illumination (at 488 nm, 3-7 W/cm^2) of a coarse diffraction grating (50, 100, 150, and 300 lines/inch), referred to as a Ronchi ruling (Edmund Scientific), held in the rear image plane of an objective mounted on an epifluorescence microscope. The stripe pattern is described by the grating constant, $K = 2\pi/L$, where L is the period of the repeat pattern in the sample. After the bleaching, the laser beam is shifted to lower intensity to detect the fading, due the diffusion, of the bleached pattern without causing additional parasitic bleaching. The detection is accomplished by a modulation system (19). After the shutter that protects the RCA 7265 photomultiplier tube (PMT) during the photobleaching step reopens, the Ronchi ruling is vibrated electromechanically in a direction perpendicular to its stripes to produce a modulated signal, which is buffered by a Stanford Research Systems SR560 low-noise preamplifier and then fed to a tuned amplifier to isolate the effects of shallow (typically 5%) photobleaching from random noise. The modulation detection system not only enables shallow, weakly perturbing photobleaches to be used, but it also eliminates decay terms due to higher Fourier components of the stripe pattern. Thus, each diffuser gives rise to a single decay term (18). A zero-crossing detector/peak voltage detection circuit triggers an analog-digital card from National Instruments, (#AT-MIO-16D Part #320489-01) to acquire the peak voltage of the modulated signal, or contrast $C(t)$, which decays exponentially due to diffusion: $C(t) = baseline + C(0)e^{-K^2Dt}$.

Results and Discussion

The aggregation kinetics of the solution were measured by DLS. The settings of the apparatus were such that multiple repeat measurements were made from a small volume that may or may not have contained large aggregates during the acquisition time. Under these conditions, hydrodynamic diameters returned from the instrument represent apparent values; however, an increase in the observation frequency of large-diameter particles unmistakably reflects aggregation. Figure 1 shows the presence of small aggregates immediately after preparation of a sample. Many aggregates were bigger than the nominal size of the membrane pores (20 nm) through which the solution was filtered. The sizes of the aggregates in solution changed slowly over the course of the first 100 hours, as evidenced by the slow shift of the distribution toward larger diameters. The distribution ultimately *appears* to become narrow—i.e., the majority of the detected aggregates has a hydrodynamic diameter of 100–200 nm.

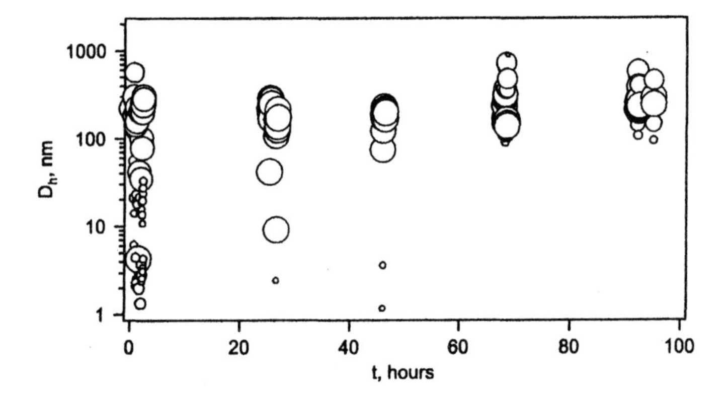

Figure 1. Growth kinetics of Aβ at a sample concentration of 50 μM $A\beta_{1\text{-}40}$ in phosphate buffer, pH 7.6 (I = 30 mM) by DLS. The graph illustrates the resulting distributions from measurements at different times following introduction of the sample into the cell. Each bubble in the plot corresponds to a mode identified during a measurement (a column of bubbles). The area of a single bubble is proportional to the area (by intensity) of the peak reported from the Zetasizer. The sum of the bubble areas for a given measurement adds to 100%. The vertical position of a bubble corresponds to the reported diameter of the particles in the peak. Thus, a small bubble near the bottom of the plot signifies a weak, fast decay mode.

The data are consistent with a report made by Kremer *et al.* (24) for an acceleration of growth kinetics during the first 100 hours. The results show much slower kinetics than reported in the literature for a solution at low pH (0.1 M HCl) (7). The data of Figure 1 and similar measurements under other conditions of pH and salt (not shown) demonstrate the limitations of DLS for characterizing dilute solutions of macromolecules that assemble into very large assemblies. The first problem is that the initial scattering may barely exceed solvent levels. Then there is almost no coherent signal above baseline (except for the solvent contribution, which occurs at very short lag times inaccessible to the correlator). This limitation applies equally to a conventional DLS instrument (using pinholes to define a coherence area at the detector, limiting the scattering to a very small volume) and to instruments, like the one used here, that rely on single-mode fiber optics to view a larger volume coherently. The scattering above solvent level is proportional to the product of mass per volume concentration, c, and the (weight average) molecular weight of the scatterer, M. The concentration remains constant during aggregation, but M increases and, with it, the scattering intensity above solvent level. After some aggregation, the DLS signal becomes easier to measure, but any small diffusers that remain are difficult to detect. It is not possible from Figure 1 to confirm or deny the presence of small aggregates (< 100 nm) at long times. Another problem arises when very large aggregates are present: the number of particles in the viewing volume may be too small. Then the magnitude of unwanted fluctuations due solely to the number of particles in the detected volume, (25) their rate being governed by transit time through a long distance defined by the illumination and detection optics, approaches that of the desired fluctuations due to diffusion on the distance scale $2\pi/q$, where q is the scattering vector magnitude. The large volumes made possible by a single-mode fiber optic instrument push the number fluctuation issue toward larger and slower diffusers, but the wide-ranging autocorrelators now commonly in use still capture these slow signal variations. One may anticipate a number fluctuation problem when slow signal variations that were not present prior to the aggregation appear. If the number of large particles in the detected volume fluctuates from zero to just a few, then spikes in the signal appear. Such spikes were visible in the present measurements at long times.

Unlike DLS, the signal intensity in FPR does not increase with aggregation. Barring quenching of dyes caused by aggregation or changes in the ease with which dyes can be bleached, the FPR signal remains constant and proportional to the number of dye molecules in the viewing volume. Small diffusers do not get hidden by large ones in FPR. After determining that the bleaching time had little effect on the diffusion values (data not shown), samples were prepared having different ratios of 5-carboxyfluorescein labeled to unlabeled $A\beta_{1\text{-}40}$. These samples were measured and compared to 100% labeled

$A\beta_{1-40}$ as a test that attachment of a probe had little effect on the sample properties (Figure 2). The diffusion coefficient of the 100% labeled sample was within error of that measured in the mixed samples, indicating that the dye moiety does not itself preferentially interact with unlabeled $A\beta_{1-40}$ molecules.

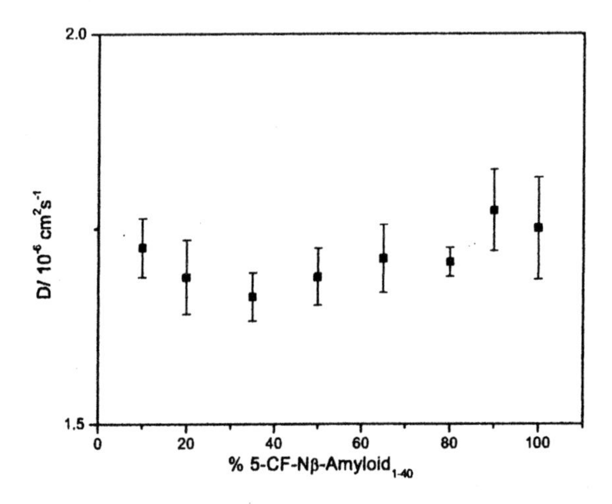

Figure 2. Diffusion as a function of dyed to undyed Aβ, expressed as percentage of 50 μM 5-carboxyfluorescein $A\beta_{1-40}$ mixed with 50 μM $A\beta_{1-40}$ Conditions: 50 mM PBS, 150 mM NaCl, pH 7.4. Error bars are standard deviations from triplicate runs.

Subsequently, FPR measurements were made on samples containing 100% 5-carboxyfluorescein-labeled peptide at different pH values. In Figure 3 are shown results for labeled $A\beta_{1-40}$ peptide at three different pH values (2.7, 6.9, 11), spanning the range where the peptide is believed to exist in the high to low oligomeric states. The diffusion coefficient values at pH 6.7 and 11 were almost identical and remained constant over a period of two weeks. The diffusion values are consistent with theoretical predictions (26), as are those in Figure 2. They also match experimental values for monomeric $A\beta_{1-40}$ measured elsewhere using diffusion-ordered NMR spectroscopy (DOSY) (27). This gives us confidence that the attachment of dye does not radically alter the hydrodynamic size of the peptide; however, these preliminary results are perplexing as the $A\beta_{1-40}$ peptide was widely thought to aggregate within the experimental time frame at neutral pH

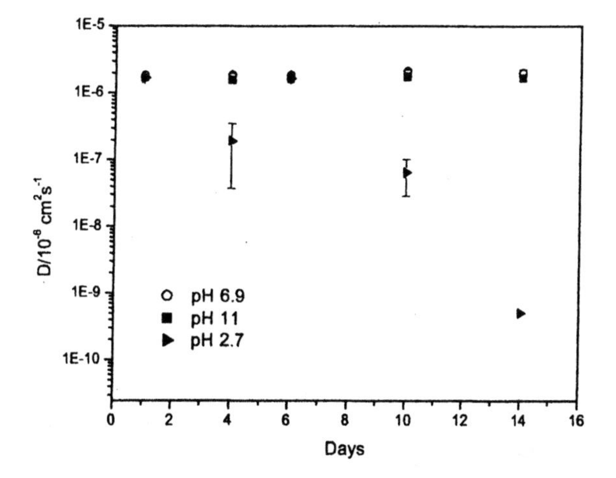

Figure 3. Diffusion of 100 μM 5-carboxyfluorescein $A\beta_{1\text{-}40}$ in 50 mM PBS, 150 mM NaCl, pH values ○ 6.9, ■ 11, ▶ 2.7. Error bars are standard deviation of triplicate runs.

values. Thus, smaller values of diffusion coefficient might have been expected at long times. At pH 2.7, the diffusion coefficient did decrease significantly with time. When this result is converted to diameter via the Stokes-Einstein equation ($D_h = kT/3\pi\eta D$, where η represents the viscosity), one finds an enormous increase in hydrodynamic diameter from about 2.6 nm to about 860 nm. This observation correctly indicates the power of FPR. Such a large change in size is difficult to follow by DLS, which struggles with the small-diameter diffusers at low concentration while requiring a particle form factor correction of unknown nature for the large aggregates, not to mention the number fluctuation issues already discussed. DOSY NMR spectroscopy could probably succeed during the initial stages, despite the low concentrations, but the larger aggregates might pose problems. Fluorescence correlation spectroscopy can handle this range of diffusers, but in addition to a more delicate setup and the need to arrange for a vanishingly small number of molecules labeled with a nonbleachable dye, adequate acquisition time must be allowed to sample the full population of diffusing entities.

The decreases in diffusion in Figure 3 do not fully reflect the presence of macroscopically large aggregates that were visible immediately after lowering the pH. An fluorescence microscopy image of the aggregated peptide (100 μM 5-carboxyfluorescein $A\beta_{1\text{-}40}$, 50 mM PBS, 150 mM NaCl, pH 2.7), is shown in Figure 4a, accompanied by an image of the stripe pattern in Figure 4b. Visible aggregates are present throughout, but they are not uniformly distributed; thus, the measured diffusion can vary within the sample cell, depending a little on the position chosen for measurement. One advantage of FPR is that one may choose visually to measure regions that do not possess extraordinarily large aggregates. If the striped pattern were to illuminate regions with many of these very large aggregates, the recovery of the signal would be incomplete on the time scale of observations in this study. Very long observations might then reveal the rate of exchange of molecules into and out of the very large aggregates and/or very slow diffusion along the fibrils.

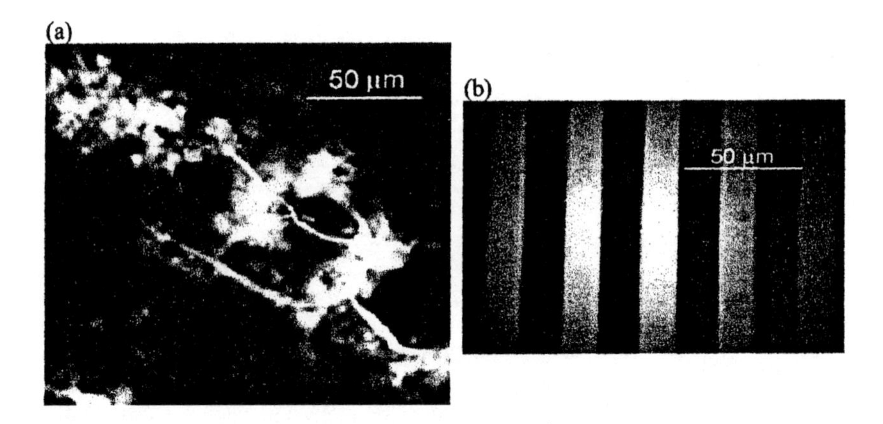

Figure 4. (a) Fluorescence microscopy image of 100 μM 5-carboxyfluorescein $A\beta_{1\text{-}40}$ in 50 mM PBS, 150 mM NaCl, pH 2.7. (b) microscopy image of Ronchi ruling stripe pattern in labeled gelatin.

The log-log curves of Figure 5 demonstrate the effectiveness of FPR as a tool to detect simultaneously both the large, slow diffusers and the small, fast ones within a given sample. Over 99% of the contrast was relaxed in these measurements, indicating that immobile fragments were avoided. Even in cases where a large fragment is illuminated, such that the recovery levels out after some time, it is possible to determine the size of the mobile fraction by treating the immobile fraction as a baseline term during analysis. The FPR experimenter is not easily confounded.

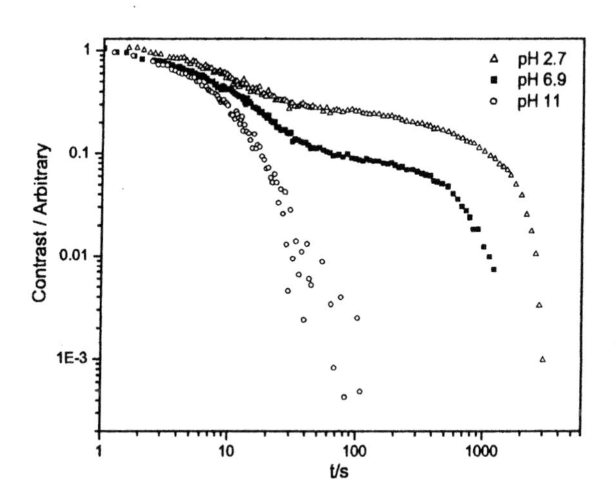

Figure 5. Log-log FPR traces for three pH values. (25/75 % Mixture 5-carboxyfluorescein $A\beta_{1\text{-}40}$ and $A\beta_{1\text{-}40}$).

Conclusions

For the first time, an FPR instrument using a modulated detection scheme has been applied to the study of Aβ aggregation. Attachment of 5-carboxyfluorescein does not cause a specific interaction with unlabeled peptide materials, and the resulting diffusion coefficients are similar to those measured by DOSY NMR and from theoretical expectations. Incorporation of labeled material into large fibrils occurs. FPR proves more sensitive than DLS for detection of low oligomer aggregates of $A\beta_{1\text{-}40}$ coexisting with much larger fibrils. Ongoing studies suggest that factors other than pH and time can affect the conformation and aggregation state of $A\beta_{1\text{-}40}$. In some instances, a contraction of fluorescently labeled $A\beta_{1\text{-}40}$ at low and medium pH has been observed, with concomitant effects on stability. These observations deserve additional attention, as does the reversibility of aggregate formation under conditions of salt, pH and potential therapeutic agents. The FPR method holds much promise for characterization of self-assembling systems, both biological and synthetic, in which small and very large particles coexist.

Acknowledgment. This work was supported by the National Institutes of Aging of the National Institutes of Health (AG17983) and NSF-IGERT, Division of Graduate Education (9987603). We thank Malvern Instruments for providing the Zetasizer instrument on loan.

References

1. Esler, W.P.; Wolfe, M.S. A Portrait of Alzheimer Secretases-New Features and Familiar Faces. *Science* **2001**, *293*, 1449-1454.

2. Saitoh, T.; Mook-Jung, I. Is Understanding the Biological Function of APP Important in Understanding Alzheimer's Disease? *J. Alz. Disease* **1999**, *1*, 287-295.

3. Lansbury, P.T., Jr. A Reductionist View of Alzheimer's Disease. *Acc. Chem.Res.* **1996**, *29*, 317-321.

4. Selkoe, D.J. Alzheimer's Disease: Genotype, Phenotype, and Treatments. *Science* **1997**, *275*, 630-631.

5. Jarrett, J.T.; Berger, E.P.; Lansbury, P.T.; Jr. The Carboxyl Terminus of β-Amyloid Protein is Critical for the Seeding of Amyloid Formation: Implications for the Pathogenesis of Alzheimer's Disease. *Biochemistry* **1993**, *32*, 4693-4697.

6. Harper, James D.; Wong, Stanislaus S.; Lieber, Charles M.; Lansbury, P.T.; Jr. Assembly of Aβ Amyloid Protofibrils: An in Vitro Model for a Possible Early Event in Alzheimer's Disease. *Biochemistry* **1999**, *38*, 8972-8980.

7. Bush, A.I.; Pettingell, W.H. Jr.; Paradis, d. M.; Tanzi, R.E. Modulation of Aβ Adhesiveness and Secretase Site Cleavage by Zinc. *J. Biol. Chem.* **1994**, *269*, 12152-12158.

8. Moore, S.; El-Agnaf, O.; Davies, Y.; Allsop, D. Studying the Aggregation of Proteins Implicated in Neurodegenerative Diseases in Real Time. *American Genomic/Proteomic Technology* **2002**, 40-43.

9. Lomakin, A.; Chung, D.S.; Benedek, G.B.; Kirschner, D.A.; Teplow, D.B. On the Nucleation and Growth of Amyloid β-protein fibrils: Detection of Nuclei and Quantitation of Rate Constants. *Proc.Natl.Acad.Sci.U.S.A.* **1996**, *93*, 1125-1129.

10. Lomakin, A.; Benedek, G.B.; Templow, D.B. Monitoring Protein Assembly Using Quasielastic Light Scattering Spectroscopy. *Meth.Enzymology* **1999**, *309*, 429-459.

11. Lomakin, A.; Teplow, D.B.; Kirschner, D.A.; Benedek, G.B. Kinetic Theory of Fibrillogenesis of Amyloid β-Protein. *Proc.Natl.Acad.Sci.U.S.A.* **1997**, *94*, 7942-7947.

12. Thunecke, M.; Lobbia, A.; Kosciessa, U.; Dyrks, T.; Oakley, A.E.; Turner, J.; Saenger, W.; Georgalis, Y. Aggregation of Aβ Alzheimer's Disease-Related Peptide Studied by Dynamic Light Scattering. *J.Peptide Res.* **1998**, *52*, 509-517.

13. Pallitto, M.M.; Ghanta, J.; Heinzelman, P.; Kiessling, L.L.; Murphy. R.M. Recognition Sequence Design for Peptidyl Modulators of β-Amyloid Aggregation and Toxicity. *Biochemistry* **1999**, *38*, 3570-3578.

14. Ghanta, J.; Shen, C.; Kiessling, L.L.; Murphy, R.M. A Strategy for Designing Inhibitors of β-Amyloid Toxicity. *J.Biol.Chem.* **1996**, *272*, 29525-29528.

15. Murphy, R.M. Static and Dynamic Light Scattering of Biological Macromolecules. What Can We Learn? *Curr.Opin.Biotechnol.* **1997**, *8*, 25-30.

16. Bu, Z.; Russo, P.S. Diffusion of Dextran in Aqueous (Hydroxypropyl)cellulose. *Macromolecules* **1994**, 27, 1187-1194.

17. Bu, Z.; Russo, P.S; Tipson, D.L.; Negulescu, I.I. Self-Diffusion of Rodlike Polymers in Isotropic Solutions. *Macromolecules* **1994**, 27, 6871-6882.

18. Ware, B.R. Fluorescence Photobleaching Recovery. *Am. Lab.* **1984**, *16*, 16.

19. Lanni, F.; Ware, B. R. Modulation Detection of Fluorescence Photobleaching Recovery. *Rev. Sci. Instrum.* **1982,** *53* (6), 905.

20. Fulop, L.; Penke, B.; Zarandi, M. Synthesis and Fluorescent Labeling of Beta-Amyloid Peptides. *Journal of Peptide Science* **2001,** *7,* 397-401.

21. Prior, R.; D'Urso, D.; Frank, R.; Prikulis, I.; Cleven, S.; Ihl, R.; Pavlakovic, G. Selective Binding of Soluble Abeta 1-40 and Abeta 1-42 to a Subset of Senile Plaques. *American Journal of Pathology* **1996,** *148,* 1749-1756.

22. Aucoin, P.J. Dissertation, Louisiana State University, Baton Rouge, LA, **2003.**

23. Gorman, P.M.; Yip, C.M.; Fraser, P.E.; Chakrabartty, A. Alternate Aggregation Pathways of the Alzheimer's β-Amyloid Peptide: Aβ Association Kinetics at Endosomal pH. *J.Mol.Biol.* **2003,** *325,* 743-757.

24. Kremer, J.J.; Pallitto, M.M.; Sklansky, D.J.; Murphy, R.M. Correlation of β-amyloid aggregate size and hydrophobicity with decreased bilayer fluidity of model membranes. *Biochemistry,* **2000,** *39* (33), 10309-10318.

25. Berne, B.; Pecora, R. *Dynamic Light Scattering* (Interscience, New York, 1975).

26. Massi, F.; Peng, J.W.; Lee, J.P.; Straub, J.E. Stimulation Study of the Structure and Dynamics of the Alzheimer's Amyloid Peptide Congener in Solution. *Biophysical Journal* **2001,** *80,* 31-44.

27. Sticht, H.; Bayer, P.; Willbold, D.; Dames, S.; Hilbich, C.; Beyreuther, K.; Frank, R.; Rösch, P. Structure of Amyloid A4-(1-40)-peptide of Alzheimer's Disease. *Eur. J. Biochem.* **1995,** *233,* 293-298.

Chapter 10

Synthesis of an ABA Triblock Copolymer of Poly(DL-lactide) and Poly(ethylene glycol) and Blends with Poly(ε-caprolactone) as a Promising Material for Biomedical Application

Aleksandra Porjazoska[1,2], Oksan Karal Yilmaz[2], Nilhan Kayaman-Apohan[3], Kemal Baysal[2], Maja Cvetkovska[1,*], and Bahattin M. Baysal[2,4]

[1]Faculty of Technology and Metallurgy, The "Sv. Kiril and Metodij" University, 1000 Skopje, Republic of Macedonia
[2]TUBITAK, Research Institute for Genetic Engineering and Biotechnology, P.O. Box 21, 41470 Gebze, Turkey
[3]Department of Chemistry, Marmara University, 81040, Göztepe-Istanbul, Turkey
[4]Department of Chemical Engineering, Boğaziçi University, 34342, Bebek-Istanbul, Turkey

The aim of this work was preparation of blends based on ABA triblock copolymers [A=poly(DL-lactide); B=poly(ethylene glycol)] and poly(ε-caprolactone) (PCL), their characterization, hydrolytic degradation study, and evaluation of the possibility to use these materials in tissue engineering. GPC, FTIR, ^{1}H NMR and DSC analyses were used for block copolymers characterization, and the corresponding blends with PCL, were characterized by FTIR, TGA, DSC and contact angle measurements. Porous films were prepared from triblock copolymers, and corresponding blends using solvent casting particulate-leaching technique. Hydrolytic *in vitro* degradation was performed in phosphate buffer solution. Preliminary cell growth experiments, using L929 mouse fibroblasts, were performed to observe the attachment and growth of cells on polymer matrices.

Biodegradable and bioabsorbable aliphatic polyesters, like polylactide (PLA), polyglycolide (PGA), and polycaprolactone (PCL), as well as their copolymers and blends (*1-4*), are probably the most attractive and promising materials for biomedical applications (biodegradable sutures, artificial skin, bone reparation systems, scaffolds in the tissue regeneration, matrices for implants of drug delivery systems etc.) (*5-8*).

Poly(lactide-co-glycolide) (PLGA) have been widely used as bidegadable sutures and microparticle matrix for drug delivery owing to their safety, satisfactory mechanical properties and varied degradation rates. By changing the molecular weight and chemical composition, as well as the morphology, polymers with degradation rates ranging from weeks to months can be obtained. PCL was also used as a drug carrier because of its excellent drug permeability, but it is only suitable for long-term drug delivery system, due to its high crystallinity and low degradation rate (*2-4*).

Aliphatic polyesters are however, hydrophobic and they adsorb proteins from the blood which can cause some side effects when these are implanted (*9*). This phenomenon can be minimized by copolymerization or blending with poly(ethylene oxide) (PEO), or by adding copolymers of ethylene oxide (EO) and propylene oxide (PO), as surfactants (*10-12*). PPO is hydrophobic and thus compatible with hydrophobic bulk material, whereas the hydrophilic PEO will mitigate protein adsorption (*12*).

It was reported, for a family of tyrosine/PEG – derived polycarbonates (*13*), or for PLGA/PEG copolymers and their blends with PLGA (*14*), that a very narrow PEG concentration range (between 4 and 16 wt% PEG at the surface), had a strong regulatory effect on key cellular responses. It is believed that these observations could be generalized and used in designing of substrates in tissue engineering.

Thus, the main objectives in our study were to synthesize and characterize ABA triblock copolymers (A = poly(DL-lactide), and B = poly(ethylene glycol)), to prepare their blends with poly(ε-caprolactone), and to evaluate the efficiency of these polymer systems as supports in cell growth experiments.

Experimental

Materials

α,ω-dihydroxy terminated poly(ethylene glycol) with molar mass Mn = 2000 g mol^{-1} (PEG 2000) was supplied from Fluka (Germany).

DL-lactide (Polysciences (PA, USA)) was purified by recrystalization from dry toluene and kept in vacuum until use.

A catalyst – stannous octoate, supplied by Sigma Corp. (St. Louis, MO, USA), was used as it was received.

Dihydroxy terminated poly(ε-caprolactone), with molar mass $Mn = 4.0 \times 10^4$ g mol^{-1} was purchased from Polysciences.

Synthesis of poly(DL-lactide)-poly(ethylene glycol)-poly(DL-lactide) triblock copolymer (PDLLA-PEG-PDLLA)

ABA triblock copolymer was synthesized by ring-opening polymerization of DL-lactide, using α,ω-dihydroxy terminated poly(ethylene glycol) as macroinitiator, in the presence of stannous octoate as a catalyst.

The feed composition was DL-lactide/PEG = 97/3 and/or 90/10 wt/wt. 0.3 and/or 1 g PEG was first introduced into the polymerization tube and melted in vacuum, at 70 °C, for 1h. A predetermined amount of DL-lactide (9.7 and/or 9 g) and a solution of stannous octoate in dry chloroform (molar ratio monomer/catalyst = 1000), were added to the melted and dried PEG. Chloroform was removed from the reaction mixture using vacuum, and the tube was sealed and put in silicone oil bath at 110 °C, for 20 h. The product was dissolved in chloroform, precipited in methanol and dried in vacuum at room temperature, until a constant weight was obtained.

Preparation of PDLLA-PEG-PDLLA/PCL blends

Blends with composition of PDLLA-PEG-PDLLA/PCL = 95/5, 90/10, 85/15 and 80/20 wt/wt were prepared using solvent evaporation technique. The predetermined amounts of the copolymer and PCL were dissolved in chloroform, precipited in methanol and dried in vacuum until a constant weight was attained.

Characterization

Molar masses were determined by gel permeation chromatography (GPC) (Waters, MA, USA), with Waters styragel column HT6F and Waters 410 differential refractometer detector. The eluting solvent was THF at a flow rate of 1 ml min^{-1}.

^{1}H NMR spectra of the triblock copolymers, were taken in deuterated chloroform at 20 °C, on a Bruker AC 200L spectrometer (MA, USA) at 200 MHz.

Fourier transform infra red spectra were obtained using Perkin Elmer 983 IR spectrometer (MA, USA). PDLLA-PEG-PDLLA triblock copolymers and blends with PCL were recorded from their solution in chloroform, poured onto NaCl disk.

DuPont DSC 910 Model (DE, USA) device was used for thermal characterization of the ABA triblock copolymers, PCL, and their blends. The samples were first heated under nitrogen to +100 °C, then quenched to -140 °C, with a heating rate of 10 °C min^{-1}. This heating/cooling cycle was repeated twice. Reported thermograms were always taken from the second heating run.

TGA spectra were obtained using DuPont 951 Thermogravimetric Analyzer (DE, USA), at the temperature range from +50 to +600 °C, with a heating rate of 10 °C min^{-1}, in the nitrogen atmosphere.

The contact angle measurements were carried out using a Model G-III contact angle meter (KERNCO Instrument Co., El Paso, TX) at room temperature and ambient humidity. Experiments were performed using liquid drops deposited from a microliter syringe onto the smooth polymer substrates, freshly prepared by spin-coating technique (*15*). The contact angles of triple-distilled water, glycerol, ethylene glycol (as polar), and methylene iodide (as a non-polar solvent) drops were measured. Contact angle data were used to calculate solid surface free-energy components of the samples.

Porous film fabrication and characterization

The porous films from PDLLA-PEG-PDLLA copolymer and PDLLA-PEG-PDLLA/PCL blends were prepared by solvent casting, particulate leaching method. The weight ratios polymer/salt were 85/15 for the copolymer, and 80/20 for the blends. 1 ml of the polymer solution (3.5% *w/V*) was poured into the teflon molds (diameter 22 mm, height 1 mm). The solvent was left to evaporate, and the entrapped salt particles were removed by immersing the films in distilled water for 48 h. Sponges (1 mm thickness) were freeze dried to remove residual water. Porosity of the films were determined by mercury porosimetry (Autopore II 9220, Micromeritics).

Hydrolytic degradation

Hydrolytic *in vitro* degradation studies of the polymer sponges prepared from ABA triblock copolymer and its blends with PCL were performed in phosphate buffered saline (PBS) with pH 7.4, at 37 °C. The phosphate buffer was composed of Na_2HPO_4 (1.15 g), KH_2PO_4 (0.2 g), KCl (0.2 g), and NaCl (8 g) in distilled water (1 litre). Porous sponges were placed in 15 ml PBS and incubated at 37 °C for various periods of time (from 0 to 60 days). The buffer solution was replaced every 60 hours.

The degree of degradation was observed through the mass and molar mass loss, determined by GPC and viscosity measurements, and through the water uptake. The water uptake of the samples was calculated as the ratio $(W_w - W_d) / W_d$, where W_w and W_d were the weights of wet and dry samples, respectively.

Culture of the cells and cell growth experiments

L929 mouse fibroblasts were grown in DMEM/F12 medium containing 10% fetal calf serum, 100 IU ml^{-1} penicillin and 0.1 mg ml^{-1} streptomycin (Culture medium, Biological Industries, Israel) at 37 °C in a 5% CO_2 incubator. Experiments were performed with cells from passages 14-19.

These were performed in 6 well tissue culture polystyrene plates (Techno Plastic Products, Switzerland). The porous biopolymer films, prepared as described above, were sterilized using 100% ethanol which was subsequently washed in sterile PBS. A 100 μl of cell suspension (in trypsin-EDTA solution, Biological Industries) containing 150 000 cells (counted by haemocytometer) was placed onto the biopolymers or previously prepared glass coverslips. After 1 h incubation period, 2 ml of culture medium was added to each well and cells were grown in a CO_2 incubator for another 72 hours.

The amount of cells present on the surfaces of the materials was quantified using a neutral red uptake assay (*16*). Cells grown on the above materials were incubated in sterile PBS containing 2% FCS and 0.001% neutral red (Sigma, St.Louis, MO) for 2 hours at 37 °C in the CO_2 incubator, during which the dye is internalized into viable cells by an active process. This was followed by a brief wash with 4% formaldehyde in PBS, which enabled the cells to be fixed. The internalized dye was solubilized in 1 ml of 50% EtOH, 1% acetic acid in H_2O. The amount of entrapped dye was measured spectrophotometrically at 540 nm, the absorbance is proportional to the number of viable cells in each well.

Results and Discussions

Copolymer synthesis, preparation of blends and their FTIR and NMR analyses

The feed and copolymer composition, and some characteristics of PDLLA-PEG-PDLLA copolymers are given in Table I.

Table I. Characteristics of the copolymers

DL-LA/PEG Feed ratio (wt/wt)	*DL-LA/PEG [a] Copolymer composition (wt/wt)*	*Yield (%)*	*Mn · 10^{-4} [b] (g mol^{-1})*	*Mw/Mn [b]*	*Tg [c] (°C)*
97 / 3	92 / 8	85.4	5.38	1.167	36
90 / 10	83.5 / 16.5	55.5	1.18	1.419	12

NOTE: [a] Determined by ^{1}H NMR; [b] Determined by GPC; [c] Determined by DSC

The data in Table I indicate that when the PEG content in the feed increase from 3 to 10%, both, the yield and the molar mass decreases. During the polymerization reaction, competing depolymerization reactions, and other non-specific chain scission reactions involving contaminants in the reaction mixture, such as water, caused the formation of a product with lower molar mass. It is also possible that the higher concentration of PEG reduce the polymerization rate, thereby decreasing the molecular weight and yield (*17, 18*). Another explanation for the yield reduction is that the methanol precipitation used for purification, removed low-molecular oligomers, with a cut-off around 5.0×10^3 g mol^{-1} (*17*). The single symmetric profile observed for all the GPC chromatograms, suggested that the PDLLA-PEG-PDLLA triblock copolymers contained no oligomers or residual macromonomer.

A representative ^{1}H NMR spectrum of PDLLA-PEG-PDLLA copolymer is shown in Figure 1. The copolymers composition (Table I) was calculated from the integral intensities of methine protons at 5.18, and methylene protons at 3.64 ppm, which correspond to DL-LA and PEG units, respectively.

The higher PEG content of the copolymer resulted in a softer, more stick, and therefore, a less suitable material for processing. So, for the preparation of blends with PCL we used the copolymer with a composition of PDLLA/PEG = 92/8, having $Mn = 5.38 \times 10^4$ g mol^{-1}.

Figures 2a-d show the FTIR spectra of the PDLLA-PEG-PDLLA copolymer, PCL and PDLLA-PEG-PDLLA/PCL 90/10 and 80/20 blends.

In the spectrum of PDLLA-PEG-PDLLA triblock copolymer (Figure 2a) we could observe a peak at 3505 cm^{-1} for O-H streching vibration, the bands at 2945 and 2880 cm^{-1} are attributed to C-H streching of methyl and methylene groups, respectively. The sharp band at 1759 cm^{-1} appears due to the carbonyl groups, while peaks at 1454 and 1383 cm^{-1} correspond to the asymmetrical and symmetrical banding vibrations of CH_3 groups. C(C=O)–O streching peak and the C-H banding peak can be seen at 1180 and 758 cm^{-1}, respectively.

The characteristic peaks of PCL (Figure 2b) were observed at 3440 cm^{-1} for O-H streching vibrations, at 2945 and 2866 cm^{-1} for methylene groups, 1726 cm^{-1} for carbonyl group, and 1190 cm^{-1} for C(C=O)–O streching peak.

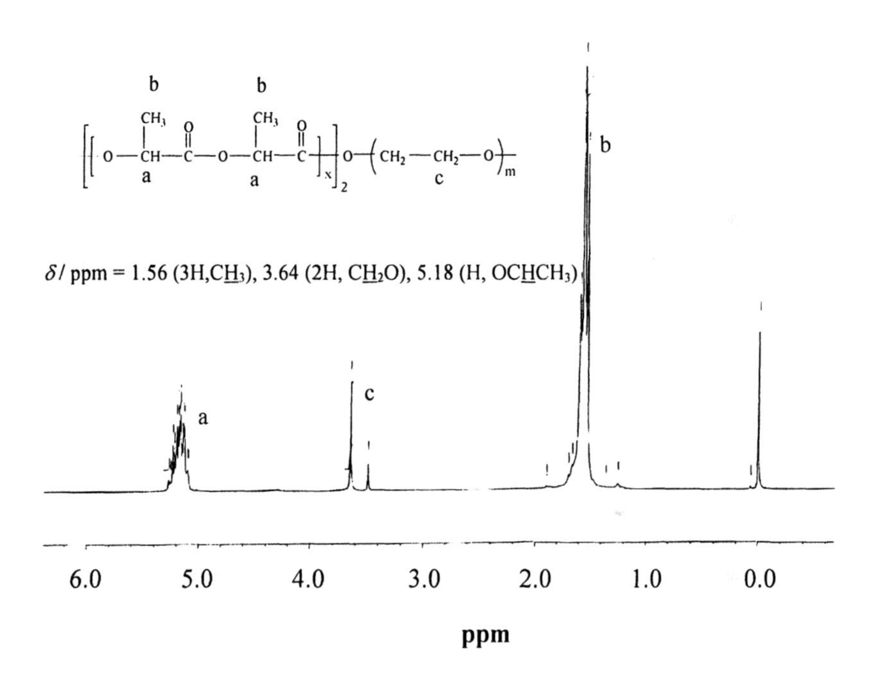

Figure 1. Representative ^1H NMR spectrum of PDLLA-PEG-PDLLA triblock copolymer.

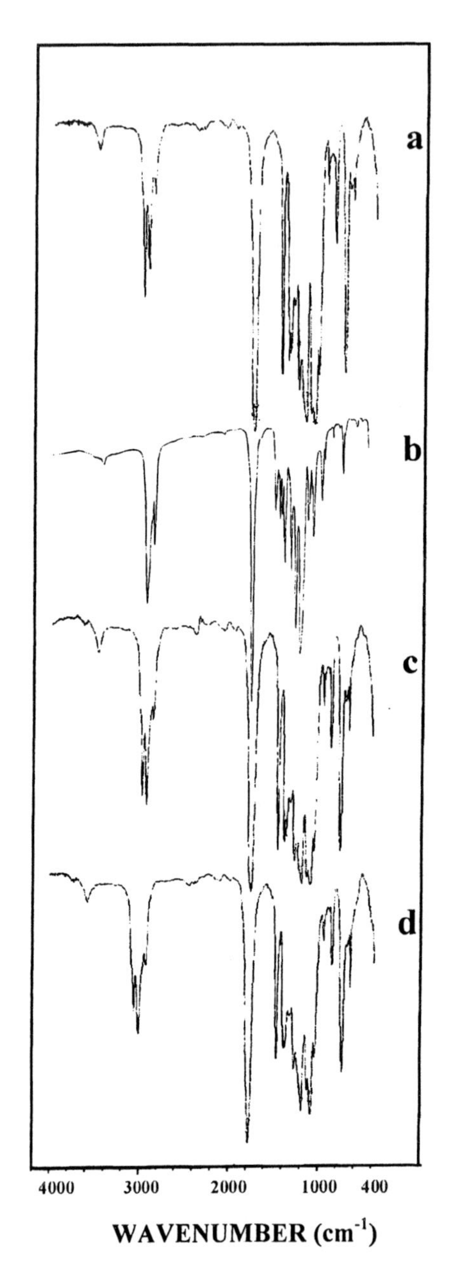

Figure 2. FTIR spectra of PDLLA-PEG-PDLLA triblock copolymer (a); PCL (b); and their blends: 90/10 (c); 80/20 (d).

FTIR spectra of PDLLA-PEG-PDLLA/PCL blends, Figuress 2c,d, have characteristic peaks of the two components. The carbonyl absorption of all blends (95/5, 90/10, 85/15 and 80/20) was observed at 1755 cm^{-1}, which is between the values of the same peak for copolymer and PCL. With the increase of the content of PCL in blends, the peak at 756 cm^{-1}, corresponding to C-H bending of lactyl units, is reduced in intensity and shifted to lower frequencies.

DSC analysis

DSC results for the PDLLA-PEG-PDLLA triblock copolymer and PDLLA-PEG-PDLLA / PCL blends with four different compositions (95/5, 90/10, 85/15 and 80/20) are summarized in Figure 3. The glass transition temperature (T_g) was taken at the onset of the corresponding heat capacity jump and the melting points are the peaks temperatures of the melting endotherms.

DSC analysis of pure PCL indicated a formation of two-phase morphology with T_g at –60 °C and melting point (T_m) at 68 °C. For pure triblock copolymer (Figure 3), T_g was observed at approximately 36 °C.

At 5 % PCL in the blend, a single transition temperature was observed, T_g = 25 °C, and as the concentration of PCL in blends increased, it was shifted to lower values. Thus, for blend with 20 wt% PCL, T_g moved to 5 °C. As far as the melting endotherm is concerned it was not noticed for triblock copolymer and blend with 5% PCL, both of the samples being amorphous.

With an increase of he amount of PDLLA-PEG-PDLLA triblock copolymer in the blends, the PCL T_m decreases. T_m depression indicate that the crystallization of PCL in blends become restricted by the presence of the PDLLA-PEG-PDLLA soft segments.

TGA analysis

Thermal stability results are summarized in Table II.

T_1 is the temperature of the inflection point, ΔW (%) is the weight loss at T_1 + cca 2°C, and T_2 is the temperature where 97.5% of the initial weight is lost.

It can be seen that the triblock copolymer exhibits lower thermal stability than dihydroxy terminated PCL. The first inflection point for ABA copolymer is at 285 °C, at which the weight loss is 95.5%, while for PCL is 390 °C, with a weight loss of 93.5%.

Temperature values for the inflection points, as well as the comparison of the percentages of volatilized material suggest that this step can be essentially ascribed to the degradation of the PDLLA-PEG-PDLLA portion in blends.

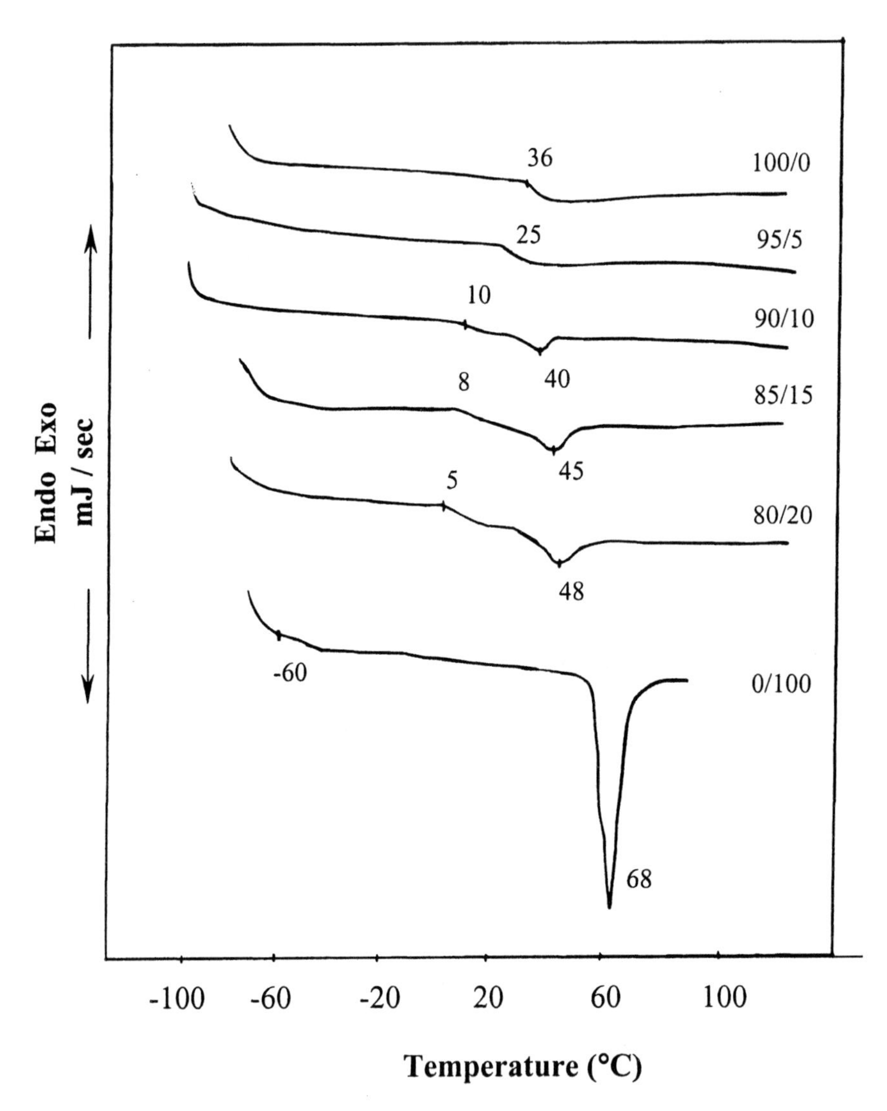

Figure 3. DSC thermograms of PDLLA-PEG-PDLLA triblock copolymer(100/0); PCL (0/100); and their blends (95/5; 90/10; 85/15; 80/20).

Table II. TGA results for PDLLA-PEG triblock copolymer and their blends

Sample	T_1 (℃)	ΔW (%)	T_2 (℃)
PCL	390	93.5	450
PDLLA-PEG-PDLLA	285	95.5	330
95/5 PDLLA-PEG-PDLLA / PCL	290	95	310
90/10 PDLLA-PEG-PDLLA / PCL	285	88.5	315
85/15 PDLLA-PEG-PDLLA / PCL	280	78.6	330
80/20 PDLLA-PEG-PDLLA / PCL	275	72.5	340

Water absorption and contact angle measurements

Water uptake of porous sponges during in vitro degradation varied considerably, depending on the blend composition. As a semicrystalline polymer, PCL is hydrophobic and its presence tends to lower water uptake of the blends. Thus, as the PCL concentration in blends increased, the initial water absorption value decreased from 0.24% for PDLLA-PEG-PDLLA/PCL (95/5) to 0.15% for PDLLA-PEG-PDLLA/PCL (80/20) blend. For both samples there is a slight increase of water absorption with the degradation time. After 50 days water absorption of the same blends was 0.40 and 0.25% respectively.

The increase of the hydrophobicity of blends with the PCL content was also shown with contact angle measurements, performed at room temperature and ambient humidity using four solvents having different polarity (water, glycerol, ethylene glycol, as polar solvents, and methylene iodide, as non-polar). Smooth polymer substrates on glass surface, were prepared by the spin-coating technique. The contact angle measurement results are given in Table III. With the increase of the PCL content in blends, and thus with the increase of the hydrophobic component in the systems, contact angle values for polar solvents increased, due to weaker interactions with the polymeric surface.

Calculated surface free-energy components of the samples according to the procedure used in ref. 15, are given in Table IV. The higher values of γ_S^- compared to γ_S^+ point out that the blend surface has slightly electron donor (basic) character as a result of oxygen atoms in the PDLLA-PEG-PDLLA and PCL chains. As known from acid-base theory, when the surface basicity of one component is high and the surface acidity of the other component is also high, then better adhesion between the components occurs (*15*).

Table III. Contact angle measurements

Content of the PCL in the blend (wt %)	θ *(degree)*			
	water	*glycerol*	*ethylene glycol*	*methylene iodide*
0	68.0	65.0	51.2	36.0
5	69.5	66.5	51.2	35.0
10	70.5	67.7	52.0	29.8
15	71.0	68.2	52.3	29.4
20	74.0	71.1	52.4	24.8

Table IV. Surface free-energy components (mJ/m²) for PDLLA-PEG-PDLLA triblock copolymer (DL-LA/PEG = 92/8 wt/wt) and PDLLA-PEG-PDLLA/PCL blends

Content of the PCL in the blend (wt %)	γ_S^{LW}	γ_S^-	γ_S^+	γ_S^{AB}	γ_S^{TOT}
0	41.6	3.130	1.045	3.62	45.22
5	42.0	3.605	0.021	0.550	42.55
10	44.3	3.152	0.0935	1.081	45.38
15	44.5	3.056	0.109	1.154	45.654
20	46.2	2.500	0.186	1.36	47.56

γ_s^{LW} - apolar Lifshitz-van der Waals (LW) component of the surface energy.
γ_s^{AB} - polar component of the surface free energy, caused by Lewis acid-base interactions, which comprises two non-additive parameters: the electron-acceptor surface free energy component (γs^+), and the electron-donor component (γs^-).

Hydrolytic degradation

The hydrolytic degradation study was carried out on porous films prepared from ABA triblock copolymer, and its blends with PCL.

Porosity of the films determined by mercury porosimetry was between 66 and 74%.

The degradation process was followed by intrinsic viscosity changes, mass and molar mass loss. The viscosity measurements showed that the 95/5 PDLLA-PEG-PDLLA/PCL blend degraded faster than the 80/20 PDLLA-PEG-PDLLA/PCL blend, and it is in agreement with the observed difference in their hydrophobicity.

Figures 4 and 5 represent the changes in mass and molar mass loss during degradation. For all observed samples, the mass loss remained constant (around 5%) during the time frame of the experiment (60 days), and it is in agreement with the results for water uptake. The molar mass chages with the degradation time (Figure 5), but the constant values of the mass loss indicate that it is not reduced below the critical value, where the samples become soluble in the aqueous medium.

Viability of cells

Mouse L929 fibroblasts were used as model cells to test the cell affinity of the copolymer and blends films. An experiment in which cell viability was

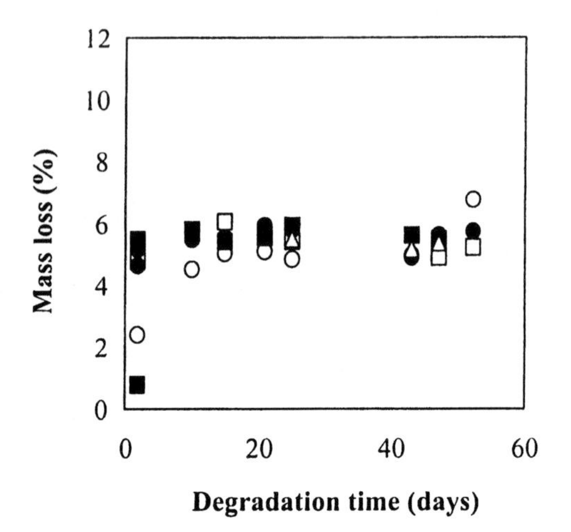

Figure 4. Mass loss for (○) PDLLA-PEG-PDLLA triblock copolymer and its blends: (□) 5% PCL; (•) 10% PCL; (■) 15% PCL and (Δ) 20% PCL in the blend.

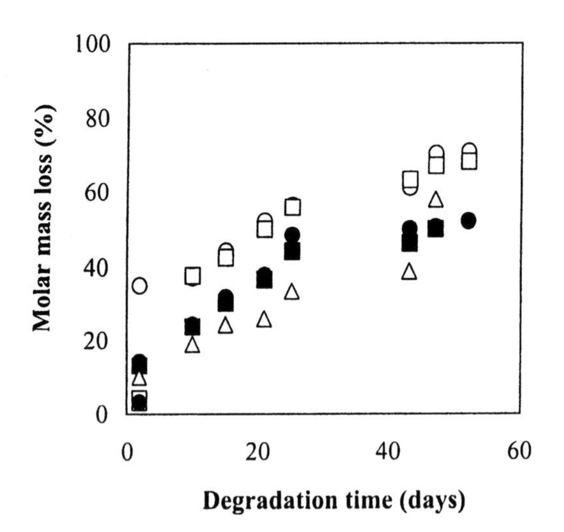

Figure 5. Molar mass loss for (○) PDLLA-PEG-PDLLA triblock copolymer and its blends: (□) 5% PCL; (•) 10% PCL; (■) 15% PCL and (Δ) 20% PCL in the blend.

measured after seeding copolymer and different blends films with L929 cells is shown in Figure 6. It is known that, among other factors, the degree of hydrophobicity, surface morphology and surface physicochemical characteristics exert profound effects on cell attachment (*19*). *In vitro* experiments showed that these films allowed the attachment and proliferation of a significant number of L929 cells. The number of cells on blends was higher than on copolymer and concidering blends, the number of cells was higher on 90/10, than on 80/20 PDLLA-PEG-PDLLA/PCL blend. It indicates that 90/10 blend has more favorable hydrophobic/hydrophilic ratio and other surface physicochemical characteristics, like surface free energy, for cell attachment.

Conclusions

In this study an ABA triblock copolymer (PDLLA-PEG-PDLLA) with a PEG content in a range with strong regulatory effect on cellular responce, was synthesized. For further modification of hydrophobic/hydrophylic characteristics its blends with PCL (copolymer/PCL=95/5, 90/10, 85/15 and 80/20 wt/wt) were prepared by solvent evaporation technique. Characterization of copolymers were performed by GPC, FTIR, NMR and DSC. The blends with PCL were characterized by FTIR, TGA, DSC and contact angle measurements. Hydrolytic degradation studies on porous films obtained from the blends of copolymers and PCL using solvent casting particulate leaching technique indicated that these materials could be used as scaffolds in tissue engineering. Preliminary cell growth experiments showed the attachment and proliferation of a significant number of L929 cells. The number of cells on blends was higher than on copolymer. Considering blends, the number of cells was higher on 90/10, than on 80/20 PDLLA-PEG-PDLLA/PCL blend indicating that 90/10 blend has more convenient surface characteristics (hydrophobic/hydrophylic ratio, surface free energy, surface morphology) for cell attachment.

Acknowledgment: This work was supported by the Turkish Scientific and Technical Research Council (TUBITAK), Turkey and the Ministry of Education and Science of the Republic of Macedonia.

References

1. Hurrell, S.; Cameron, R. E. *J. Mater. Sci. Materials in Med.*, **2001**, *12*, 811-816.

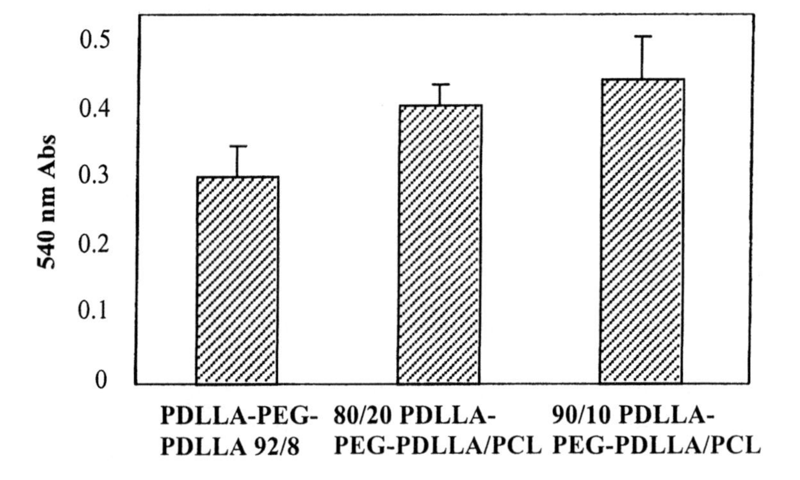

Figure 6. Copmposite graph for cell attachment and growth on PDLLA-PEG-PDLLA copolymer or PDLLA-PEG-PDLLA/PCL blends. 540 nm absorbance is a measure of the number of attached cells on the biopolymers.

2. Cai, Q.; Bei, J.; Wang, S. *J. Biomater. Sci. Polymer Ed.*, **2000**, *11*, 273-288.
3. Penco, M.; Bignotti, F.; Sartore, L.; D'Antone, S.; D'Amore, A. *J. Appl. Polym. Sci*, **2000**, *78*, 1721-1728.
4. Kim, Ch. H.; Cho, K. Y.; Choi, E. J.; Park, J. K. *J. Appl. Polym. Sci.*, **2000**, *77*, 226-231.
5. Davis, M. W.; Vacanti, J. P. *Biomaterials*, **1996** , *17*, 365-372.
6. Mooney, D. J.; Park, S.; Kaufman, P. M.; Sano, K.; Mcnamara, K.; Vacanti, J. P.; Langer, R. *J. Biomed. Mater. Res.*, **1995**, *29*, 959-965.
7. Wang, J.; Wang, B. M. Schwendeman, S. P. *J. Controll. Rel.*, **2002**, *82*, 289-307.
8. Berkland, C.; King, M.; Cox, A.; Kim, K. K.; Pack, D. W. *J. Controll. Rel.*, 2002, *82*, 137-147.
9. Chen, Ch. Ch.; Chueh, J. Y.; Tseng, H.; Huang, H. M.; Lee, Sh. Y. *Biomaterials*, **2003**, *24*, 1167-1173.
10. Lee, S. H.; Kim, S. H.; Han, Y. K.; Kim, Y. H. *J. Polym. Sci. A. Polym. Chem.*, **2002**, *40*, 2545-2555.
11. Sheth, M.; Kumar, R. A.; Dave, V.; Gross, R. A.; McCarthy, S. P. *J. Appl. Polym. Sci.*, **1997**, *66*, 1495-1505.
12. Xiong, X. Y.; Tam, K. C.; Gan, L. H. *Macromolecules*, **2003**, *36*, 9979-9985.
13. Tziampazis, E.; Kohn, J.; Moghe, P. V. *Biomaterials*, **2000**, *21*, 511-520.
14. Jeong, J. H.; Lim, D. W.; Han, D. K.; Park, T. G. *Colloids and Surfaces B. Biointerfaces*, **2000**, *18*, 371-379.
15. Dogan, M.; Eroglu, M. S.; Erbil, H. Y. *J. Appl. Polym. Sci.*, **1999**, *74*, 2848-2855.
16. Guo, Y.; Baysal, K.; Kang, B.; Yang L. J.; Williamson, J. R. *J. Biol. Chem.*, **1998**, *273*, 4027-4034.
17. Barrera, D. A.; Zylstra, E.; Lansburg, P. T.; Langer, R. *Macromolecules*, **1995**, *28*, 425-432.
18. John, G.; Tsuda, S.; Morita, M. *J.Polym.Sci.A:Polym.Chem.*, **1997**, *35*, 1901-1907.
19. Sarazin, P.; Roy, X.; Favis, B. D. *Biomaterials*, **2004**, 25, 5965-5978.

New Polymeric Materials in Nanotechnology

Chapter 11

Nanocomposites Based on Cyclic Ester Oligomers

Amiya R. Tripathy, Stephen N. Kukureka, and William J. MacKnight

Polymer Science and Engineering Department, University of Massachusetts, Amherst, MA 01003

Introduction

The term nanocomposite as applied to polymer-based materials has come to signify those based on clay particles, mainly of the montmorillonite type, exfoliated in mixtures with low viscosity precursors which are subsequently polymerized *in situ.* The montmorillonite clay consists of silicate layers. By modifying the clay appropriately, usually by treating the clay layers with a long-chain surfactant having a primary ammonium ion at one end and a carboxylic acid group on the other, cations on the surface of the clay layers (usually Na^+, K^+, Mg^{+2}, Ca^{+2}) can be replaced by the quaternary ammonium ions. This allows monomers such as ε-caprolactam to penetrate and swell the clay layer. Polymerization of the ε -caprolactam results in a composite with the clay layers dispersed at the nanometer level in a matrix of nylon 6 (Fig. 1) **(1).**

The general approach outlined above has also been applied to the synthesis of nanocomposites of poly(caprolatone), epoxy resins, and poly (lactic acid). Usuki et al **(2,3)** have reported work with other synthetic materials, which led to acrylic polymer, nitrile rubber, polyimide, polypropylene and EPDM based nanocomposites. Nylon 6-clay composites have found applications in the automotive industry. It is reported that they offer significant property enhancement with respect to nylon 6. Thus the modulus of the nanocomposite is 1.5 times that of nylon 6 without clay; the heat distortion temperature increases from 65° to 140°C and the gas barrier effect is doubled at a 2wt% loading of clay **(4-6).**

This laboratory and others have been investigating the synthesis and polymerization of cyclic oligomers of bis-phenol-A polycarbonates (BPACY), and cyclic ester oligomers (CEO) for a number of different applications, including reaction injection molding (RIM), reinforced reaction injection molding (RRIM), and resin transfer molding (RTM). Both the BPACY and the

CEO are low viscosity liquids (viscosity about 20 centipoises), either above the glass transition temperature (T_g) in the case of the BPACY or above the melting point (T_m), in the case of the CEO.

In the following sections, we shall discuss the synthesis and properties of two different families of nanocomposites, one based on PBACY and the other based on CEO.

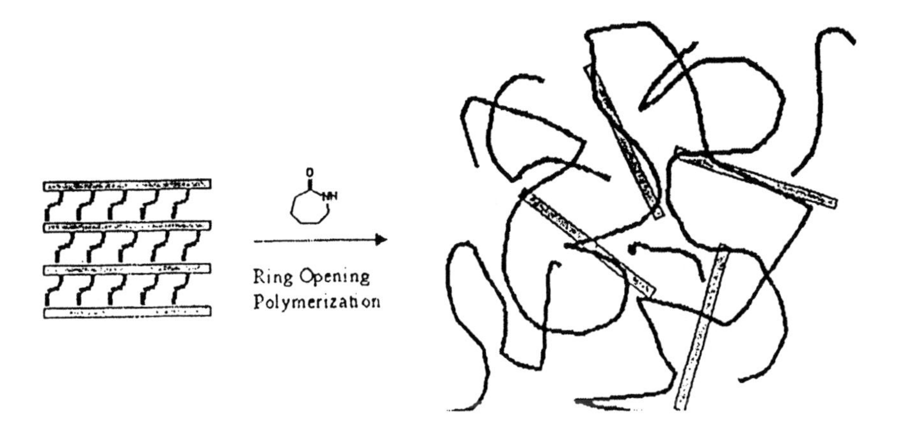

Figure 1. Schematic representation of exfoliation of nylon 6 and modified nanoclay

a. Nancocomposites from nanophase-separated blends of BPACY

It is possible to form nanocomposites by liquid-liquid phase separation during polymerization of cyclic oligomers initially dissolved in a polymer with which they do not react. We will exemplify this with cyclic carbonate oligomers blended with styrene-acrylonitrile copolymer (SAN). The thermodynamics of blending and kinetics of phase separation will be seen to produce unique nanostructures, unobtainable with blends based on conventional linear polymers. Furthermore the morphology of the resulting nanocomposites will be shown to have interesting effects on mechanical properties with potential applications. We also consider the more general effects of the topology of cyclic polymers on blend miscibility and hence on the production of nanocomposites.

Cyclic carbonate oligomers

Considerable work has been done on the synthesis, polymerization and properties of cyclic oligomers of bisphenol-A carbonate (BPACY) **(7-11).** The synthesis of BPACY is by an interfacial hydrolysis/condensation of bisphenol-A-bis(chloroformate). The resulting cyclic oligomers can be isolated as an amorphous material with a glass transition temperature (T_g) of 147°C and a density of 1197 kg m^{-3}. Their absolute molecular weight M_w has been determined as 1529 by high-performance liquid chromatography (HPLC), and they contain less than 0.05% linear contaminants. There are from 1 to 20 repeat units in the oligomers. These BPACY cyclic oligomers can be rapidly polymerized to linear, high-molecular-weight polycarbonate at, say, 260 °C in the presence of a catalyst. Typical catalysts used include tetrabutylammonium tetraphenylborate $(Bu)_4NH(Ph)_4$. Further details of starting materials and preparation methods are given in **(10)**. The polymerization proceeds by a ring-opening polymerization mechanism **(9-12)** and produces linear polycarbonate indistinguishable from that produced by conventional polymerization **(10)**.

In situ blending with SAN - thermodynamics and phase Behavior

Conventional blending usually involves mixing of high-molecular-weight polymers and results in miscible or immiscible blends, which may be of commercial interest in either form **(13)**. However, it is also possible to undertake *in situ* blend formation by polymerizing a monomer in the presence of another polymer. This has been widely used for rubber-toughened polymers **(13,14)** and in previous work on cyclic polymers **(10)**. *In situ* formation of multicomponent systems is also used for interpenetrating polymer networks (IPN), usually formed by synthesizing or cross-linking one network in the immediate vicinity of another **(14)**. In the present work, BPACY cyclic oligomers are used as novel building blocks for polymer blends. The *in situ* polymerization of these reactive oligomers from an initially homogeneous mixture of styrene-acrylonitrile copolymer enabled the preparation of materials with variable but controlled morphologies through liquid-liquid phase separation **(10)**.

Successful control of phase dispersion in a system such as this, to produce nanocomposite structures, is dependent on four key factors. First the initial mixture of cyclic oligomer and polymer to be blended must be miscible. Second, the polymerization kinetics of the cyclic oligomers, dispersed through the host polymer, must be more rapid than the dynamics of phase separation. Third, phase separation of the polymers must occur in a sensible timescale to produce immiscible structures. Fourth, phase coarsening must be restricted if the nanostructures are to be preserved.

With the BPACY-SAN system described above, it was found that polymerization of the cyclic oligomers took a little over 60 s. The resulting polycarbonate had a weight-average molecular weight M_w of 39000 and a polydispersity M_w/M_n of 2.6 with less than 0.25% unreacted oligomers remaining after polymerization **(10)**. This compares favorably with conventional polycarbonate resins. Even when the proportion of BPACY in the mixture was varied over the range 10 to 90% it was found that polymerization always took place in less than 80 s.

In order to examine progress of the *in situ* polymerization, a series of samples taken from 50/50 blends was studied by heat treatment followed by quenching and transmission electron microscopy (TEM). This was complemented by differential scanning calorimetry (DSC). Micrographs of a series of 50/50 samples held at 260°C (the polymerization temperature) for times varying from 30s to 300s are shown in Fig 2. After the heat treatment period the samples were quenched to below T_g before TEM examination. Samples held for 30s, are homogeneous but after 60 s, phase separation occurs. The structure begins as an SAN-rich phase in a PC matrix. As the holding time increases, a phase inversion occurs, after all the cyclic oligomers have been converted to polycarbonate. For holding times of 150s and greater there is a dispersion of a PC-rich phase in a SAN matrix. Further holding times only have the effect of coarsening the PC domains. This morphology is approximately 100 times finer than that produced by conventional melt-blending of PC and SAN. DSC confirms that for times over 60s, two T_g's are seen and there is no further change in the overall miscibility **(10)**. Thus it can be seen that both *in situ* polymerization from cyclic oligomers and conventional melt-blending result in complete immiscibility of PC and SAN and a resulting two-phase structure. The significant difference is the morphology of the blend with domains produced on a much finer nanoscale from the cyclic oligomers. Furthermore, the coarsening of these domains and its limitations leads to further interesting features.

Kinetics of phase coarsening

The initial phase-separation processes in these immiscible blends are nucleation and growth or spinodal decomposition. However, once the equilibrium phase composition is reached, coarsening processes are independent of the initial phase-separation mechanism. Nachlis, Kambour and MacKnight **(10)**, investigated a series of *in situ* polymerized BPACY/SAN blends with compositions of 90/10, 70/30, 50/50, 10/90. They measured the number-average radius (R) of dispersed domains as a function of time (t) and found that $R = t^{\alpha}$ where the scaling exponent α varied with blend composition. It was found that α was highest for 70/30 mixtures ($\alpha = 1$) or 50/50 mixtures ($\alpha = 0.55$) whereas for 10/90 mixtures α was 0.33 and for 90/10 no coarsening was observed at all ($\alpha = 0.00$).

Predictions of scaling exponents depend on the proposed mechanism of coarsening. For Ostwald ripening, an evaporation-condensation mechanism in which larger droplets grow at the expense of smaller ones, scaling exponents of 0.33 are predicted **(15)**. An alternative diffusion- reaction process due to Binder and Stauffer **(16)** predicts scaling exponents of between 0.33 and 1. However these mechanisms were not devised specifically for polymers and there are some special features of polymer phase separation which lead to pinning of domain growth, especially for *in situ* polymerized cyclic systems. These effects were first noted by Hashimoto *et al* **(17)** and the ideas subsequently developed by

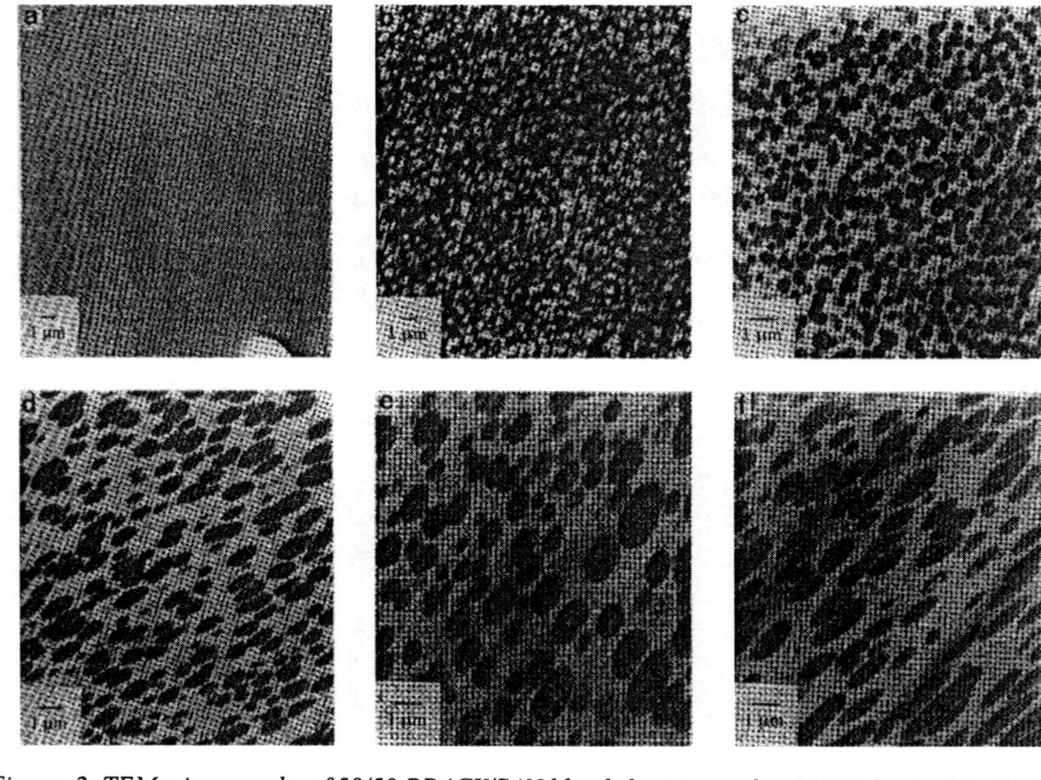

Figure 2. TEM micrographs of 50/50 BPACY/SAN blends heat-treated at 260°C for (a) 30s (b) 60s (c) 90s (d) 150s (e) 210s (f) 300s.

Muthukumar et al **(18).** They are particularly relevant to the production of nanocomposites from cyclic oligomers.

In polymers there is a barrier to coarsening by diffusive mechanisms because a large entropy barrier exists to diffusion across a sharp interface. This means that for material to leave one domain and diffuse across the matrix to join a larger growing domain, these entropy barriers must be overcome which is very difficult in practice. This phenomenon can lead to 'domain pinning' whereby coarsening may cease under certain conditions. The entropic barriers are particularly effective at restricting coarsening for systems of independent dispersed domains - either because there is a low volume fraction of the second phase or because the dispersed domains are small. In both these cases the important point is that the second-phase domains are not touching and so in order for material transport to take place from one domain to another the entropic barriers must be overcome. However, for a dispersed phase with larger or clustered domains, which are touching then material transport is much easier **(10).**

This mechanism can explain the pinning of domain growth in the 10/90 and 90/10 systems of BPACY/SAN. In both cases there is a fine dispersion of second-phase domains which do not form local clusters. These domains do not coarsen significantly with time. (The slight coarsening of the 10/90 BPACY/SAN system may be due to the much lower viscosity of the SAN matrix compared with polycarbonate making material transport easier when the polycarbonate is the second phase.) The process can be optimized using the unique ability of cyclic oligomers to polymerize *in situ* giving very fine dispersions of small second phase domains. Domain pinning and the so-called 'frozen morphology' **(18)** thus achieved point the way to a novel method of producing nanocomposites from cyclic ester oligomers which are unobtainable in any other way. This has interesting consequences for mechanical properties and more general features of topological effects on blend miscibility.

Effects of nanophase separation on mechanical properties

Domain size effects have important effects on mechanical properties and these were investigated further by Nachlis, Kambour and MacKnight **(10).** A series of 70/30 BPACY/SAN25 blends were prepared with the polycarbonate polymerised *in situ* as above. They were heat-treated at 260°C for various times (60s, 240s, 420s, 720s) before rapid quenching to below T_g. All of the cyclic BPACY oligomers were completely converted to polycarbonate in less than 60s and the effects of extended heat-treatment times were purely to produce greater degrees of domain coarsening with time. Mechanical tests in tension were conducted on both the blend samples and on pure polycarbonate and pure SAN. As expected, the pure PC was ductile and the pure SAN was brittle in all tests.

The blend samples exhibited strong effects of domain size. Blends consisted of dispersed second-phase particles of SAN in a PC matrix. The samples which had been heat-treated for only 60s (and therefore had the smallest SAN domain size) were ductile at a strain rate of 2 x 10^{-4} s though brittle at higher rates. As the domain size increased, fewer specimens were ductile until for heat-treatment times of greater than 720s, all specimens were brittle. The ductile behavior, both of the matrix *and* the second-phase domains, was confirmed by extensive electron microscopy and morphological studies. Clear yielding was seen with neck formation and no evidence of whitening or craze formation. At greater domain sizes, crazing and cracking was seen in the second phase domains.

The very non-linear effects of increasing ductility with decreasing domain size are summarized in Fig 3 **(10).** Previous work by Berger and Kramer has demonstrated a brittle-to-ductile transition as brittle particles become more finely dispersed in a ductile matrix. They proposed that the volume of an inclusion must be at least an order of magnitude greater than the typical craze-fibril spacing of 20 to 100 nm for craze formation to be feasible. It is also true that the stress concentration caused by a crack in a second-phase particle is

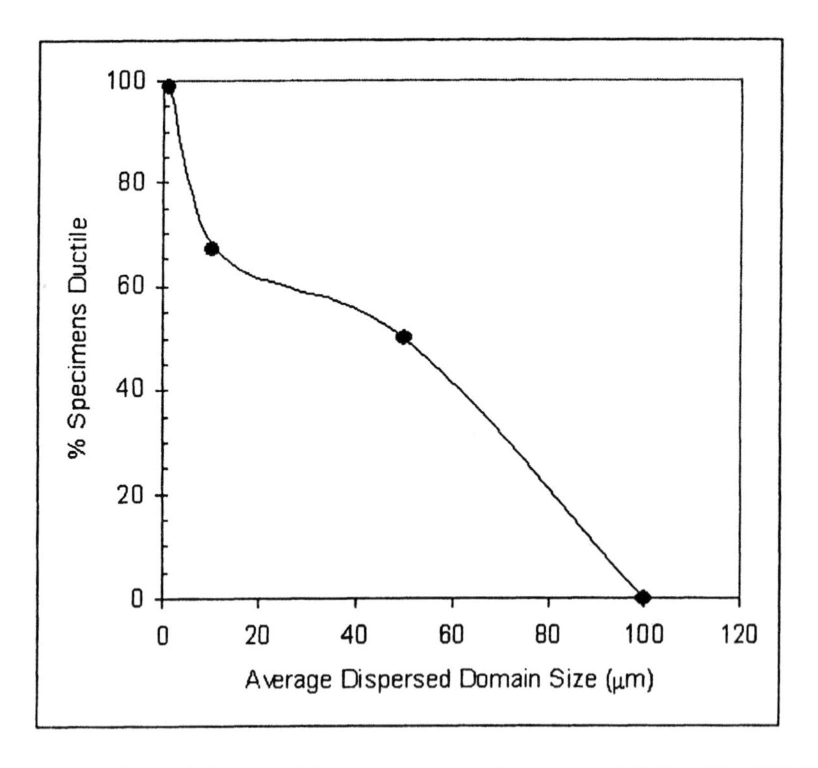

*Figure 3. Effects of dispersed domain size on deformation of 70/30 BPACY/SAN blends (*after *10).*

smaller for smaller particles, limiting the ability of cracks to continue into the matrix **(19)**. The extreme versatility of cyclic systems in producing finely dispersed composites on a micro or nanoscale by *in situ* polymerization can have tremendous benefits in producing ductility with otherwise brittle materials. It is likely that further developments in blend systems and domain pinning will help to provide a new generation of nanocomposites from cyclic materials.

Topological effects of cyclic oligomers on blend miscibility

It is apparent that blending of cyclic oligomers with a wide variety of polymer systems is possible, but are there more general considerations which favor such blending over that of linear systems? Nachlis, Bendler, Kambour and MacKnight **(11)** set out to examine topological effects on the thermodynamics of polymer mixing with cyclic oligomers. They compared the miscibility of bisphenol-A carbonate cyclic oligomers (BPACY) with a variety of linear polymers and compared the results with those from blends of equivalent linear oligomers. Detailed experimental studies were performed on blends with a series of polystyrenes of narrow molecular-weight distribution complemented by theoretical studies based on the Flory-Huggins theory of polymer mixing **(20)**.

It was found that the BPACY oligomers were only miscible with polystyrenes of molecular weight below 20,400 and so systems in this range were studied in more detail. Similar work was also performed with linear carbonate oligomers. In each case, miscibility was determined by the presence of one T_g peak in a trace obtained by differential scanning calorimetry (DSC).

Flory-Huggins solution theory, applied to polymer miscibility, combines the enthalpy of mixing with the combinatorial entropy of mixing (together with other non-combinatorial entropy terms in various modifications to the original theory) **(13,14)**. An important parameter is, the Flory-Huggins interaction parameter. For miscibility χ is negative. Careful experimental work exploring the 'miscibility gap' in polystyrenes, which was found at certain molecular weights, enabled the phase boundaries to be determined experimentally. This included binodals and spinodals for binary systems (as well as consideration of pseudo-binary systems and cloud point curves). The critical condition for miscibility can also be expressed from the Flory-Huggins theory in terms of the molecular weight distribution of the components **(11)** and so theory and experiment can be compared in providing bounds to χ for both blends of linear polystyrene/cyclic polycarbonate (χ PS_L/PC_C) and linear polystyrene/linear polycarbonate (χ PS_L/PC_L).

It was found that the interaction parameter χ was significantly lower for linear polystyrene/cyclic polycarbonate blends than for linear polystyrene/linear polycarbonate blends i.e. χ $PS_L/PC_C < \chi$ PS_L/PC_L in all cases. This was a fundamental result which is considered further below, together with a proposed model to explain the behavior. It did not depend, for instance, on the presence of end groups in linear polymers since it was insensitive to changes in end groups and the difference between interaction parameters was too large to be accounted for by end groups alone.

Differences in interaction parameter as well as experimental miscibility data indicate that blends of cyclic polycarbonate oligomers are miscible over a wider range of polystyrene molecular weights than corresponding linear oligomers. It was proposed that this is because the interaction parameter depends strongly on topology **(11)**. These topological effects appear to be due to intermolecular repulsion effects with cyclic polymers which do not exist with linear polymers. Rings which are not catenated during synthesis cannot, topologically, exist in a linked or threaded state so that there is effectively an excluded volume for intermolecular interactions. Further molecular modeling by Nachlis et al determined that the mean-square radius of gyration of a cyclic molecule is smaller than that of a linear molecule - in fact half the value. This smaller size corresponds to a higher density, a higher probability of self-contacts and a reduced number of intermolecular contacts with other cyclics.

This work has produced a result of considerable general significance. The Flory-Huggins mean field theory predicts the shape of phase boundaries well for blends of both linear PS/linear PC and linear PS/cyclic PC. However, the interaction parameter is significantly lower for blends of cyclic oligomers and this has been explained in terms of a topological effects which favor blends from cyclic systems. This enhances our ability to design systems of blends and nanocomposites which utilize the versatility of cyclic ester oligomers.

b. Nanocomposites based on CEO's and organoclays

In this section polyester nanocomposites produced via the *in situ* polymerization of cyclic poly(butylene terepthalate) oligomers (c-PBT) and modified organoclays in the presence of a stannoxane catalyst are discussed. *In situ* polymerization was conducted with various c-PBT and organoctay feed ratios, and nanocomposites were also synthesized based on copolymers of *in situ* polymerized c-PBT oligomers and ε-caprolactone together with a modified organoclay.

Macrocyclic polyester oligomers have unique properties such as low melt viscosity (0.017 Pa.s, water like) and the capacity for rapid polymerization to c-PBT virtually isothermally and without the evolution of low molecular-weight by-products **(21-23)**. These properties make them attractive matrix resin for engineering thermoplastic composites. Here we assess the production of high performance nanocomposites of c-PBT and organoclay. It is possible for polymer/clay nanocomposites to exist in two idealized forms **(24-27)**: intercalated and exfoliated. Intercalation involves polymer chains penetrating the silicate layers and expanding them, increasing the *d* spacing. Exfoliation involves more extensive polymer penetration into the silicate, resulting in a random dispersion of clay layers in the polymer matrix. This random dispersion of clay layers determines property enhancement. Clearly, whether a mixture of polymer and organoclay produces intercalated or exfoliated nanocomposites depends on the characteristics of the polymer matrix and the organoclay. The improved properties of nylon 6 - clay nanocomposites are believed to arise from the exfoliated nature of the organoclay in the nylon matrix helped by the polarity of the nylon **(4, 29)**. The precursors for the c-PBT polymer are CEO oligomers capable of *in situ* polymerization in the presence of a transesterification catalyst such as stannoxane within a few minutes. CEO oligomers can successfully be intercalated in the clay layers prior to the *in situ* polymerization because of their low molecular weight [M_w = (220), n = 2-7]. Subsequent polymerization can result in an exfoliated nanocomposite by tailoring the interactions between c-PBT oligomers, organoclays, and the clay surface.

Nanocomposites of c-PBT/clay **(28)** were first made by dissolving CEO, organoclay, and catalyst in a suitable solvent so that the clay layers would be delaminated prior to polymerization, resulting in an exfoliated structure. Unlike other systems this is a demonstration of *in situ* polymerization where the organoclay was swollen in a catalyst solution first (usually catalyst/CEO (wt/wt) 2.6/1000) rather than a monomer (CEO) solution so that catalyst could be intercalated within clay layers for the polymerization reaction to occur inside the clay galleries. The intercalated-exfoliated PBT/clay nanocomposite was produced by *in situ* polymerization in the presence of cyclic stannoxane catalyst and sodium montmorillonite modified by dimethyl dehydrogenated tallow ammonium ions. Also, we found that by adding 2wt% of modified organoclay, the onset of degradation on TGA of the c-PBT polymer nanocomposites was improved by 12 - 14^{0}C compared with c-PBT polymer without clay **(28)**. Subsequently, we also made *in situ* polymerized c-PBT oligomer and ε-caprolactone copolymer nanocomposites with an in-house modified organoclay to avoid the heat instability of the organic modifier of the organoclay and the excessive crystallinity of c-PBT polymer (more than 70%) at the processing temperature. Higher thermal stability and better mechanical properties are obtained in the copolymer nanocomposites than in the corresponding $CaCO_3$ filled composites. The experimental details are given elsewhere **(28)**. A variety of characterization techniques were used to investigate the blends and nanocomposites and the results are detailed below.

Blend results

Figure 4 shows X-ray diffraction curves of the 2-6 wt% organoclay/CEO mixtures, both before and after polymerization. There is a clear diffraction peak at $2\theta = 4.7°$ for the 20A clay but none for the nanocomposites between $2\theta = 2°$ and 10°. This indicates the presence of exfoliated silicate layers of organoclay dispersed in the CEO matrix. However, when the mixture is polymerized, a small peak at $2\theta = 2.3°$ appears for the 20A6%P sample, corresponding to a d-spacing (d_{001}) of 3.7 nm, which is higher than that for the organoclay 20A (d_{001} =

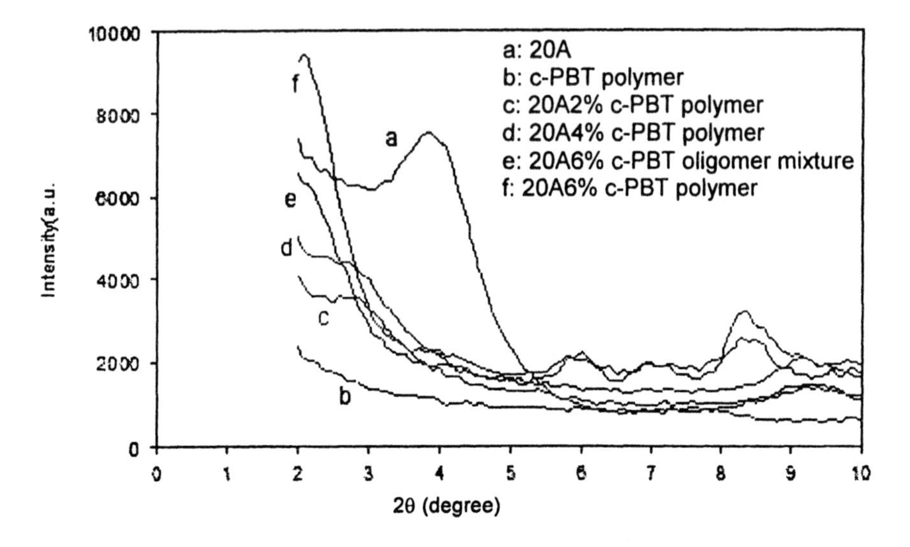

Figure 4. X-ray diffraction of c-PBT homopolymer nanocomposites

2.2 nm). This suggests that a small amount of the organoclay is not exfoliated in the PBT but exists in the form of an intercalated layered structure. We also observed that XRD along the edge of the compressed polymeric film of the 20A6%P does not show the diffraction pattern of a crystalline material, indicating no further orientation of the clay layer but an exfoliated-intercalated structure.

Further evidence of the nanometer-scale dispersion of silicate layers in the PBT/clay is provided by the TEM micrographs, as shown in Figure 5. Individual silicate layers, occasionally two, three or more layers stacked together, are observed to be well dispersed in the PBT host matrix before polymerization in Figure 5a. Here, the darker lines are clay layers and the polymer matrix is brighter. A few multilayer tactoids also exist; but mostly single clay layers are observed, and there are no large, undispersed clay tactoids. This indicates that the organoclay is well dispersed and exfoliated in the c-PBT host before polymerization and that is why the X-ray diffraction shows no peak in 2θ corresponding to dispersed clay (Figure 4). The individual silicate layers have an original thickness of ~1 nm and average length of ~100-150 nm, **(31)** connected through their edge and changed to an average ~5 nm thickness and a length of ~100-200 nm in 20A6%M, exhibiting an orientation in the same direction throughout the matrix (Figure 5a). Figure 5b shows elements of both exfoliation and intercalation of the clay layers. The clay layers are observed to be better oriented than before polymerization. However, clay stacks are still visible in abundance. There are a few aggregates but mostly well-dispersed clay layers due to the disruption of clay tactoids (originally present in the 20A6%M) located with the polymer matrix in 20A6%P, and reflected in the X-ray diffraction of the

Figure 5a: TEM micrograph of CEO & 20A(6%) mixture (before polymerization); b: TEM micrograph of CEO & 15A(6%) nanocomposites (after polymerization)

increment of the d-spacing in comparison to the original clay (Figure 4). The average thickness of the clay stacks is ~7 nm and the length is ~200-250 nm as observed after polymerization.

Thermogravimetric analysis of nanocomposites (Figure 6) indicates better thermal stability than of the corresponding polymer without clay. There is an almost 12-14 °C increment in the onset temperature of degradation of the 2wt% nanocomposite compared with the c-PBT polymer without clay. This observation is very much consistent with that for other reported nanocomposite systems **(32)**. It means that a mere 2wt% of nanoclay is capable of improving the thermal stability of PBT. In fact, with an increase in the organoclay percentage in the nanocomposite the thermal stability decreases gradually, when in a nitrogen atmosphere (Figure 6). There are two possibilities for ordering of polymer chains inside the silicate galleries: as the temperature increases to around 160°C (the softening range of CEO), CEO oligomers tend to melt and the already-dispersed clay layers have enough time to aggregate prior to polymerization. Once the polymerization starts, the movement of clay layers will be restricted and, possibly that is the reason for the presence of aggregates and thicker stacks in the 20A6%P than in 20A6%M. There is another possibility: that the long hydrocarbon chain of surfactant (dihydrogenated tallow ammonium ions) could be decomposed at the temperature of polymerization (190 °C) so that adjacent clay layers have an opportunity for crystalline growth on cooling. However, aggregation can be prevented during the processing if clay is modified with some heat-stable organic modifier. Another difficulty is excessive crystallinity of the c-PBT polymer, which makes it very brittle. Under these circumstances, incorporation of organoclay into the polymer matrix further increases its brittleness.

To overcome these problems Na-montmorrilonite clay was modified with 1,2 dimethylimidazole and bromohexadecane, described in detail elsewhere **(33),** c-PBT/CL copolymer was made via *in situ* copolymerization of CEO and ε-caprolactone using a stannoxane catalyst (I) and thus copolymer nanocomposites were produced.

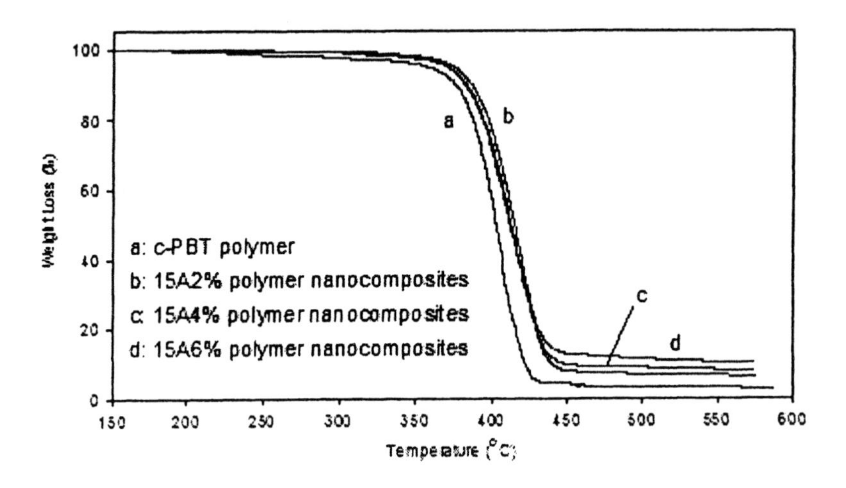

Figure 6: TGA of c-PBT polymer clay nanocomposites in N_2

The reduction of the crystallinity of the copolymer from that in CEO and CL has been discussed **(34).** Also, it was found that the polymers obtained are random copolymers. The failure characteristics **(34),** shown in the stress-strain curve of the copolymer of (70/30 wt/wt) c-PBT/CL are typical of thermoplastic elastomers. Similarly, a copolymer nanocomposite was synthesized using 15A clay and also a copolymer composite using $CaCO_3$ was synthesized for comparison with the C16-MMT copolymer nanocomposites.

X-ray diffraction data were taken on these nanocomposites. Figure 7a shows a small peak at 2θ = 2.1 ° corresponding to a 4.2 nm d-spacing from the 15A clay in the 15A4% 70/30 cPBT/CL copolymer nanocomposite. Clay 15A originally had a d-spacing at 3.04 nm. Therefore, there is an increase of 1.16 nm in d-spacing of the clay layers in the nanocomposite. The curve corresponding to c-P BT/CL (70/30) copolymer in Figure 4a reveals the characteristic peaks of the polymer along the d_{002} and d_{011} planes, and they are also visible in the 15A4% copolymer nanocomposite. The X-ray diffraction curves of the C16-MMT-modified nanoclay and the 4 wt% C16-MMT (70/30) c-PBT/CL copolymer nanocomposite are shown in Figure 7b. C16-MMT nanoclay shows a peak at 2θ = 4.9 ° in comparison with the diffraction peak at 2θ = 2.9 ° for the

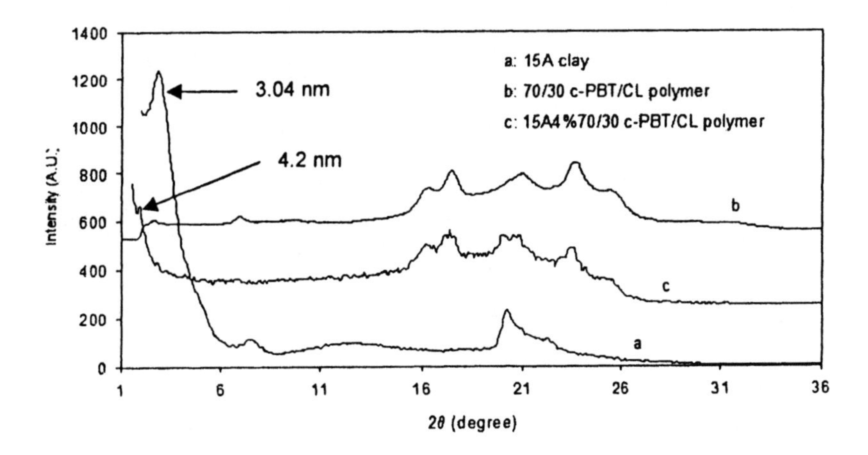

Figure 7a: WAXD of 15A nanoclay c-PBT/CL copolymer nanocomposites

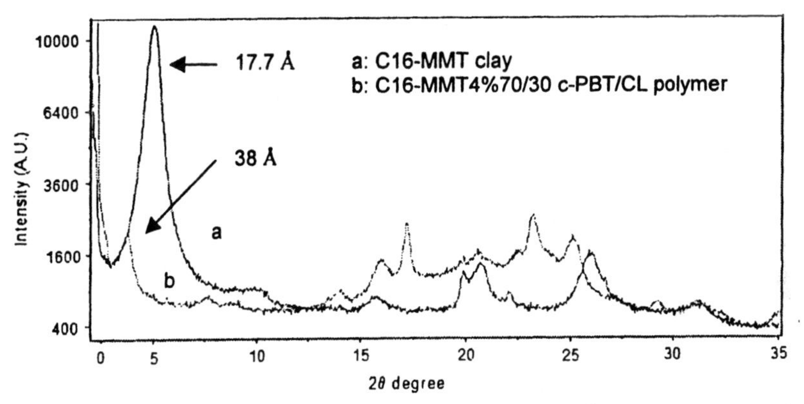

Figure 7b: WAXD of C16-MMT nanoclay c-PBT/CL copolymer nanocomposites

15A clay. The d-spacing also increases to 3.8 nm corresponding to 2θ = 2.3 ° in the 4 wt% C16-MMT copolymer nanocomposites. This indicates that there is at least a 2.03 nm increase in d-spacing for the C16-MMT4% 70/30 c-PBT/CL copolymer nanocomposite in comparison with an increase of 1.16 Å for the 4 wt%15A 70/30 c-PBT/CL copolymer nanocomposites. This implies that modified C16-MMT organoclay is more intercalated than the 15A. It is interesting that the intensity of the diffraction peak is drastically reduced in both cases (Figures 7a 7b), suggesting the breaking up of the original clay stacks in the copolymer nanocomposites. We also observed that XRD along the edge of the compressed copolymeric film does not show the diffraction pattern of a

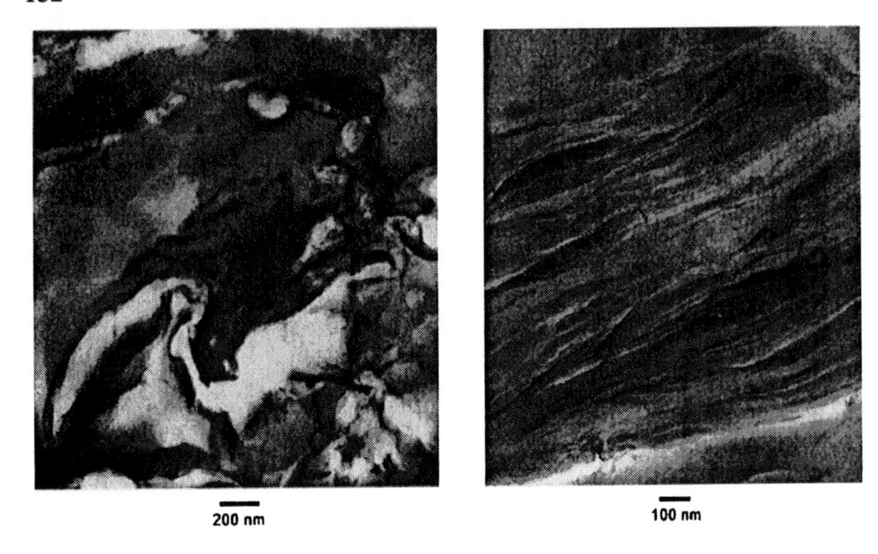

Figure 8a: TEM micrograph of c-PBT & 15A(4%) c-PBT/CL (70/30) copolymer; b: TEM micrograph of c-PBT & C16-MMT(4%) c-PBT/CL (70/30) copolymer

crystalline material, indicating no further orientation of the clay layer but suggesting an exfoliated-intercalated structure. Further details of the nanostructure can be understood from the TEM micrographs of the respective samples.

In Figure 8a, we observe an aggregation of 15A clay in the matrix of c-PBT/CL copolymer. Here, the darker lines are clay layers and the polymer matrix is brighter. A few multilayer tactoids also exist but mostly undispersed clay tactoids are observed. This indicates that the 15A organoclay is not well dispersed in the c-PBT/CL copolymer host, which is also reflected in the X-ray diffraction. Figure 5b shows both exfoliation and intercalation of the clay layers of 4 wt% C16-MMT (70/30) c-PBT/CL copolymer nanocomposites. The clay layers are observed to be better oriented than those in the 15A4% (70/30) c-PBT/CL copolymer nanocomposite. Individual clay stacks are still visible in abundance. There are very few aggregates but mostly well dispersed clay layers due to disruption of clay tactoids (originally present in the C16-MMT clay) in the polymer matrix. The average thickness of the clay stacks is ~5-6 nm and the length is ~100-200 nm in the 4 wt% C16-MMT (70/30) c-PBT/CL copolymer nanocomposites. As a whole the nanoclay is dispersed in the polymer matrix with some orientation.

The GPC traces for (70/30) c-PBT/CL copolymer and the corresponding copolymer composites, shown in Figure 9, exhibit a single peak, which differs in magnitude and retention time between samples. It is observed that the maximum

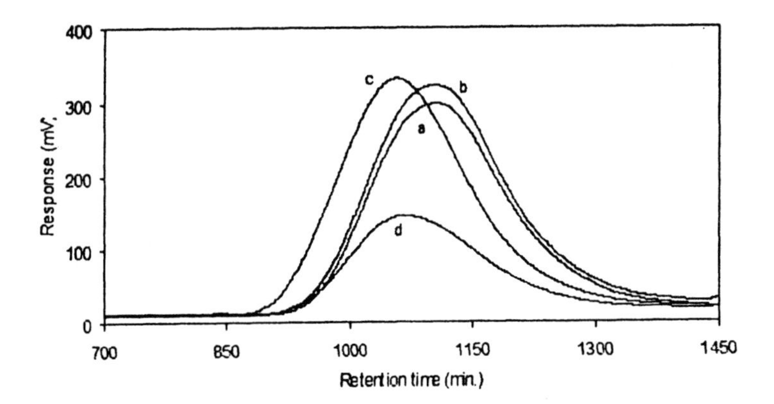

a: 70/30 c-PBT/CL copolymer
b: 15A4% 70/30 c-PBT/CL copolymer nanocomposites
c: C16-MMT4% 70/30 c-PBT/CL copolymer nanocomposites
d: CaCO34% 70/30 c-PBT/CL copolymer composites

Figure 9: GPC traces of c-PBT/CL (70/30) copolymer composites

Table 1. GPC data of copolymer nanocomposites

Sample	M_n [g/mol]*10^4	M_w/M_n
70/30 c-PBT/CL copolymer	5.58	1.7
15A4% 70/30 c-PBT/CL nanocomposite	6.12	1.6
C16-MMT 4% 70/30 c_PBT/CL nanocomposite	8.49	1.6
CaCO3 4% 70/30 c-PBT/CL composite	7.43	1.6

shifts to higher retention time from 4 wt%C16-MMT (70/30) c-PBT/CL copolymer nanocomposites to (70/30) c-PBT/CL copolymer. The weight average molecular weights roughly estimated through a polystyrene calibration curve, are shown in Table 1. The number-average molecular weight varies between 85,000 and 55,000 and the polydispersity index remains approximately 1.6 in all cases. The maximum molecular weight is obtained in the 4 wt%C16-MMT (70/30) c-PBT/CL copolymer nanocomposites.

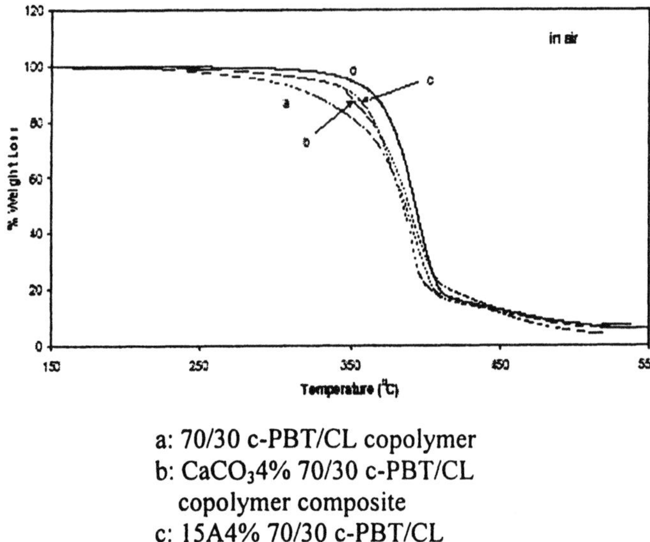

a: 70/30 c-PBT/CL copolymer
b: $CaCO_3$4% 70/30 c-PBT/CL copolymer composite
c: 15A4% 70/30 c-PBT/CL copolymer nanocomposite
d: C16-MMT4% 70/30 c-PBT/CL copolymer nanocomposite

Figure 10: TGA of c-PBT/CL (70/30) copolymer composites

Thermogravimetric analysis of the nanocomposites (Figure 10) indicates better thermal stability than the corresponding copolymer without clay. There is an almost 35 °C increase in the onset temperature for degradation of both 15A4% (70/30) c-PBT/CL copolymer and $CaCO_3$4% (70/30) c-PBT/CL copolymer composites compared with the (70/30) c-PBT/CL copolymer without clay. In fact, with the 4 wt% C16-MMT nanoclay the onset of degradation has increased at least 75 °C (Figure 10) in air compared with the copolymer without clay.

Mechanical testing was carried out on melt compressed test samples with an average thickness of 1 mm and width 6.5 mm. The stress-strain behavior of (70/30) c-PBT/CL copolymer (Figure 11) is typical of ductile materials - a yield point, necking and strain-hardening until fracture. However, copolymer nanocomposites failed in a brittle manner with yielding and a very low elongation at break. Cold drawing with limited strain hardening was observed in (70/30) c-PBT/CL copolymer and $CaCO_3$4% (70/30) c-PBT/CL copolymer composite. The Young's modulus (E), yield stress (σ_Y) and the stress at break (σ_B) exhibit an increase both in C16-MMT4% (70/30) c-PBT/CL and 15A4% (70/30) c-PBT/CL copolymer nanocomposites. Copolymer nanocomposites made from nanoclay exhibit higher moduli than the respective $CaCO_3$ composites for the same loading. The highest modulus was observed in C16-

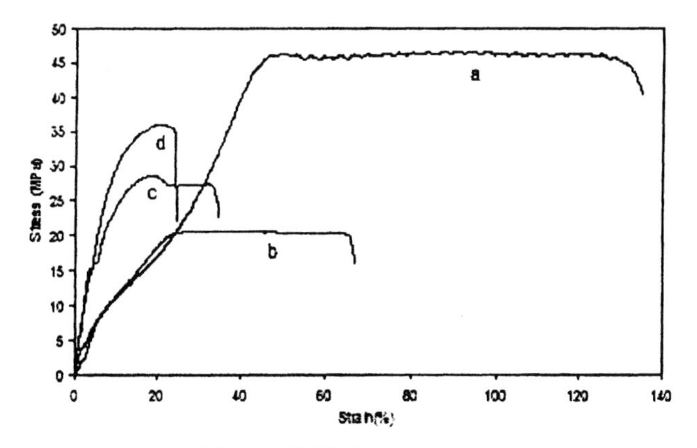

a: 70/30 c-PBT/CL copolymer
b: $CaCO_3$4% 70/30 c-PBT/CL copolymer composite
c: 15A4% 70/30 c-PBT/CL copolymer nanocomposite
d: C16-MMT4% 70/30 c-PBT/CL copolymer nanocomposite

Figure 11: Stress strain curves of c-PBT/CL (70/30) copolymer composites

MMT4% (70/30) c-PBT/CL, which exhibits an almost three-fold increase compared with the (70/30) c-PBT/CL copolymer without clay. Overall, the copolyester nanocomposites enhance the onset of degradation by at least 75 °C compared with c-PBT/CL copolymers without clay. The modulus of the copolyester nanocomposite using 4 wt% in-house modified nanoclay is improved by more than three times compared with that of the corresponding $CaCO_3$-filled copolymer composites.

Conclusions

Cyclic oligomers of bis-phenol A polycarbonates (BPACY), and cyclic ester oligomers (CEO) provide routes to nanocomposite materials with interesting properties. The rather wide range of miscibility of BPACY and CEO with polymers affords the possibility of morphology control by *in situ* polymerization of the oligomers. Phase separation taking place during each polymerization can be controlled to produce nano-scale particles in a matrix. In the case of *in situ* polymerization of BPACY in SAN, the resulting nanocomposites show enhanced tensile strength compared with typical blends of glassy polymers. The polymerization of CEO's with modified mortonorillonite

24. Pinnavia, T. J.; Beal, C. W., Eds. *Polymer Clay Nanocomposites;* John Wiley & Sons: New York, **2001**.
25. Kojima, Y.; Usuki, A.; Kawasumi, M.; Okada, A.; Kurauchi, T.; Kamigaito, O. *J. Polym. Sci., Part A: Polym. Chem.* **1993,** *31,* 983.
26. Kim, B. H.; Jung, J. H.; Hong, S. H.; Joo, J.; Epstein, A. J.; Mizoguchi, K.; Kim, J. W.; Choi, H. J. *Macromolecules* **2002**, 35, 1419.
27. Fornes, T. D.; Hunter, D. L.; and Paul, D. R. *Macromolecules,* **2004**, 37, 1793.
28. Tripathy, A. R.; Burgaz, E.; Kukureka, S. N.; MacKnight, W. J. *Macromolecules,* **2003**, 36, 8593.
29. Gilman, J. W.; Awad, W. H.; Davis, R. D.; Shields, J.; Harris, H. R.; Davis, C.; Morgan, A. B.; Sutto, T. E.; Callahan, J.; Trulove, P. C.; and Delong, H. C. *Chem. Mater* **2002**, 14, 3776.
30. Ray, S. S.; Maiti, P.; Okamoto, M.; Yamada, K.; Ueda, K. *Macromolecules* **2002**, 35, 3104.
31. Shi, H.; Lan, T.; Pinnavaia, T. *J. Chem. Mater.* **1996**, 8,1584.
32. Wang, Z.; Pinnavaia, T. *J. Chem. Mater.* **1998**, *10,* 1820.
33. Takekoshi, T.; Khouri, F. F. ; Campbell, J. FR.; Jordon, T. C.; Dai, K. H. U.S. Patent 5,707,439, **1998**.
24. Tripathy, A. R.; MacKnight, W. J.; Kukureka, S. N.; *Macromolecules,* **2004**, *37,* 6793.

Chapter 12

Selective Solvent-Induced Reversible Surface Reconstruction of Diblock Copolymer Thin Films

Ting Xu[1], Matthew J. Misner[1], Seunghyun Kim[1], James D. Sievert[1], Oleg Gang[2], Ben Ocko[2], and Thomas P. Russell[1,*]

[1]Department of Polymer Science and Engineering, University of Massachusetts, Amherst, MA 01003
[2]Department of Physics, Brookhaven National Laboratory, Upton, NY 11973

Through the use of a selective solvent a reversible surface reconstruction of diblock copolymer thin films was observed. The solvent selectivity and solubility of the minor component block were found to be crucial to generate nanoporous films with pores that penetrate through entire film thickness. The process was shown to be reversible by thermal annealing and was easily monitored using in-situ grazing incidence small angle x-ray scattering and scanning force microscopy. At temperatures of 60-90°C, only a small fraction of the nanopores relaxed to regenerate the original nanotemplate. However, by heating to 90-100°C, the original nanotemplate was completely regenerated. Even though the bulk mobility of PS and PMMA is low at these temperatures, the local mobility required to regenerate the template was sufficient.

We dedicate this contribution to Frank E. Karasz who has been a leader in the field of polymer science for many years and an inspiration to many.

Introduction

Block copolymers comprised of chemically distinct polymers covalently joined at one end, self-assemble into well-defined, ordered arrays of nanoscopic domains ranging from spheres to cylinders to lamellae, depending on the volume fraction of the components.[1] Solvent has been shown as an effective, strong and highly directional method to manipulate the microdmain orientation by controlling the solvent selectivity, and evaporation rate.[2-12]For example, the rate of solvent evaporation and the evaporation-induced flow were used to highly orient cylindrical domains of a diblock copolymer parallel to the surface over very large surface areas. Also, it was shown that solvent annealing could be used to markedly improve long-range order of the copolymer thin films, since solvent imparts the mobility to the copolymer that enables a rapid removal of defects.

Recently, we reported an approach to generate nanoporous films of polystyrene-block-poly(methyl methacrylate) PS-b-PMMA diblock copolymer by swelling the minor component cylindrical PMMA microdomains with a preferential solvent, acetic acid.[8,13] A solvent-induced reconstruction of the film was shown to be responsible for the formation of the pores. Upon exposure to acetic acid the PMMA was solubilized, drawn to the surface, and subsequently trapped at the surface upon drying. Grazing incidence small angle x-ray scattering (GISAXS) and x-ray reflectivity studies have shown that the pores penetrated through the entire film and the lateral spacing of the pore was not affected by the solvent swelling. The process was shown to be fully reversible, since neither block was degraded or removed to generate the pores. Heating above the glass transition temperature of the block copolymer resulted in a recovery of the initial film morphology i.e., nanoscopic cylindrical PMMA domains oriented normal to the surface.

Here, we report on this solvent induced surface reconstruction, emphasizing the role of solvent selectivity and solubility, the role of the random copolymer brush used to balance interfacial interactions, and the mechanism of the regeneration of the nanotemplate. The regeneration of the initial nanotemplate by thermal annealing was studied using *in-situ* grazing incidence small angle x-ray scattering (GISAXS) and scanning force microscopy (SFM). At temperatures of 60-90°C, only a small fraction of the pores close to the bottom of the film were re-filled. The pores were not fully filled to re-generate the nanotemplate until being heated up to 90-100°C indicating that majority of the PMMA block did not have mobility at temperatures below 90-100°C.

Experimental

Two PS-b-PMMA diblock copolymers were prepared by conventional anionic block co-polymerization and were characterized by size exclusion chromatography that was calibrated against polystyrene standards. One has molecular weight of 7.1×10^4 g/mol with polydispersity index of 1.08 with a PS volume fraction of 0.5 that formed lamellar microdomains in the bulk. This will be referred to as LSM71k. The other PS-b-PMMA had a molecular weight of 6.9×10^4 g/mol with polydispersity index of 1.06 and a PS volume fraction of 0.29 that formed cylindrical microdomains in the bulk. This copolymer will be refereed to as CSM69k. Narrow molecular weight distribution random copolymers of styrene and methylmethacrylate, P(S-r-MMA), with 58 mol% styrene and benzyl alcohol end-groups, were prepared by a "living" free-radical polymerization. The P(S-r-MMA) as anchored to the native oxide layer on silicon substrates to generate a surface that exhibits balanced interfacial interactions with respect to PS and PMMA.[14-16] Another method to balance the interfacial interactions between each block with the substrate is to passivate the silicon substrate by dipping it into 5% HF solution for 3 minutes and rinsing with deionized water for 3 minutes.[17] This etches the native oxide and silanizes the surface, which has been found to produce a surface with balanced interactions. Films of PS-*b*-PMMA with a thickness near the repeat period were spin coated onto these substrates, annealed at 170°C under vacuum for 40hrs, soaked in acetic acid, and thoroughly rinsed with deionized water.

Samples for TEM were prepared on a silicon wafer that had a 1μm thick layer of thermally grown silicon oxide. After solvent swelling and rinsing, the diblock copolymer film was floated off the substrate by dipping into an aqueous 5% HF solution and then transferred to copper grids. TEM was performed using a JEOL 100CX electron microscope operating at 100 kV.In order to investigate the temperature effects on the regeneration process of the initial nanotemplate, one sample was cut into several pieces. Each piece was then annealed at a desired temperature under N_2 for 1 hr and quenched to room temperature. Tapping mode SFM was then performed on each sample with a Dimension 3100, Nanoscope III from Digital Instruments Corp.

The *in-situ* GISAXS experiments were performed on beamline X22B at the National Synchrotron Light Source at the Brookhaven National Laboratory, using x-rays with a wavelength of 1.567Å.[18] The transversal and longitudinal coherence length is ~1-2μm. A heating stage controlled the temperature with an accuracy of 0.1°C. The sample was heated to the desired temperature and annealed at that temperature under N_2 for 6 minutes before beginning the GISAXS measurements.

clay leads to exfoliated clay nanocomposites. The crystallinity of the CEO's and hence the mechanical properties of the nanocomposites can be controlled by using mixtures of different CEO's such as cyclic butylene terephthalate oligomers and ε-caprolactone, leading to copolymers.

Acknowledgements

We wish to thank the National Environmental Technology Institute (NETI) for their support and the Cyclics Corporation of Schenectady, NY for their continued interest in, and samples for, this work.

References

1. Kawasumi, M. SAE Technical Paper, **1991**, 910584.
2. Kawasumi, M.; Hasegawa, N.; Kato, M.; Usuki, A.; Okada, A. Macromolecules, **1997**, 30, 6333.
3. Usuki, A.; Tukigase, A.; Kato, M. Polymer, **2002**, 43, 2185.
4. Brunelle, D. J. U.S. Patent 5,498,651, **1996**.
5. Takekoshi, T. ; Pearce, E. J. U.S. Patent 5,386,037, **1995**.
6. Brunelle, D. J. ; Serth-Guzzo, J. A. U.S. Patent 5,661,214, **1997**.
7. Brunelle, D. J.; Bowden, E. P.; Shannon, T. C. *J. Am. Chem. Soc.,* 117, 2399 (**1990**).
8. Brunelle, D. J. and Shannon T. C. *Macromolecules,* 24, 3035 (**1991**).
9. Brunelle, D. J.; *Ring Opening Polymerization* Ch 11 (ed D. J. Brunelle, Hanser 1993).
10. Nachlis, W. L.; Kambour, R. P.; MacKnight, W. J. *Polymer,* 35, 3643-3657 (**1994**).
11. Nachlis, W. L.; Bendler, J. T.; Kambour, R. P.; MacKnight, W. J. *Macromolecules,* 28, 7869-7878 (**1995**).
12. Semlyen, J. A. *Cyclic Polymers,* (2nd ed, Kluwer, 2000).
13. Paul, D. R.; Bucknall, C. B. *Polymer Blends,* (Wiley 2000).
14. Sperling, L. H. *Polymeric Multicomponent Materials,* (Wiley 1997).
15. Furukawa, H. *Adv. Phys.,* 34, 703 (**1985**).
16. Binder, K. and Stauffer, D. *Phys. Rev. Lett.,* 33, 1006 (**1974**).
17. Hashimoto, T.; Takenaka, M.; Izumitani, T, *J. Chem. Phys.* 97, 679 (**1992**).
18. Kotnis, M. A. and Muthukumar, M. *Macromolecules,* 25, 1716 (**1992**).
19. Berger, L. L. and Kramer, E. J. *J. Mat. Sci.,* 22, 2739 (**1987**).
20. Flory, P. J. *Principles of Polymer Chemistry,* (Cornell 1953).
21. Katahira, S., Tamura, T., Yasue, K., and Watanabe, M. US Pat. US5414042 (**2000**).
22. Yasue, K.; Tamura, T.; Katahira. S.; Watanabe, M., Eur. Pat. EP0605005 (**1995**).
23. Giannelis, E. P. *Adv. Mater.* **1996,** *8,* 29.

Results and Discussions

Figure 1a shows the TEM image of an annealed LSM71k thin film on a silicon substrate modified with random copolymer where the surface is neutral to both blocks. Lamellar microdomains oriented normal to the surface are seen. Figure 1b is the TEM image after dipping the film into acetic acid, a selective solvent for the PMMA blocks. The bright white lines in this image have the same intensity as a section without film, suggesting that the lines are hollow and extend completely through the film. This is identical to what was previously seen in films with cylindrical morphology where the minority component PMMA blocks were drawn to the surface by the acetic acid and were trapped at the surface upon drying. With the lamellar morphology, the same behavior can be produced with the PS block when a selective solvent for the PS block is used. Shown in Figure 1c is a TEM image of an LSM71k film after a dipping into cyclohexane, a selective solvent for the PS blocks. Here, the hollow lines are similar to those in the films swollen by acetic acid. However, is evident that the PS was not completely drawn to the surface, and as a consequence, the definition of the lines is not as sharp as when PMMA was drawn to the surface by acetic acid. The observed difference can be attributed to the fact that cyclohexane is not as good a solvent for PS as acetic acid is for PMMA. Thus, it is important to choose a solvent with higher selectivity/solubility to generate lines or pores that span the entire film.

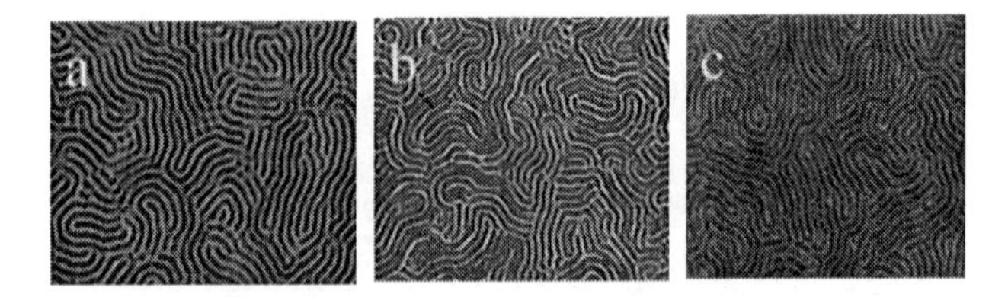

Figure 1 TEM images of an LSM71k diblock copolymer thin film (a) after being annealed and then (b) dipped into acetic acid, (c) cyclohexane.

The TEM results were confirmed with GISAXS, which is very sensitive to the formation of pores due to the high electron density difference between the copolymer and air. Figure 2a shows the GISAXS pattern of an LSM71k film after acetic acid exposure at a grazing incidence angle of $\alpha \approx 0.2^{o}$, which is the same sample preparation as in Figure 1b. The pattern shows multiple higher order reflection in q_y (in the plane of the film) resulting from the lateral ordering of the lamellar microdomains normal to the surface. There are also multiple oscillations along the first diffuse rod (in the q_z direction) with a frequency corresponding to a depth of the nanoscopic channels. The observation of fringes out to 0.15 $Å^{-1}$ suggests that the termination of the channels is sharp. Figure 2b is the GISAXS pattern of an LSM71k film after dipping into cyclohexane (same condition as Figure 1c). In comparison to the data in Figure 2a, the intensity of scattering is much lower, indicating the channel formation is not as complete and that some of the PS remains in the channels. . Figure 2c shows the q_z scans along the first diffuse rod of both GISAXS patterns. The oscillations in the frequency of the scans are inversely related to the depth of the channels. As can be seen, the oscillations along q_z for the film after dipping into cyclohexane have lower frequency than for the film after dipping in acetic acid confirming that the channels are shallower as indicated by the TEM images.

Previous studies have shown that, with exposure to acetic acid, nanoporous films with pores penetrating through the entire film could be achieved[13]. In addition, electron density profiles obtained by fitting the x-ray reflectivity profiles indicated that the thickness of the random copolymer brush layer increased after swelling. It was argued that the solubility of the random copolymer in acetic acid and the diffusion of PMMA blocks into the brush layer could explain the penetration of the pores through the entire film. As stated previously, passivation of the silicon substrate is an alternate method to balance interfacial interactions between the PS and PMMA blocks with the substrate and to orient the PS-b-MMA microdomains normal to the surface. Therefore, passivation can be used to clarify the role of the brush layer during the selective solvent swelling.

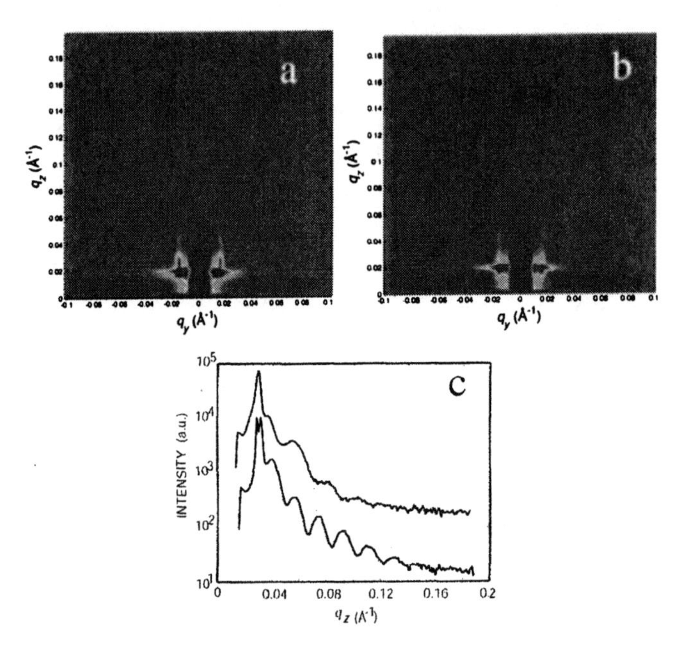

Figure 2 Grazing incidence small angle x-ray scattering pattern of LSM71k diblock copolymer thin films prepared on a substrate modified with random coplymer after being dipped into (a) acetic acid and (b) cyclohexane. (c) The qz scans qy=0.0178 Å-1 of both GISAXS patterns. (See page 1 in color insert.)

Two thin films of LSM71k on different substrates were prepared; one on a neutral brush modified substrate, the other on a passivated substrate. After annealing, SFM (not shown here) shows that, in both cases, the lamellar microdomains orient normal to the surface. After exposure to the acetic acid, comparable SFM images were observed for both substrates. Figure 3a is the GISAXS pattern of an LSM71k film on a passivated surface after dipping into acetic acid at a grazing incidence angle, $\alpha \approx 0.2^{o}$ (same conditions as that used to obtain the data in Figures 1b and 2a). The pattern is similar to Figure 2a, showing higher order diffractions along q_y and oscillations along q_z direction. Figure 3b shows the q_z scans along the first diffuse rod for both GISAXS patterns (Figure 2a and 3a).

The fringes decayed more rapidly for the copolymer on the passivated surfaces (at q_z = 0.08 $Å^{-1}$) in comparison to the copolymer film on a neutral brush (at q_z = 0.15 $Å^{-1}$) indicating a rough termination of the empty channel. Though it is difficult to determine the frequency of the fringes accurately, the wavelength for the copolymer on the passivated surface is slightly higher than that of the copolymer film on the neutral surface, implying slightly shallower channels on the neutral brush. Therefore, the neutral brush plays a critical role in forming channels that penetrate though the entire film. This agrees with previous x-ray reflectivity observations where it was argued that the PMMA blocks diffuse into the underlining brush layer. For the copolymer thin films on the passivated substrate, the PMMA can only go to the top of the film, and those chains close to the copolymer/substrate interface cannot reach the top surface during the rinsing process due to the connectivity with the PS bock, which produces rough termination of the channels.

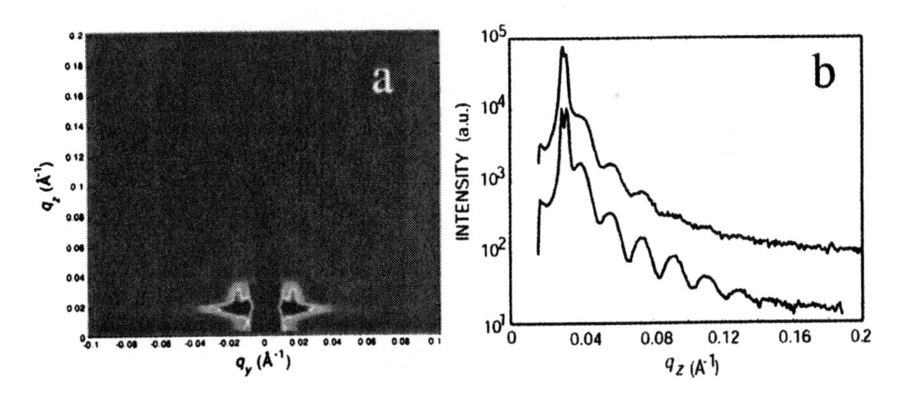

Figure 3(a) Grazing incidence small angle x-ray scattering pattern of an LSM71k diblock copolymer thin film prepared on a passivated substrate after being dipped into acetic acid. (b) The qz scans qy=0.0178 Å-1 of the GISAXS patterns of Figure 2a and 3a. (See page 1 in color insert.)

The re-annealing process of the cylindrical microdomains, CSM69k on a neutral brush modified Si substrate was monitored using *in-situ* grazing incidence small angle x-ray scattering as well.

With the formation of pores, the scattering intensity should be intense due to the large electron density difference between both blocks and air. However, when the pores are refilled with PMMA blocks, the scattering intensity decreases due to the smaller electron density between PMMA and PS. Thus, the refilling process can be studied by simply monitoring the scattering intensity. Figure 4 shows a series of GISAXS patterns captured during the re-annealing process. The sample was first put on the hot stage at 60°C. The GISAXS pattern shows multi-order diffraction along q_y and q_z, resulting from the nanoporous film. This result indicates that the nanoporous structure was still stable at this temperature. From 60 to 90 °C, there is only a small decrease in the scattering intensity, indicating that only a small portion of the PMMA blocks has mobility. The majority of the PMMA chains could not relax back into the pores until being

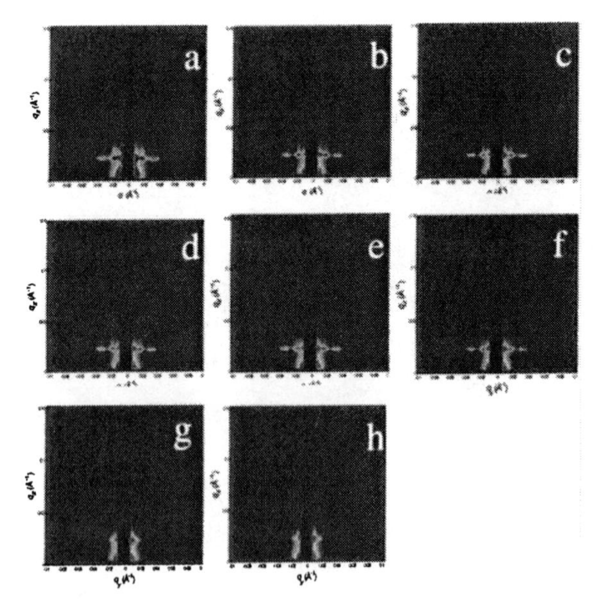

Figure 4 Grazing incidence small angle x-ray scattering patterns of the re-annealing process of a CSM69k diblock copolymer thin film swollen with acetic acid. (a) 60°C, (b) 70°C, (c) 80°C, (d) 85°C, (e) 90°C, (f) 100°C. The film was annealed at each temperature for 6 minutes. (g) was taken after annealing the sample at 100°C for 15 minutes. (See page 2 in color insert.)

heated to 90°C. The small fractions of the PMMA chains that relaxed initially are, most likely, those chains close to the substrate interface, where the polymer chains are stretched the most during solvent swelling. With increasing temperature, the scattering intensity decreased with a dramatic drop between 90-100 °C, as shown in Figure 5a, where the qy scans at q_z = 0.022 Å^{-1} for all the GISAXS patterns plotted. As described before, the decrease in scattering intensity results from the refilling of the pores. It should be noted that at 100 °C, the scattering nearly vanished and with further annealing (15 mins), the scattering intensity increased again (note the intensity of the first order diffraction in Figure 5). This suggests that the PMMA blocks were not closely packed within the pores initially, thus reducing the electron density contrast between the PMMA and PS blocks. With further annealing, the improved packing increased the density in the PMMA domain and electron density contrast.

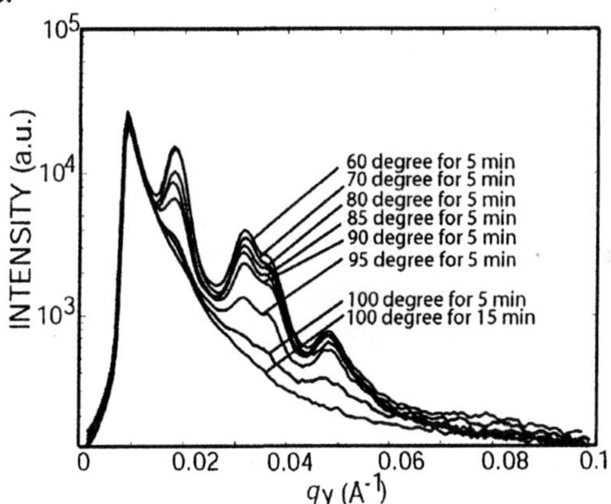

Figure 5a The q_y scan at q_z=0.022 Å^{-1} of the GISAXS patterns in Figure 4.

A series of SFM images of the swollen diblock copolymer thin films after annealing at different temperatures is shown in Figure 6. From both height and phase images, almost no difference was seen on the film surface after annealing at 90°C. After annealing at 100°C for 1 hr, the film surface changed dramatically. The surface roughness decreased, the phase contrast decreased dramatically and some pores were filled. After annealing at 110°C for 1 hr, no deep pores were seen on the surface. However, the pores were not completely filled until the film was annealed at 130°C (SFM image is not shown here). The SFM images further show that the pores were not refilled until heating to 100°C. These results are consistent with the arguments made previously from the GISAXS, where it was argued that the decrease in the scattering intensity was due to the filling of the channels from the bottom portion of the pores at the lower temperature.

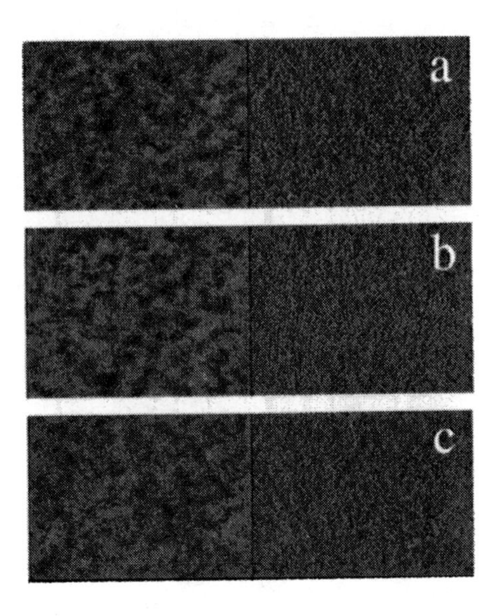

Figure 6 AFM images of CSM69k diblock copolymer thin films after being (a) swollen with acetic acid at room temperature and annealed at (b) 80°C, (c) 90°C, (d) 100°C and (e) 110°C for an hour. The height scale is 15 nm and the phase scale is 15° for all the images. (See page 2 in color insert.)

In conclusion, a reversible selective solvent induced surface reconstruction in diblock block copolymer thin films was investigated. A random copolymer brush layer, solvent selectivity, and very good solvent selectivity are crucial to obtain the nanoporous film with pores penetrating through entire film. The re-annealing process to regenerate the nanotemplate was monitored using *in-situ* GISAXS and SFM. From 60-90°C, only a small fraction of the pores close to the substrate interface were refilled. The template was not fully regenerated until the annealing temperature was close to the glass transition temperature of the film.

Acknowledgment

This work is funded by the Army Research Laboratory, the National Science Foundation and the Department of Energy, Office of Basic Energy Sciences (DE-FG02-96ER45612), Material Research Science and Engineering Center (MRSEC) at University of Massachusetts, Amherst (DMR-0213695). Work at Brookhaven National Laboratory was supported through the Department of Energy, Office of Basic Energy Sciences (DE-AC02-98CH10886) and from the Nanoscale Science, Engineering, and Technology Program.

References

(1) Bates, F. S.; Fredrickson, G. H. Annual Reviews of Physics and Chemistry 1990, 41, 5252.
(2) Kim, G.; Libera, M. Macromolecules 1998, 31, 2569.
(3) Kim, S. H.; Misner, M. J.; Xu, T.; Kimura, M.; Russell, T. P. Adv Mater 2004, 16, 226.
(4) Kimura, M.; Misner, M. J.; Xu, T.; Kim, S. H.; Russell, T. P. Langmuir 2003, 19, 9910.
(5) Lin, Z.; Kim, D. H.; Wu, X.; Boosahda, L.; Stone, D.; LaRose, L.; Russell, T. P. Adv. Mater. 2002, 14, 1373.
(6) Rehse, N.; Knoll, A.; Magerle, R.; Krausch, G. Macromolecules 2003, 36, 3261.

(7) Temple, K.; Kulbaba, K.; Power-Billard, K. N.; Manners, I.; Leach, K. A.; Xu, T.; Russell, T. P.; Hawker, C. J. Adv Mater 2003, 15, 297.
(8) Xu, T.; Stevens, J.; Villa, J. A.; Goldbach, J. T.; Guarim, K. W.; Black, C. T.; Hawker, C. J.; Russell, T. R. Adv Funct Mater 2003, 13, 698.
(9) Huang, H. Y.; Hu, Z. J.; Chen, Y. Z.; Zhang, F. J.; Gong, Y. M.; He, T. B.; Wu, C. Macromolecules 2004, 37, 6523.
(10) Mori, H.; Hirao, A.; Nakahama, S.; K. Senshu. Macromolecules 1994, 27, 4093.
(11) Zhang, Q.; Tsui, O. K. C.; Du, B.; Zhang, F.; Tang, T.; He, T. Macromolecules 2000, 33, 9561.
(12) Knoll, A.; Horvat, A.; Lyakhova, K. S.; Krausch, G.; Sevink, G. J. A.; Zvelindovsky, A. V.; R. Magerle, ,. Phys. Rev. Lett. 2002, 89, 035501.
(13) Xu, T.; Goldbach, J. T.; Misner, M. J.; Kim, S.; Gibaud, A.; Gang, O.; Ocko, B.; Guarini, K. W.; Black, C. T.; Hawker, C. J.; Russell, T. P. Macromolecules 2004, 37, 2972.
(14) Benoit, D.; Chaplinski, V.; Braslau, R.; Hawker, C. J. J. Am. Chem. Soc., 1999, 121, 3904.
(15) Huang, E.; Rockford, L.; Russell, T. P.; Hawker, C. J. Nature 1998, 395, 757.
(16) Mansky, P.; Liu, Y.; Huang, E.; Russell, T. P.; Hawker, C. Science 1997, 275, 1458.
(17) Xu, T.; Kim, H. C.; DeRouchey, J.; Seney, C.; Levesque, C.; Martin, P.; Stafford, C. M.; Russell, T. P. Polymer 2001, 42, 9091.
(18) Doshi, D. A.; Gibaud, A.; Goletto, V.; Lu, M. C.; Gerung, H.; Ocko, B.; Han, S. M.; Brinker, C. J. J. of Am. Chem. Soc. 2003, 125, 11646.

Chapter 13

Thermoplastic Molecular Sieves: New Polymeric Materials for Molecular Packaging

Giuseppe Milano[1], Christophe Daniel[1], Vincenzo Venditto[1], Paola Rizzo[1], Gaetano Guerra[1,*], Pellegrino Musto[2], Giuseppe Mensitieri[3]

[1]Department of Chemistry, University of Salerno, Via S. Allende, 84081 Baronissi (SA), Italy
[2]Institute of Chemistry and Technology of Polymers, National Research Council of Italy, Via Campi Flegrei 34, 80078 Pozzuoli (NA), Italy
[3]Department of Materials and Production Engineering, University of Napoli Federico II, P.le Tecchio 80, 80125 Napoli, Italy

New polymeric materials based on syndiotactic polystyrene (s-PS) are presented. These materials are able to absorb rapidly and efficiently volatile organic compounds from water and air, also when present at very low concentrations and can be considered as the first example of polymeric molecular sieves, as they display a high sorption selectivity similar to zeolites. Moreover these new molecular sieves are hydrophobic and hence seem particularly suitable for water and moist air purification. Different aspects relative to the structure, absorption properties, and host-guest interactions of these new polymeric materials are described.

Introduction

Syndiotactic polystyrene (s-PS), whose synthesis was reported about two decades ago (*1*, *2*), is a easily crystallizable and high melting (~270 °C) stereoregular polymer presenting a very complex polymorphic behavior (*3*).

In addition to four different crystalline forms, several clathrate structures, mainly including halogenated or aromatic hydrocarbons as guest molecules, have been described (*4*).

The crystalline δ-form is nanoporous and can be obtained by removal of guest molecules from clathrate samples, by suitable solvent treatments (5). The corresponding crystal structure (6) is shown in Figure 1A. Sorption studies from liquid and gas phases, into s-PS samples (mainly films) being in this nanoporous δ-form, have shown that this thermoplastic material is able to absorb selectively some organic substances from different environments also when present at low concentrations (*5, 7, 8*).

To our knowledge, this is the first case of polymeric semicrystalline material whose sorption ability is higher for the crystalline phase than for the amorphous phase.

These polymeric molecular sieves present narrow distributions of nanopore volumes (of the order of 120-170 $Å^3$) and hence, like zeolites, good molecular selectivities. In fact, in both cases, the nanopores correspond to cavities with well defined positions and shapes inside crystalline lattices (*9*).

Sorption studies from liquid and gas phases, into s-PS samples being in the nanoporous δ-form (*7*), have shown that this thermoplastic material is able to absorb selectively some volatile organic compounds (VOCs) (mainly halogenated or aromatic hydrocarbons) also when present at low concentrations (*5, 7-9*), forming the corresponding clathrate forms. These sorption studies have suggested that this material is promising for applications in chemical separations as well as in water and air purification. In particular, thin polymeric films have been suggested as sensing elements of molecular sensors (*10*). Moreover, these materials are hydrophobic (their water absorption as well as adsorption are negligible) while absorb rapidly and efficiently all VOCs mostly presented in industrial waste waters (benzene, toluene, chloroform, tetrachloroethylene, trichloroethylene, etc.) and hence seem particularly suitable for water and moist air purification.

It is worth noting that these new molecular sieves are based on thermoplastic polymers and as a consequence could be used not only as powders or grains (as carbon adsorbents) but also processed to obtain morphologies and manufacts possibly useful for molecular separation processes (e.g., aerogels or membranes or macroporous beads).

In the following sections, different aspects relative to the structure, absorption properties, and host-guest interactions of syndiotactic polystyrene crystalline δ-form will be discussed.

Shape and Volume of the Cavities

Empty Volume Calculation

The monoclinic structure of the δ-form of s-PS (*6*) (space group P21/a, a = 17.5 Å, b = 11.9 Å, c = 7.7 Å, and γ = 117°) has per unit cell two identical cavities centered on the center of symmetry, bounded by 10 phenyl rings (Figure 1) (*9*).

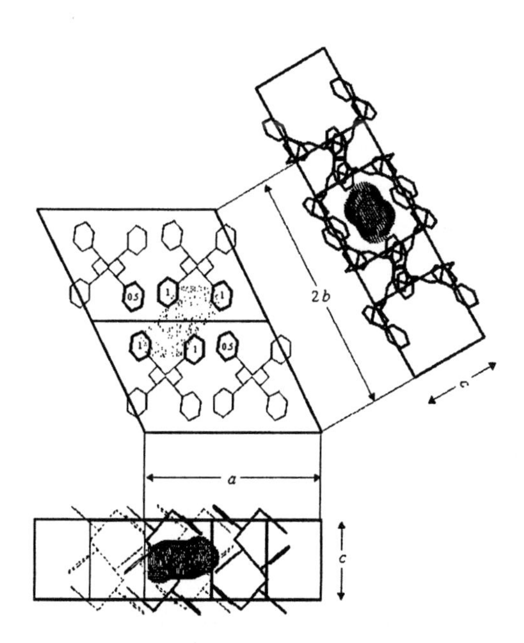

Figure 1: Region of empty space (dotted) calculated for the δ-form, by assuming r= 1.8 Å (that is a typical van der Waals radius of chlorine atoms or methyl groups) shown for three different views of two unit cell: along c and perpendicular to ac and bc planes. Along the c view, the phenyl rings that define the boundary of the cavity are bolded and labelled by heights (expressed as fractions of c=7.7 Å)

A complete filling of the cavities by guest molecules generates clathrate structures generally presenting an organization of the polymeric host very similar to that observed for the δ-form.

Crystal structure data contain all the information necessary to evaluate the empty volume fraction as well as the shapes and sizes of possible empty spaces (cavities and/or channels). The fraction of empty volume (Φ_e) can be defined on the basis of the volume that can be filled by a hypothetical sphere of radius *r* (*9*).

The region of empty space calculated for the δ-form, by assuming r = 1.8 Å, that is a typical van der Waals radius of chlorine atoms or methyl groups is shown as a dotted region for three different views of the unit cell, along *c* and perpendicular to *ac* and *bc* planes, in Figure 1. The figure shows that this empty space corresponds to finite cavities (two per unit cell) centered on the center of symmetry of the crystal structure, whose boundary is essentially defined by 10 phenyl rings. By considering, for instance, the view along the *c* axis, there are four phenyl rings below the cavity (average height 0), four phenyl rings above the cavity (average height c = 7.7 Å), and two phenyl rings whose average height is equal to the average height of the cavity ($c/2$) (bolded rings in Figure 1). For a probe sphere with r = 1.8 Å, the cavity has a volume of nearly 115 $Å^3$ and its maximum dimension is nearly 8.1 Å (essentially along the *a-b* direction) while its minimum dimension is nearly 3.4 Å (essentially along the *c* axis).

Following the same procedure, the shapes and volumes of the hypothetical cavities generated by ignoring the presence of the guest molecules have been evaluated for the clathrate structures[4] and the relative cavity volumes are reported in the last column of Table I.

Table I : Crystal Structure Parameters, Guest Volume Fraction, Guest Volume, and Volume of the Cavity for the δ–form and Three Different Clathrates of s-PS

Crystal structure	*a (Å)*	*b(Å)*	*c(Å)*	*γ*	*Guest volume fraction*[a]	*Guest volume ($Å^3$)*[a]	*Cavity volume ($Å^3$)*[b]
δ -form	17.5[c]	11.8[c]	7.7[c]	117[c]			115
s-PS/DCE	17.1[d]	12.1[d]	7.7[d]	120[d]	0.22	91	125
s-PS/iodine	17.3[e]	12.9[e]	7.7[e]	120[e]	0.29	124	151
s-PS/toluene	17.6[f]	13.3[f]	7.7[f]	121[f]	0.30	132	161

NOTE: [a]Guest volume fraction = $V_{guest}/(4 \times V_{styrenic\ unit})$. Volume calculations are based on the van der Waal radii. [b]Calculated with the assumption of r = 1.8 Å. [c]From Reference 6. [d]From Reference 4c. [e]From Reference 4b. [f]From Reference 4a.

Nitrogen Sorption Experiments

Following standard procedures for porosity evaluation of powders (*11*), by nitrogen sorption experiments at low temperature (77 K) on amorphous and semicrystalline s-PS samples, it is possible to give an experimental evaluation of the δ-form cavity size.

Sorption and desorption isotherms of δ- and γ- (not nanoporous) form samples are compared in Figure 2, where the sorption is expressed as cm^3 of nitrogen in normal conditions (1 atm, 0 C) per gram of polymer.

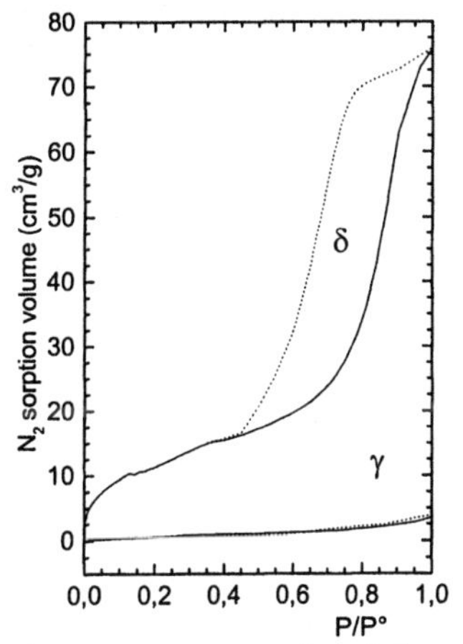

Figure 2: Sorption (continuous curves) and desorption (dashed curves) isotherms of nitrogen at 77 K into s-PS powder samples, including δ and γ crystalline phases.

Moreover, starting from p/p_0 0.6, there is a steeper increase of sorption of the δ-form up to $p/p_0 = 0.9$. This makes the sorption difference between the δ -form and the other semicrystalline samples particularly large (nearly 70 cm^3/g) at $p/p_0 = 0.98$. It is also worth noting that for the δ-form powder there is a remarkable hysteresis in the sorption-desorption phenomenon.

Because the two samples present similar crystallinity, it has been assumed that the sorption increase observed for the δ-form sample is essentially associated with condensation of nitrogen molecules into the crystalline nanoporous phase and that the observed hysteresis is due to the formation of polymer-gas intercalates.

In the assumption of formation of s-PS/nitrogen intercalates, the number of N_2 molecules per crystalline cavity can be approximately evaluated. In fact, if n_{N2} and n_{cavity} are moles of nitrogen and cavities, respectively, for a given mass of polymer M_{pol}, then:

$$\frac{n_{N_2}}{n_{cavity}} = \frac{\left(PV_{N_2}M_{pol}\right)/RT}{\left(M_{pol}X_c\right)/4M_{styr}} = \frac{4PV_{N_2}M_{styr}}{RTX_c} = 0.0186V_{N_2}/X_c \qquad (1)$$

where V_{N_2} is the volume of nitrogen (for P = 1 atm and T = 273 K) sorbed in the crystalline phase per gram of polymer, M_{styr} is the molecular mass of the styrenic unit, X_c is the crystalline fraction of the δ-form polymer sample, and 4 is the number of styrene units per cavity in the δ crystalline phase.

On the basis of eq 1, by taking V_{N_2} = 70 cm^3/g (from Figure 2), and X_c = 0.43 (determined from the X-ray diffraction pattern (not reported here)), the number of nitrogen molecules per crystalline cavity is calculated to be close to 3.

Two possible limit evaluations of the space occupied by the molecule of N_2 can be obtained considering the space occupied by three molecules in the two possible nitrogen crystalline structures (cubic and hexagonal) (*12*): 137 and 163 Å^3, respectively (*13*). In our assumptions these volumes can be taken as a rough evaluation of the volume of the cavity of the crystalline phase. This independent evaluation is in satisfactory agreement with the calculated volumes of column 8 in Table 1.

Absorption of VOCs from Water

Chlorinated and aromatic hydrocarbons are readily sorbed by the nanoporous crystalline form of s-PS. Just as an example, sorption kinetics of 1,2-dichloroethane (DCE) from saturated (8100 ppm) or diluted (100 ppm) aqueous solutions by powder s-PS samples, (*14*) characterized by surface areas of nearly 4 m^2/g, are shown in Figure 3. In addition to powders in δ-form, powders in the orthorhombic β crystalline form (a = 8.81 Å, b = 28.82 Å, c = 5.1 Å) (*15*) which absorb low molecular mass compounds only into the amorphous phase (*16*), are also considered. The stability of the crystalline β-form of s-PS in the presence of organic compounds can be attributed to a much lower fraction of empty volume than for the δ–form (9) which is due to a larger density (1.078 g/cm^3 for the β-form vs. 0.997 g/cm^3 for the δ-form).

It is apparent that for δ-form powders, few minutes are sufficient to obtain substantial sorption of DCE from water solutions, whereas β-form powders sorb DCE more slowly and to a lower extent. Sorption from s-PS samples, which are amorphous or in crystalline forms other than δ, is negligible for low DCE activities as occurs for β -form samples. For the considered δ-form sample, the equilibrium sorption of DCE from 100 and 10 ppm aqueous solutions is of 7 and 4.5%, respectively. These results hence indicate the occurrence of high partition coefficients between the polymeric phase and the aqueous phase. The reported results indicate that nanoporous crystalline samples of s-PS can be suitable for water purification from some chlorinated hydrocarbons. As will be discussed in the next sections, the occurrence of different conformational equilibria of some guest molecules, into amorphous and clathrate phases of s-PS, allows evaluation of the guest contents into both phases. On this basis, it is established that sorption involves preferentially the clathrate phase and that desorption from clathrate phase is much slower than desorption from amorphous phase *(8)*.

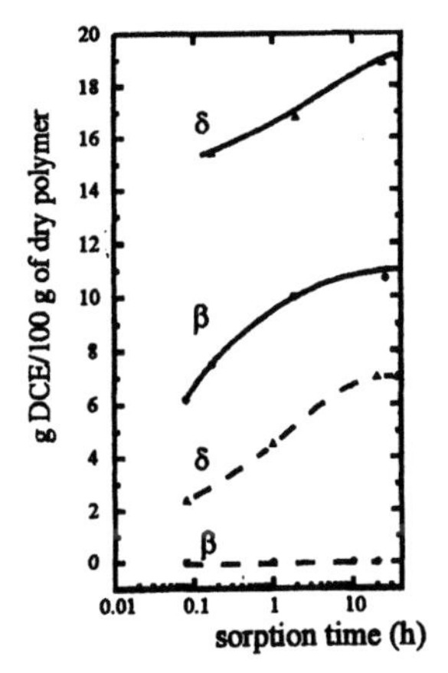

Figure 3: Sorption kinetics of DCE from saturated (8100 ppm; solid lines) or 100 ppm (dashed lines) aqueous solutions by semicrystalline powder samples:( ) δ-form and ( ) β-form

Nature of the Host-Guest Interactions

FTIR study of conformational selectivity

For the conformational studies, 1,2-dichloroethane (DCE), 1,2-dichloropropane (DCP) and 1-chloropropane (CP) have been chosen as guest molecules, since they present similar and simple conformational equilibria and the different conformations are reasonably populated and readily detectable by spectroscopic means. In fact, the FTIR spectra of these chlorinated compounds in the wavenumber range 1500-450 cm^{-1}, present a number of well-resolved peaks. Moreover, due to the relative simplicity of these molecules a complete normal vibrational analysis is feasible, which has allowed unambiguous assignment of the various absorptions to the normal modes of the different conformers (*17-19*).

A comparison between the conformational equilibria of the three considered chlorinated compounds into s-PS films is presented in Figure 4, as a plot of the fraction of the trans conformers (see Scheme 1) versus the molar concentration of the chlorinated compounds in the polymer samples.

Scheme 1

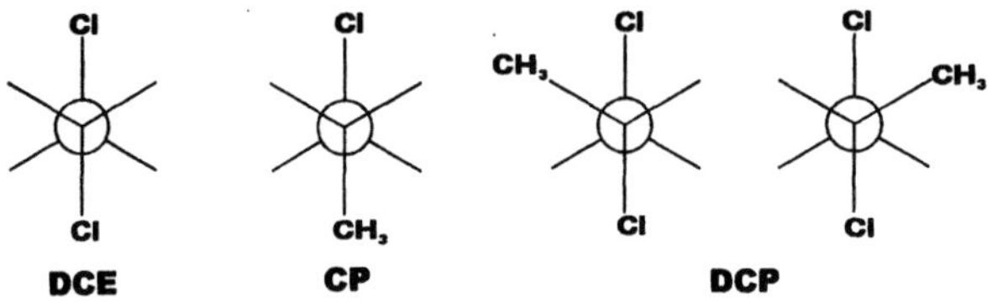

DCE CP DCP

For each set of data, the maximum value of the guest molar concentration (the experimental point on the extreme right of each curve of Figure 4) corresponds to that observed immediately after sorption. The other experimental points correspond to the same samples after different desorption procedures.

Data corresponding to DCE sorption from a 0.5 wt % aqueous solution followed by isothermal desorption at 40 °C (empty circles), 60 °C (filled circles), and 80 °C (squares) relative to β- and δ-form s-PS samples are labeled A and B, respectively, in Figure 4. It is apparent that, the trans and gauche conformations are nearly equally populated for DCE molecules sorbed in the amorphous phase of samples including the dense and impermeable β crystalline phase (*16*) (data set A in Figure 4).

The observed fractions of trans conformer (Xt) are intermediate between those observed for pure liquid (Xt = 0.35) and for vapor (Xt = 0.75) DCE (*17*). For equal sorption conditions, the DCE concentration is higher and the fraction of its trans conformer is larger when DCE is sorbed into δ-form crystalline samples (cf., e.g., data points on the extreme right of curves B and A of Figure 4). Moreover, the Xt value increases up to 0.94 as DCE concentration in the polymer is reduced by desorption (curve B in Figure 4), that is to the same value obtained for DCE sorption into nanoporous polymer samples from very diluted aqueous solutions (10 ppm) (*8*).

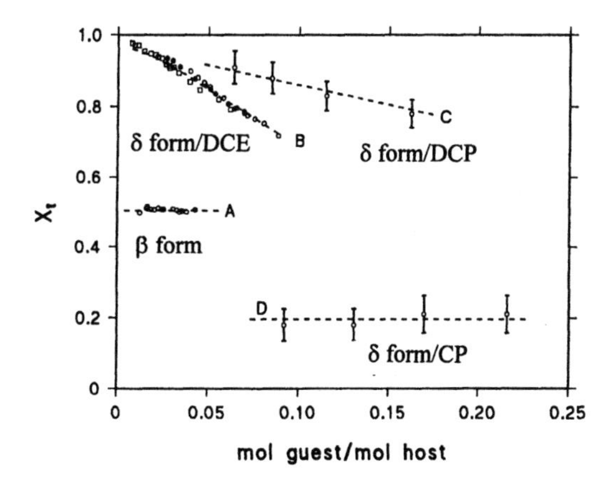

Figure 4: Plot of the fraction of the trans conformers for the three chlorinated hydrocarbons considered (see scheme 1) as a function of molar concentration of the chlorinated compounds in s-PS semicrystalline films (expressed as mole of guest per mole of styrene monomeric units): (A) DCE absorbed from a 0.5 wt% aqueous solution by a β-form film; (B) DCE absorbed from a 0.5 wt% aqueous solution by a nanoporous δ-form film; (C) DCP absorbed from pure liquid by a nanoporous δ-form film; and (D) CP absorbed from pure liquid by a nanoporous δ-form film.

A similar behavior is observed for DCP. For instance, sorption experiments into a nanoporous s-PS film from pure liquid, followed by desorption at room temperature (data set C in Figure 4), show large Xt values, larger than those measured for diluted DCP solutions in benzene (Xt = 0.65) (*18*) or for liquid DCP (Xt = 0.62) (*23*), which increase as desorption proceeds.

In contrast the CP behavior is completely different, in fact the population of the trans conformer is not increased by clathration into s-PS. Take for example CP sorbed from pure liquid at room temperature by a nanoporous s-PS sample: although the clathration phenomenon is confirmed by the typical changes in the X-ray diffraction patterns (*6*), Xt remains close to 0.2 (data set D of Figure 4), which is not far from the value observed for liquid (0.27) and vapor (0.38) CP (*19*).

These results, therefore, indicate that the conformational equilibria of the chlorinated compounds are poorly affected by sorption into the amorphous polymer phase. Moreover, the conformational equilibrium of CP is poorly affected also by inclusion into the s-PS nanoporous phase. On the contrary, for DCE and DCP, the conformers presenting two trans chlorine atoms are largely prevailing into the clathrate phases.

The observed increases of the Xt values during guest desorption procedures can be explained by a faster desorption kinetics from the amorphous phase. In particular, by assuming Xt values for both phases independent of guest concentration, guest contents into clathrate and amorphous phases can be evaluated from the above-described FTIR peak intensities. For instance, Xt for DCE sorbed into the amorphous phase can be reasonably assumed equal to 0.51, which is equal to the experimental value observed for different amorphous and semicrystalline samples sorbing DCE only into the amorphous phase (see, e.g. curve A of Figure 4 and ref. 8). As for the clathrate phase, Xt can be assumed close to 0.94, i.e., the highest experimental value observed for δ-form s-PS samples after substantial guest desorption (see curve B in Figure 4) as well as after DCE absorption from diluted solutions (*8*).

In these assumptions, separate desorption kinetics from the amorphous and clathrate phases of s-PS can be evaluated (*8b*). Just as an example, desorption kinetics at 40 °C from amorphous and clathrate phases of a s-PS film after DCE sorption from a 0.5 wt % aqueous solution (based on the Xt values corresponding to the empty circles of curve B of Figure 4) are shown in Figure 5.

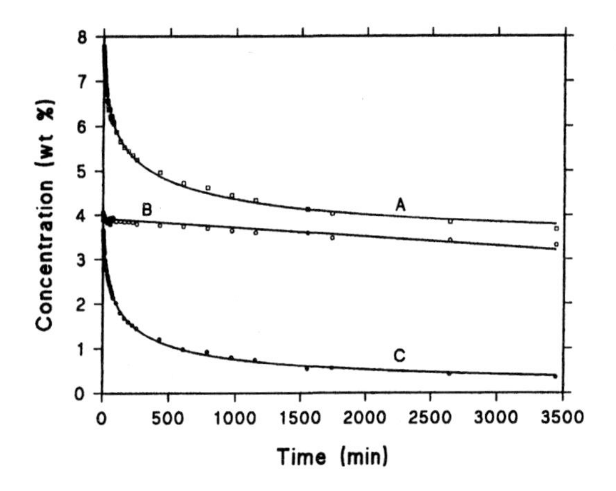

Figure 5:Desorption kinetic of DCE at 40°C from a s-PS film, after sorption from a 0.5 wt% aqueous solution: (A) overall; (B) for the clathrate phase; (B) for the amorphous phase. Data of curves B and C have been calculated on the basis of Xt variations (see text).

Initially, in the considered sample, DCE is partitioned almost evenly in the amorphous and the crystalline phases, while the difference in the desorption rates between the two phases is dramatic. For instance after about 60 h the DCE content in the amorphous phase has decreased by about 90% of its initial value, while, in the crystalline phase, the reduction is close to 10%.

Analogous conformational studies, applied to sorption kinetics relative to semi-crystalline δ-form s-PS samples, have indicated that the absorption of the considered chlorinated compounds from aqueous solutions occurs preferentially into the crystalline phase and essentially only into the crystalline phase for low guest concentrations (*8*). This sorption mechanism based on the formation of a clathrate crystalline phase involves a high potential selectivity, typical of host-guest compounds.

Gas sorption (CO_2 and CH_4) and transport into a polymeric crystalline phase have already been observed for isotactic poly(4-methyl-1-pentene) (*21, 22*). However in that case the molecules do not form clathrate structures but dissolve in the crystals at about 1/3 to 1/4 the level they do in the amorphous phase.

The Role of Electrostatic Interactions: Molecular Modelling Results

The nature of host-guest interactions and the role of electrostatic ones in s-PS clathrates have been subject of several theoretical studies (*14, 23, 24*).

The occurrence for DCE and DCP (but not for CP) of conformational selectivity, favouring the trans conformers in the cavities of the δ-form of s-PS, indicates the presence of specific attractive, rather than van der Waals repulsive, interactions involving the trans chlorine atoms. In fact, since chlorine atoms and the methyl groups present similar van der Waals radii (close to 1.8 Å), if the conformational selectivity had been due to repulsive non-bonded interactions, it would have been observed also for CP (see Scheme 1).

It is reasonable to assume that the specific locations of the phenyl rings delimiting the cavities of the nanoporous structure would influence the location of the two chlorine atoms of DCE and DCP, thus determining the observed conformational selectivities.

A molecular mechanics approach is used in the following to rationalize the previously described experimental conformational results. This can contribute to elucidating the nature of the host-guest interactions into s-PS clathrate compounds.

As for the polymeric host, an electrostatic potential map, relative to the cavity region of the unit cell of the nanoporous δ-form of s-PS indicated (dotted) in Figure 1, at a quote of 1/2c, is shown in Figure 6A.

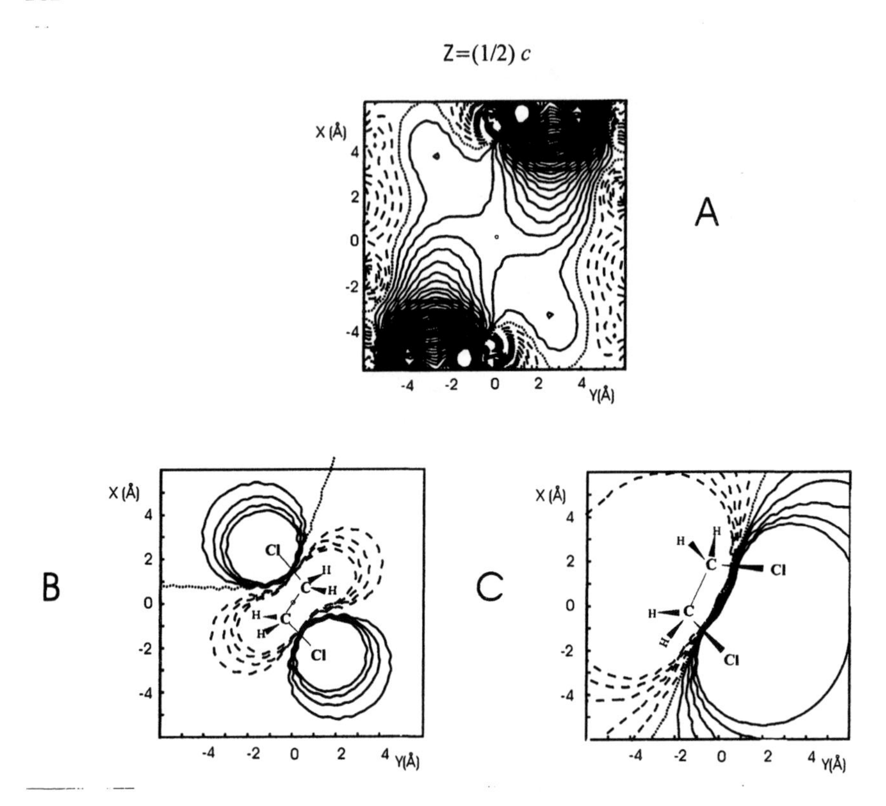

Figure 6: Potential maps, relative to the cavity region of the unit cell of the nanoporous δ-form of s-PS of indicated (dotted) in Figure 1, at a quote of 1/2c: (A) electrostatic potential of the cavity; electrostatic potential maps relative to the trans (B) and gauche (C) conformers of DCE positioned with respect to the cavity in a way suitable to minimize the overall potential energy. The zero energy curves is dotted and positive and negative are dashed and continuous, respectively

The map of Figure 6A shows that the electrostatic potential is essentially negative in the central part of the cavity while it is prevailingly positive in some regions that are external to the cavity.

As for the chlorinated guests, trans and gauche conformers of DCE, positioned with respect to the cavity in the way suitable to minimize overall potential energy, are shown in Figure 6, parts B and C, respectively. For comparison with electrostatic potential of the cavity (Figure 6A), the electrostatic potential of both conformers of DCE is also shown.

It is apparent on inspection that there is a good electrostatic fit between the quadrupolar trans conformer and the substantially quadrupolar cavity. Presence of a center of symmetry in the geometric center of the cavity determines quadrupolar character of the electrostatic potential. Furthermore, negative regions of electrostatic potential (left-bottom and right-upper corners in figure 6A) are determined by the presence of phenyl rings at a quote ½ c in the cavity boundary.

The positive electrostatic region of the DCE trans conformer, corresponding to the carbon and hydrogen atoms (Figure 6B), presents a good superposition with the negative electrostatic nature of the cavity (Figure 6A).

The electrostatic fit between the cavity and the substantially dipolar gauche conformer (cf. Figure 6C with Figure 6A) is instead poor. Strictly analogous considerations hold for DCP. On the contrary, CP is dipolar and its electrostatic field is essentially independent of its conformation, as a consequence its electrostatic interaction with the cavity is similar for all conformers.

Results of minimizations of the host-guest interaction energies for the different guest conformers, into the cavity of Figure 6, parts B and C, are reported in Table II.

Table II. Host-guest Interaction Energies[a]

	DCE		*DCP*		*CP*	
	T	*G*	*T*	*G*	*T*	*G*
Van der Waals	-12.8	-12.8	-16.0	-16.1	-13.5	-13.6
Electrostatic	-1.1	0.0	-0.8	0.2	0.2	-0.6
Total	-13.9	-12.8	-16.8	-15.9	-14.3	-14.2

NOTE: [a]Results of minimization of the interaction energies (kcal/mol) with the cavity of figure 1 for *trans* and *gauche* conformers of chlorinated guests.

For the sake of simplicity, only the absolute energy minimum situations of both conformers are listed. It is apparent that for all the considered chlorinated guests the main attractive contribution (up to 16 kcal/mol) is of van der Waals type, being however poorly dependent on the guest conformation. The electrostatic attractive contribution is smaller for all the considered chlorinated compounds; however, it is nearly 1 kcal/mol larger for the trans conformers, with respect to the gauche conformers, for both DCE and DCP. On the other hand, the electrostatic contribution to the minimum energy values is poorly dependent on the CP conformation. Hence, the calculations of Figure 6 and Table II are able to rationalize the absence of conformational selectivity for CP and the preference toward trans conformers for DCE and DCP, in the cavities of the δ-form s-PS.

Concluding Remarks

In this contribution different aspects relative to the structure, absorption properties, and host-guest interactions of syndiotactic polystyrene crystalline nanoporous δ-form have been presented.

Semi-crystalline δ-form samples are able to absorb, mainly in the crystalline phase, most of the volatile organic compounds present in industrial wastes from liquid or gaseous mixtures (i.e. to form clathrate), also when those are present at low concentrations. Hence, these materials provide possible industrial uses for environment pollution control or for water and moist air purification. In particular semi-crystalline δ-form polymer films have been tested as sensing probes for the detection of chemical pollutants (for example in resonant sensors) (*10*) since they display a good rigidity and at the same time they present a higher sensitivity and selectivity than other polymeric sensing films used so far.

Furthermore, it is worth noting that the crystalline δ-form could provide an innovative approach in the area of polymeric materials with optoelectronic and photonic properties. Indeed, the δ-form offers the possibility to arrange the photoactive molecules with a high degree of positional order while polymeric materials used so far are usually based on the dispersion of photoactive molecules in an amorphous polymer matrix, on the chemical bonding of photoactive groups to the polymer backbone (functionalization), or on the inclusion of the photoactive groups in the polymer chain as monomers. All these traditional procedures lead to the formation of amorphous phases are characterized by a disordered distribution of the photoactive groups. It is also worth emphasizing that it has been, recently, established the possibility to obtain three different planar orientation of the crystalline phase in δ-form semi-crystalline film samples (*25*) allowing thus a tuning of the orientation of the guest molecules absorbed in the crystalline phase through the control of the orientation of the host crystalline phase. This possibility is particularly relevant in the case of photoactive guest molecules (for example fluorescent, photoreactive, photochromic, with nonlinear activity, ...) as it is possible to distribute the molecules in an isolated way with a high orientational and positional order.

References

(1) Ishihara, N.; Seimiya, T.; Kuramoto, M.; Uoi, M. *Macromolecules* **1986**, *19*, 2465.

(2) Zambelli, A.; Longo, P.; Pellecchia, C.; Grassi, A. *Macromolecules* **1987**, *20*, 2035.

(3) Guerra, G.; Vitagliano, V. M.; De Rosa, C.; Petraccone, V.; Corradini, P. *Macromolecules* **1990**, *23*, 1539.
(4) (a) Chatani, Y.; Shimane, Y.; Inoue, Y.; Inagaki, T.; Ishioka, T.; Ijitsu, T.; Yukinari, T. *Polymer* **1993**, *34*, 1620. (b) Chatani, Y.; Inagaki, T ; Shimane, Y.; Shikuma, H. *Polymer* **1993**, *34*, 4841. (c) De Rosa, C.; Rizzo, P.; Ruiz de Ballesteros, O.; Petraccone, V.; Guerra, G. *Polymer* **1999**, *40*, 2103.
(5) (a) Guerra, G.; Manfredi, C.; Rapacciuolo, M.; Corradini, P.; Mensitieri, G.; Del Nobile, M. A. Ital. Pat. **1994** (C.N.R.). (b) Reverchon, E.; Guerra, G.; Venditto, V. *J. Appl. Polym Sci.* **1999**, *74*, 2077.
(6) De Rosa, C.; Guerra, G.; Petraccone, V.; Pirozzi, B. *Macromolecules* **1997**, *30*, 4147.
(7) Manfredi, C.; Del Nobile, M. A.; Mensitieri, G.; Guerra, G.; Rapacciuolo, M. *J. Polym. Sci., Polym. Phys. Ed.* **1997**, *35*, 133.
(8) (a) Guerra, G.; Manfredi, C.; Musto, P.; Tavone, S. *Macromolecules* **1998**, *31*, 1329. (b) Musto, P; Manzari, M; Guerra, G. *Macromolecules* **1999**, *32*, 2770. (c) Musto, P.; Mensitieri, G.; Cotugno, S.; Guerra, G.; Venditto, V. *Macromolecules* **2002**, *35*, 2296. (d) Larobina, D.; Sanguigno, L.; Venditto, V.; Guerra, G.; Mensitieri, G. *Polymer* **2004**, *45*, 429.
(9) Milano, G.; Venditto, V.; Guerra, G.; Cavallo, L.; Ciambelli, P.; Sannino D. *Chem. Mater.* **2001**, *13*, 1506.
(10) Guerra, G.; Venditto, V.; Mensitieri, G. It. Pat. SA00A23; Eur. Pat. Appl. EP1217360A2. (b) Mensitieri, G.; Venditto, V.; Guerra, G. *Sens. Actuators* **2003**, *92*, 255.
(11) Gregg, S. J.; Sing, K. S. W. *In Adsorption, Surface Area and Porosity;* Academic Press: London, **1982**.
(12) Wyckoff, R. W. G. *In Crystal Structures*; J. Wiley & Sons: New York, **1963**.
(13) The space occupied by three N_2 molecules is considered as three-halves and three-fourths of the volumes of the dimolecular cubic and tetramolecular hexagonal unit cells, respectively.
(14) Guerra, G.; Milano, G.; Venditto, V.; Musto, P.; De Rosa, C.; Cavallo, L. *Chem. Mater.* **2000**, *12*, 363.
(15) Chatani, Y.; Fujii, Y.; Stimane, Y.; Ijitsu, T. *Polymer Preprints*, Japan, (Eng. Ed.), **1988**, *37*, E428.
(16) Mensitieri, G.; Rapacciuolo, M.; De Rosa, C.; Apicella, A.; Del Nobile, M. A.; Guerra, G. *J. Mater. Sci. Lett.* **1991**, *24*, 5645.
(17) Tanabe, K. *Spectrochim. Acta* **1972**, *28A*, 407.
(18) Thorbjornsrud, J.; Ellestad, O. H.; Klaboe, P.; Torgrimsen, T. *J. Mol. Struct.* **1973**, *15*, 45.
(19) Ogawa, Y.; Imazeki, S.; Yamaguchi, H.; Matsuora, H.; Harada, I.; Shimanouchi, T. *Bull. Chem. Soc. Jpn.* **1978**, *51*, 748.
(20) Musto, P.; Manzari, M. G.; Guerra, G. *Macromolecules* **2000**, *33*, 143

(21) Puleo, A. C.; Paul, D. R.; Wong, P. K. *Polymer* **1989**, *30*, 1357.
(22) Muller-Plathe, F. *J. Chem. Phys.* **1995**, *103*, 4346.
(23) Milano, G.; Guerra, G.; Cavallo, L. *Eur. J. Inorg. Chem.* **1998**, *10*, 1513
(24) Milano, G.; Guerra, G.; Cavallo, L. *Macromol. Theory Simul.* **2001**, *10*, 349.
(25) (a) Rizzo, P; Lamberti, M.; Albunia, R. A.; Ruiz de Ballesteros, O.; Guerra, G. *Macromolecules* **2002**, *35*, 5854.(b) Guerra, G.; Rizzo, P.; Mensitieri, G.; Venditto, V., It. Pat. SA2003A000014. (c) Rizzo, P; Costabile A.; Guerra, G. *Macromolecules* **2004**, *37*, 3071.

Chapter 14

Modeling Transport Properties in High Free Volume Glassy Polymers

Xiao-Yan Wang, Kenneth M. Lee, Ying Lu, Matthew T. Stone, I. C. Sanchez, and B. D. Freeman

Department of Chemical Engineering, University of Texas at Austin, Austin, TX 78712

Molecular modeling techniques are applied to study the cavity size distributions and transport properties of two very permeable polymers, poly (1-trimethylsilyl-1-propyne) (PTMSP) and a random copolymer of tetrafluoroethylene and 2,2-bis(trifluoromethyl)-4,5-difluoro-1,3-dioxole (TFE/BDD), which have very similar and large fractional free volumes, but very different permeabilities. Using atomistic models, cavity size (free volume) distributions determined by a combination of molecular dynamics and Monte Carlo methods are consistent with the observation that PTMSP is more permeable than TFE/BDD. The average spherical cavity size in PTMSP is 11.2 Å, whereas it is only 8.2 Å in TFE/BDD. These cavity size distributions determined by simulation are also consistent with free volume distributions determined by positron annihilation lifetime spectroscopy (PALS). The diffusivity, solubility and permeability of CO_2 in these polymers were also obtained through molecular simulation. The diffusivity and permeability of CO_2 in PTMSP are higher than in TFE/BDD. Good agreement is observed between the simulation and experimental data.

Introduction

Polymer membranes with controlled permeability have found wide application in industrial gas separations, food packaging, water and air purification and biomedical engineering.[1] Poly(1-trimethylsilyl-1-propyne) (PTMSP) and the random copolymer of tetrafluoroethylene and 2,2-bis(trifluoromethyl)-4,5-difluoro-1,3-dioxole (TFE/BDD) are known to have very high permeability coefficients for glassy polymers and are potentially very useful for membrane separation technology.[2]

PTMSP has the highest permeability coefficients among all polymeric materials.[3,4] Moreover, PTMSP exhibits very unusual gas and vapor transport properties. It is more permeable to large organic vapors, such as *n*-butane, than to small, permanent gases, such as nitrogen.[4,5] TFE/BDD is a glassy random copolymer with very high permeability and excellent chemical resistance.[3,6] There are two commercially available TFE/BDD copolymers containing 65 or 87 mol % BDD from Du Pont (Wilmington, DE) under the trade names Teflon AF1600 (TFE/BDD65) and AF2400 (TFE/BDD87).[3] We chose TFE/BDD87 for study because of its higher fractional free volume (FFV) and also very high permeability. Table 1 presents chemical structures of PTMSP and TFE/BDD87, and Table 2 gives some physical properties of PTMSP and TFE/BDD87. PTMSP is a substituted polyacetylene that contains double bonds in the backbone and a bulky trimethylsilyl [$Si(CH_3)_3$] side group. PTMSP's density of 0.75 g/cm^3 is rather low relative to the 1.0 g/cm^3 or higher density typical of other glassy polymers. TFE/BDD87's apparently "higher" density is a result of fluorine replacing every possible hydrogen in the structure. Both PTMSP and TFE/BDD87 are loosely packed glassy polymers with stiff chain backbones. Notice that both polymers have relatively high glass transition temperatures, even though they have large FFV values.

The high permeability in these polymers is partially associated with the extremely high fractional free volumes (FFV). In general, higher FFV polymers are more permeable.[7-10] Group contribution, zero pressure PVT data and experimental methods can be used to determine free volumes.[7-13]

Based on density and group contribution estimates of occupied volume, the FFV values of TFE/BDD87 and PTMSP are 0.32 and 0.34, respectively.[3,11] These values were calculated using the van der Waals volumes of repeat units and group contribution increments. These values are almost twice the values for conventional, low-free-volume glassy polymers such as polysulfone, which has a FFV of 0.156 by the Bondi method[12] and 0.133 from zero pressure PVT data.[13] The FFV value of TFE/BDD87 is also obtained using experimental low pressure PVT data.[7] Although both TFE/BDD87 and PTMSP have comparable FFV's, PTMSP is much more permeable than TFE/BDD87. This suggests that FFV is only part of the story behind their permeation properties and that the distribution of free volume may also be important.

Table 1. Structure of the TFE/BDD87 and PTMSP

Polymer	Formula	
TFE/BDD87 (AF2400)	$-[CF_2-CF_2]_{0.13}-$ TFE	$-[CF-CF]_{0.87}-$ (ring: O–C(CF_3)(CF_3)–O) BDD
PTMSP	$-[C(CH_3)=C(Si(CH_3)_3)]_{0.50}-$ trans	$-[C(CH_3)=C(Si(CH_3)_3)]_{0.50}-$ cis

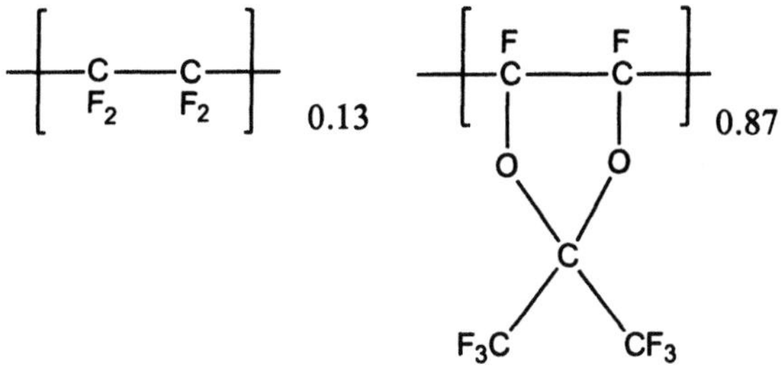

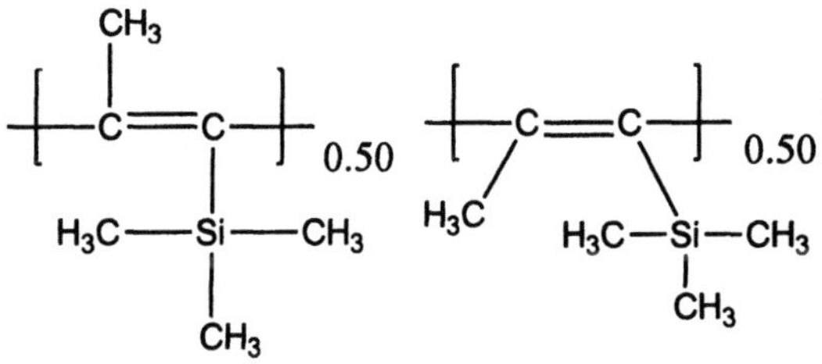

Table 2. Some properties of TFE/BDD87 and PTMSP at T = 298 K

Polymer	*Density* *(g/cm^3)*	T_g (°C)	*FFV%*[3]	*FFV%* *(PVT data*[7]*)*
TFE/BDD87	1.74	240	32	31.4
PTMSP	0.75	>280	34	NA

FFV = fractional free volume
T_g = glass transition temperature.

Positron annihilation lifetime spectroscopy (PALS) measurements support this view. They show,[2,3] on the time scale of the PALS experiment, that the largest accessible free volume elements in TFE/BDD87 are smaller and in much lower concentration than in PTMSP. In addition to PALS, there are several other experimental methods that can be used to characterize the cavity size distribution of a polymer, such as photochromic[14] and spin probe[15] methods.

Although tremendous progress has been made, the cavity size distribution of a material is still very difficult to measure. Molecular modeling provides an alternative method and its utility is demonstrated herein. Hofmann *et al.*[16] have calculated the free volume distribution in high free volume PTMSP and two lower free volume polymers using a geometric method. In our work[7], cavity size distributions (CSDs) for TFE/BDD87 and PTMSP were calculated using a combination of molecular dynamics and Monte Carlo techniques. The algorithm used to determine CSDs is based on energetic rather than geometric considerations and is described in detail in reference 17.

In this work, the permeability of CO_2 in PTMSP and TFE/BDD87 is obtained from molecular simulation.

Methodology

Permeability

The permeability coefficient, P, of a polymer film to a penetrant is given by the relationship:[18,19]

$$P = \frac{Nl}{p_2 - p_1} \tag{1}$$

where N is the steady-state penetrant flux $cm^3(\text{STP})/(cm^2 \cdot s)$, l is the film thickness (cm), p_1 and p_2 are the downstream (permeate) and upstream (feed) pressure, respectively, and also $\Delta p = p_2 - p_1$. The permeability is often expressed in barrers, and $1\,\text{barrer} = 10^{-10} cm^3(\text{STP}) \cdot cm/(cm^2 \cdot s \cdot cm\text{Hg})$, where STP represents standard temperature (273.15 K) and pressure (1 atm).

The permeability coefficient defined in eq. (1) can also be expressed as a product of diffusivity and solubility, that is:

$$P = SD \tag{2}$$

where S is the solubility coefficient and D is an average diffusion coefficient. Permeability is often measured experimentally by the constant pressure/variable volume method. For comparison, P_0, which is obtained by extrapolating permeability coefficient data to zero pressure difference across the film ($\Delta p = 0$), are presented in Table 3 for PTMSP and TFE/BDD87. From Table 3, CO_2 has extremely high permeability in both TFE/BDD87 and PTMSP. In this study, we will calculate the diffusivity and solubility of CO_2 in TFE/BDD87 and PTMSP through molecular modeling. Then, we will obtain the permeability of CO_2 in PTMSP and TFE/BDD87 based on eq. (2).

Table 3. Comparison of experimental values of permeability of TFE/BDD87 and PTMSP to a series of gases and vapors at $t = 35°C$ and $\Delta p = 0$ [18, 19]

Penetrant	P_0 (barrer)	
	TFE/BDD87[18]	*PTMSP*[19]
H_2	2100	15000
O_2	960	9000
N_2	480	6600
CO_2	2200	27000
CH_4	390	15000
C_2H_6	210	31000
CF_4	66	3100
C_2F_6	13	2400

Solubility

Experimentally, the sorption[19] of penetrants into a glassy polymer is usually described by the dual mode model:

$$C = k_D p + \frac{C'_H bp}{1 + bp} \tag{3}$$

where C is the equilibrium penetrant concentration in the polymer at pressure p(atm); it has units of (volume of penetrant (cm^3)/volume of polymer (cm^3)). k_D is the Henry's law parameter which describes penetrant dissolution into the equilibrium densified polymer matrix. C'_H is the Langmuir capacity of the glass. b is the Langmuir affinity parameter describing the affinity of a penetrant for a Langmuir site. The Langmuir capacity is equivalent to the maximum concentration of solute molecules in the unrelaxed (Langmuir) matrix of a glassy polymer. It can be viewed as a measure of excess free volume of a polymer.

The solubility of a penetrant in a polymer is defined as the ratio of equilibrium penetrant concentration to penetrant pressure:

$$S = \frac{C}{p} = k_D + C'_H \frac{b}{1 + bp} \tag{4}$$

Infinite dilute solubility coefficient, S_0, is calculated as follows:

$$S_0 \equiv \lim_{p \to 0} \frac{C}{p} = k_D + C_H^{'} b \tag{5}$$

S_0, also called the Henry's law solubility coefficient, can be calculated using the Widom insertion method[20] through molecular simulation. In this method, we calculate the energy Δu of randomly inserting a particle into a molecular configuration. Δu is the interaction energy between the inserted particle and the rest of the particles in the system. The dimensionless solubility, or Widom insertion factor B, is expressed as:

$$B = \langle \exp(-\Delta u / kT) \rangle \tag{6}$$

where $\langle \cdots \rangle$ represents an ensemble average. The Widom insertion method is most effective for systems with low to moderate densities; it may give poor estimates for low free volume systems. The dimensionless insertion factor B is related to experimentally determined Henry's law solubility coefficient S_0 by:[21-23]

$$B = \frac{T}{273.15} S_0 = \frac{T}{273.15} (k_D + C_H^{'} b)(1\,\text{atm}) \tag{7}$$

Diffusion

Experimentally, diffusion coefficients are usually calculated from permeability and solubility data using eq. (2). Permeability and solubility can be measured easily.

In molecular simulation, diffusion coefficients are calculated from the Einstein relationship[20]:

$$D_i = \lim_{t \to \infty} \frac{1}{6t} \left\langle [\mathbf{r}_i(t) - \mathbf{r}_i(0)]^2 \right\rangle \tag{8}$$

where $\mathbf{r}_i$ is the position vector of atom i, $\left\langle [\mathbf{r}_i(t) - \mathbf{r}_i(0)]^2 \right\rangle$ represents the ensemble average of the mean-square displacement of the inserted gas molecule trajectories; $\mathbf{r}_i(t)$ and $\mathbf{r}_i(0)$ are the final and initial positions of the center of mass of the gas molecules over the time interval t. Diffusion coefficient can also be obtained from the velocity autocorrelation function[20].

Modeling details

Building Amorphous Cells

The Material Studio[24] software of Accelrys Inc. was utilized to construct the amorphous packing structure. The COMPASS force field[25] was used in all simulations. The nonbonded interactions of COMPASS force field include a Lennard-Jones 9-6 function for the van der Waals interaction and a coulombic function for the electrostatic interaction.

For PTMSP, the initial polymer chain was constructed including 50 repeat units with a 50:50 probability for the occurrence of *cis* and *trans* monomers, mimicking what is believed to be the structure of PTMSP material polymerized in the presence of $TaCl_5$ catalyst.[5,16] Two PTMSP chains (each with 50 repeat units) were folded in the Amorphous Cell at a density of 0.75 g/cm^3 which corresponds to the experimental density. The resulting cell length was 29.2 Å. 60 initial states were constructed and followed by 5000 steps of energy minimization to eliminate the "hot spots". Afterwards, a 10 *ps* NVT MD run at 298 K was performed for each of the 60 states to equilibrate structures.

For TFE/BDD87, the initial polymer chain was constructed from 100 repeat units with 13 mol % TFE and 87 mol % BDD monomers and a density of 1.74 g/cm^3, which corresponds to the experimental density. The resulting cell length was around 28 Å. 60 initial states were constructed and followed by 10,000 energy minimization steps and a 10 *ps* MD run in the NVT ensemble. The temperature was also set to 298 K.

It is assumed that the resulting equilibrated structures are representative of the glassy polymers.

Simulation of Cavity Size Distribution

A cavity-sizing algorithm[17] was then applied to each of the above equilibrated configurations. The following is a quick review of cavity-sizing algorithm.

i. A polymer structure is generated by MD (or MC) simulation.
ii. The force field used to create the above structure is replaced with a pure repulsive force field. All atoms remain in fixed locations.
iii. A trial repulsive particle is then randomly inserted into the repulsive polymer structure and a local energy minimum is located in the repulsive force field.
iv. After the minimum is determined, attractive interactions are turned on, and the size of the test particle is adjusted until its potential interaction with all other atoms becomes zero. This size is taken as the diameter of a spherical cavity.

v A check is then made to determine whether the initial random inserting point is inside the cavity or not. The cavity is only accepted if the initial point is inside the cavity. This procedure leads to volume distribution rather than a number distribution of cavities.

vi Steps iii to v are repeated enough times to get a representative distribution of cavity sizes for a given structure.

Simulation of Diffusion, Solubility and Permeability Coefficients

Diffusion coefficients were determined by adding 5 CO_2 molecules to each of 10 independent states. After each state was equilibrated for 60ps, the diffusion constants were calculated by molecular center of mass displacement over a 100*ps* interval for PTMSP and TFE/BDD87. Finally, an average value of D was calculated from ten independent states.

The Widom insertion method[20] was used to simulate the solubility of CO_2 in PTMSP and TFE/BDD87. For each of the polymer's 60 configurations, we performed 20,000 CO_2 configurations.

The permeability is then obtained using eq. (2).

Results and Discussion

Free volume

Figure 1 presents a typical structure of PTMSP. Large cavities are observed in the structure. Table 4 presents the fractional cavity volume (FCV) and the average cavity size from Reference 7. The fractional cavity volume is the fraction of space occupied by spherical cavities defined by our cavity size algorithm. In other words, not all of the free volume is in the form of well-defined spherical cavities. FCV is yet another measure of free volume. Also included in Table 4 are the radii R_4 (Å) of the larger free volume elements of the polymers from the PALS measurements.[3] The average cavity size of PTMSP and TFE/BDD87 from our simulations are consistent with the PALS measurements.

The fractional free volume from Reference 3, fractional cavity volume, and the average cavity size of PTMSP are larger than that of TFE/BDD87. From Table 3, we see that the permeation coefficients of PTMSP are much larger than those of TFE/BDD87, even though PTMSP and TFE/BDD87 have comparable free volume values. This suggests that free volume can only partly explain the permeation properties. Figures 2 and 3 will show that the cavity size distribution has a big influence on the permeability.

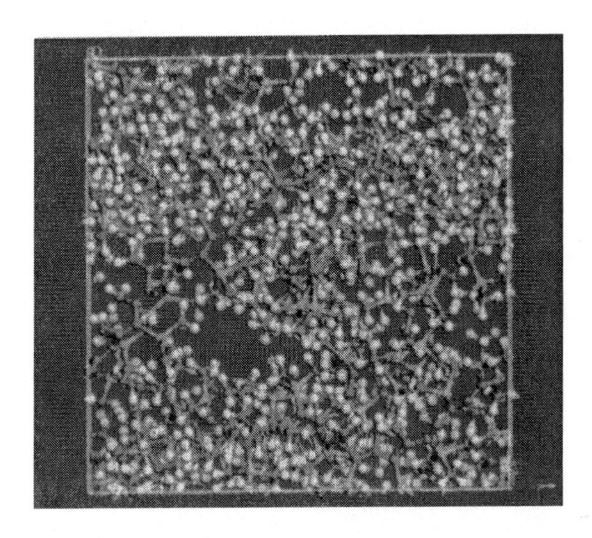

Figure 1. A typical packing model for PTMSP. Large cavities are observed. (See page 3 of color inserts.)

Table 4. Comparison of fractional cavity volume (FCV), average cavity size, PALS data of PTMSP and TFE/BDD87

Polymer	*FCV%*[7]	*Average Cavity Size (Å)*[7]	$2 \times R_4$ *(Å) (from PALS*[3]*)*
TFE/BDD87	13.2	8.2	11.9
PTMSP	15.6	11.2	13.6

Cavity Size Distribution

Cavity size distribution provides an alternative means to examine this problem. Hofmann *et al.*[16] have studied the free volume distribution in ultrahigh free volume PTMSP and two lower free volume polymers by computer simulation. PTMSP shows a broader free volume distribution than the low free volume polymers.

The cavity size distributions of PTMSP and TFE/BDD87 determined based on the algorithm described before are presented in Figures 2 and 3. The distribution in Figure 2 (PTMSP) is shifted towards larger cavity sizes relative to that of TFE/BDD87. The largest cavity diameter found in PTMSP is about 16 Å; in contrast, the largest cavity diameter found in TFE/BDD87 is 12 Å.

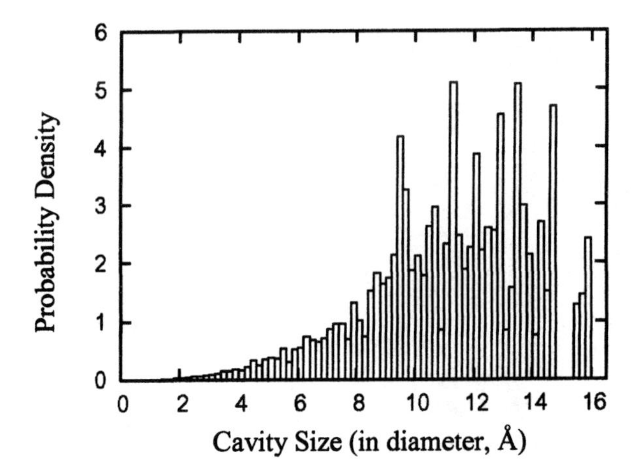

Figure 2. Cavity size distribution in PTMSP at T = 298 K and $\rho = 0.75 g/cm^3$ from molecular simulation. The average cavity size is 11.2 Å, and the fractional cavity volume is 15.6%. PTMSP has higher permeability than TFE/BDD87 (see Table 3). PTMSP has a smaller fraction of smaller cavities than TFE/BDD87 (see Figure 3).

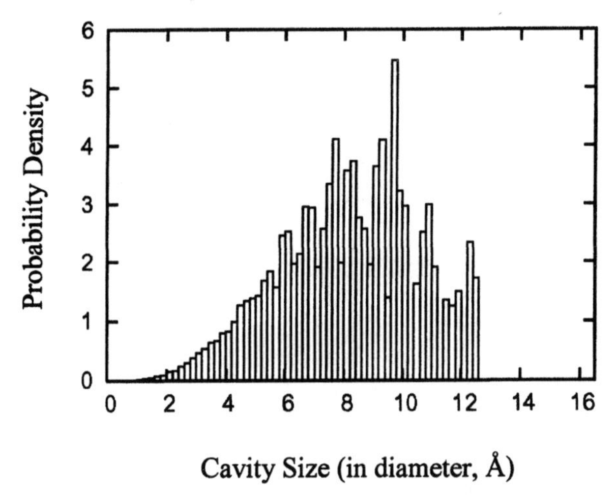

Figure 3. Cavity size distribution in TFE/BDD87 at T = 298 K and $\rho = 1.74 g/cm^3$ from molecular simulation. The average cavity size is 8.2 Å, and the fractional cavity volume is 13.2% (see Table 4). TFE/BDD87 has a higher fraction of smaller cavities than PTMSP.

This result is also consistent with PALS data that show larger free volume elements in PTMSP than in TFE/BDD87.[2, 3] PALS free volume distributions are bimodal.[3] The larger of the two cavities in PTMSP is 13.6 Å, while in TFE/BDD87 it is 11.9 Å (see Table 4).

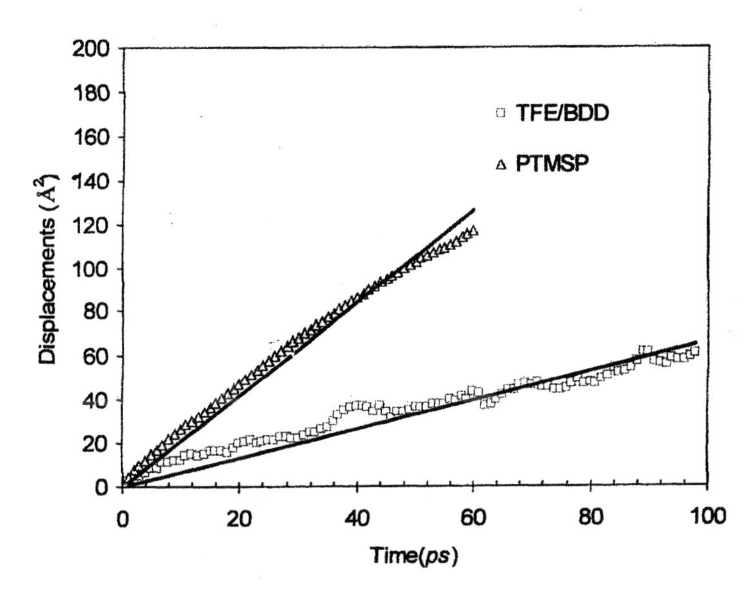

Figure 4. Mean-square displacement of CO_2 in PTMSP and TFE/BDD87 at 298K during NVT molecular dynamics. Lines represent linear fit of data. (See page 3 of color insert.)

Table 5. Comparison of diffusion, solubility and permeability of CO_2 in PTMSP and TFE/BDD87 at T=298K (Experimental data from Ref. 26,27)

Polymer	*Solubility*		*Diffusion* $(10^{-5} cm^2/s)$		*Permeability (Barrers)*	
	Simul.	*Exp.*	*Simul.*	*Exp.*	*Simul.*	*Exp.*
PTMSP[26]	10.7±1.6	9.6	3.5±0.05	3.0	35000	35200
TFE/BDD[27]	12.9±4.5	4.5	1.1±0.05	0.56	16000	3000

Diffusivity

Diffusivity of CO_2 in PTMSP and TFE/BDD87 at T = 298 K was calculated by eq. (8). Figure 4 compares the displacement of CO_2 in these two polymers as a function of time. CO_2 diffuses much faster in PTMSP than in TFE/BDD87, which may be due to the larger cavities in PTMSP. The cumulative distribution of the cavity size distribution of TFE/BDD87 and PTMSP from Reference 7 shows that 50% of the cavities in PTMSP exceed 11.3 Å, whereas 50% of the cavities in TFE/BDD87 exceed 8.1 Å. The diffusion coefficients we obtained from molecular dynamics simulations are listed in Table 5 along with the experimental data. We observe very good agreement between simulations and experiments.

Srinivasan *et al.* [26] concluded that the fast diffusion instead of high solubility is responsible for the large gas permeability observed for PTMSP. Our simulation results appear to be in agreement with their conclusions. The diffusion coefficient of CO_2 in PTMSP is about $3.0 \times 10^{-5} cm^2/s$, but is only about $0.754 \times 10^{-8} cm^2/s$ in polysulfone.[28] Larger cavities in PTMSP may contribute to its faster diffusion coefficient.

Solubility and Permeability

The solubility of CO_2 in PTMSP and TFE/BDD87 measured by the Widom insertion method is given in Table 5. We also present the experimental solubility of CO_2 in this same table. Note that the solubility values presented in Table 5 are dimensionless Widom insertion factors.

Table 5 shows that the simulation solubility of CO_2 in PTMSP compares quite well to the experimentally measured solubility. On the other hand, simulations predict a much higher solubility of CO_2 in TFE/BDD87. The substantial error in the simulation based solubility makes it unclear whether the two polymers follow the experimentally observed trend that CO_2 is more soluble in PTMSP than in TFE/BDD87. The over-predicted solubility of CO_2 in the fluorinated copolymer TFE/BDD87 may suggest a weaker-than-anticipated cross interaction between carbon and fluorine. A recent study has shown that the hydrogen-fluorine interaction is weaker than the expected geometric mean.[29]

With simulation based values for the solubility and diffusion coefficients, we can use eq. (2) to find the permeability of CO_2 in each polymer. Table 5 shows that the permeability predicted by simulation compares well with the experimental permeability for PTMSP. The simulation results over-predict the experimental results for TFE/BDD87. Nevertheless, we find that computer simulations predict the trend that the permeability of CO_2 in PTMSP will be higher than in TFE/BDD87.

Conclusions

Although PTMSP and TFE/BDD87 polymers have large and similar fractional free volumes, PTMSP is much more permeable to gases (see Table 2). As an explanation, it has been suggested that PTMSP has larger free volume elements and more connected regions of free volume than TFE/BDD87.[2, 6] Our cavity size distribution results from molecular simulation support the idea that PTMSP has, on average, larger cavities (see Figures 2-3). However, our algorithm does not address the issue of cavity connectivity. It is also been demonstrated that molecular simulations of glassy structures can be produced and probed with a cavity size algorithm to obtain results that are consistent with existing PALS measurements.

Our molecular dynamics simulation results show that the diffusion coefficient of CO_2 in PTMSP is larger than in TFE/BDD87, and they agree well with the experimental data. The simulation solubility of CO_2 in PTMSP compares quite well to the experimentally measured solubility. We find that simulation measured permeability predicts the trend that the permeability of CO_2 in PTMSP is higher than in TFE/BDD87. Larger cavities in PTMSP may account for the faster diffusion.

Acknowledgements

The authors thank A. J. Hill for her stimulating seminar at UT. This material is based upon work supported in part by the STC Program of the National Science Foundation under Agreement No. CHE-9876674.

References:

1. Nakagawa, T. In *Polymeric Gas separation Membranes*, 1st ed.; Paul, D. R.; Yampol'skii, Y. P. Eds.; CRC: Boca Raton, FL, 1994; p155.
2. Singh, A.; Bondar, S.; Dixon, S.; Freeman, B. D.; Hill, A. J. *Proc Am Chem Soc Div Polym Mater Sci Eng* **1997**, 77, 316-317.
3. Shantarovich, V. P.; Kevdina, I. B.; Yampolskii, Y. P.; Alentiev, A. Y. *Macromolecules* **2000**, 33, 7453-7466.
4. Merkel, T. C.; Bondar, V.; Nagai, K.; Freeman, B. D. *Macromolecules* **1999**, 32, 370-374.

5. Nagai, K.; Masuda, T.; Nakagawa, T.; Freeman, B. D.; Pinnau, Z. *Prog. Polym. Sci.* **2001**, 26, 721-798.
6. Alentiev, A. Y.; Yampolskii, Y. P.; Shantarovich, V. P.; Nemser, S. M.; Plate, N. A. *J. Mem. Sci.* **1997**, 126, 123-132.
7. Wang, X. Y.; Lee, K. M.; Lu, Y.; Stone, M. T.; Sanchez, I. C.; and Freeman, B. D. *Polymer* **2004**, 45, 3907-3912.
8. McHattie, J. S.; Koros, W. J.; Paul, D. R. *Polym. Sci. Polym. Phys.* **1991**, 29, 731-746.
9. McHattie, J. S.; Koros, W. J.; Paul, D. R. *Polymer* **1992**, 33, 1701-1711.
10. Niemelä, S.; Leppänen, J.; Sundholm, F. *Polymer* **1996**, 37, 4155-4165.
11. van Krevelen, W. D. *Properties of Polymers*, 3rd ed.; Elsevier: Amsterdam, 1997, p71.
12. Aitken, C. L.; Koros, W. J.; Paul, D. R. *Macromolecules* **1992**, 25, 3424-3434.
13. Sanchez, I. C.; Cho, J. *Polymer* **1995**, 36, 2929-2939.
14. Victor, J. G; Torkelson, J. M. *Macromolecules* **1987,** 20, 2241-2250.
15. Wasserman, A. M.; Kovarskii, A. L. *Spin Probes and Lables in Physical Chemistry of Polymers*; Nauka: Moskow, 1986.
16. Hofmann, D.; Heuchel, M.; Yampolskii, Y. P.; Khotimskii, V.; Shantarovich, V. *Macromolecules* **2002**, 35, 2129-2140.
17. in 't Veld, P. J.; Stone, M. T.; Truskett, T. M.; Sanchez, I. C. *J. Phys. Chem. B* **2000**, 104, 12028-12034.
18. Merkel, T. C.; Bondar, V.; Nagai, K.; Freeman, B. D.; Yampolskii, Y. P. *Macromolecules* **1999**, 32, 8427-8400.
19. Merkel, T. C.; Bondar, V.; Nagai, K.; Freeman, B. D. J. *Polym. Sci.: Part B: Polym. Phys.* **2000**, 38, 273-296.
20. Frenkel, D; Smit, B. *Understanding Molecular Simulation*, 2nd ed.; Academic Press, 1996.
21. Stone, M. T.; in 't Veld, P. J.; Lu, Y.; Sanchez, I. C. *Mol. Phys.* **2002**, 100, 2773-2782.
22. Sanchez, I. C.; Rodgers, P. A. *Pure Appl. Chem.* **1990**, 62, 2107-2114.
23. Rodgers, P. A.; Sanchez, I. C. *J. Polym. Sci., Part B: Polym. Phys.* **1993**, 31, 273-277.
24. Materials Studio is a package developed by Accelrys Inc., http://www.accelrys.com/, San Diego, CA, 2002.
25. Sun, H. *J. Phys. Chem. B* **1998**, 102, 7338-7364.
26. Srinivasan, R.; Auvil, S. R.; Burban, P. M. *J. Mem. Sci.* **1994**, 86, 67-86.
27. Pinnau, I.; Toy, L. G. *J. Mem. Sci.* **1996**, 109, 125-133.
28. Erb, A. J.; Paul, D. R. *J. Mem. Sci.* **1981**, 11, 8.
29. Song, W.; Rossky, P. J.; Maroncelli, M. *J. Chem. Phys.* **2003**, 119, 9145-9162.

Chapter 15

New Hyperbranched Urethane Acrylates

Branislav Bozic[1], Srba Tasic[1], Radomir Matovic[2], Radomir N. Saicic[3], and Branko Dunjic[1,*]

[1]DugaNova Ltd, Viline Vode 6, 11000 Belgrade, Serbia
[2]ICTM-Centre for Chemistry, Njegoseva 12, 11000 Belgrade, Serbia
[3]Faculty of Chemistry, University of Belgrade, Studentski Trg 16, P.O. Box 158, 11000 Belgrade, Serbia

Novel type of crosslinkable polymeric photoinitiators with hyperbranched architecture were synthesized and characterized. This type of UV curable hyperbranched urethane acrylates utilizes xanthate groups for photoinitiation and controlled crosslinking. A series of polymers were synthesized with different types of functionalization by varying type of the hydroxyl alkyl (meta)acrylates. Obtained polymers have low viscosities and high molecular weights. Glass transition temperatures of crosslinked samples were surprisingly high for their chemical compositions.

1. Introduction

During the last two decades, the field of hyperbranched polymers has been well established with a variety of studies on synthesis, structure, properties and possible applications of these unique materials (1). Hyperbranched polymers are highly branched macromolecules with a large number of end groups, which can be exploited to determine the physical and chemical properties of the resulting products (2). These properties make them attractive in many application fields, especially in coatings, where high degree of functionalization and high molecular weights can give enhanced coatings properties (3). One of the most promising fields of application of hyperbranched polymers in coating technologies is in UV curable coatings. These coatings use photoinitiated polymerization of (meta)acrylates for rapid production of polymeric cross-linked materials with defined properties. The fastness and efficiency of this process are reasons for the wide employment of UV initiated crosslinking in applications where emphasis is on the mechanical and optical properties of materials.

The use of acrylated hyperbranched polyesters in UV curable systems has already been described in several papers (4,5). It was shown that they have lower viscosity and higher curing rates than the linear polymers of similar molecular weights. The mechanical properties of cured films can be controlled by varying a number of acrylate groups in hyperbranched polymer precursor or by introducing non-reactive groups. However, because of the high level of functionalization of these oligomers, the obtained films are very hard and brittle, and have poor flexibility.

Urethane-acrylate oligomers in UV-curable coatings give films with excellent mechanical properties and chemical resistance. Our group has developed new hyperbranched urethane-acrylates with high functionalities and molecular weights, and with acceptable viscosities (6).

The large number of end groups in hyperbranched polymers can be used for incorporation of crosslinkable groups (such as acrylate) and photactive groups in the same macromolecule. In this way, one can obtain photoinitiators with high molecular weight and with reactive sites that can participate in polymerization. In UV curing applications, the advantages expected from the reactive polymeric photoinitiators include good compatibility and low migration, which reduces problems associated with the low molar mass photoinitiators.

Here we describe the use of xanthate mediated controlled radical initiation for crosslining of xanthate urethane-(mata)acrylate terminated hyperbranched polymers. The process can be represented as shown in Scheme 1(7-10).

Under UV light this photoinitiators decompose and give two unsymmetrical free radicals with distinctly different reactivities toward monomers. One of them acts as initiator, while the other is stable radical species with a high reversible termination efficiency (radicals I and II in Scheme 1, respectively). In this work, we have applied this type of initiation and controlled photopolymerization in the synthesis of a novel type of hyperbranched urethane acrylate.

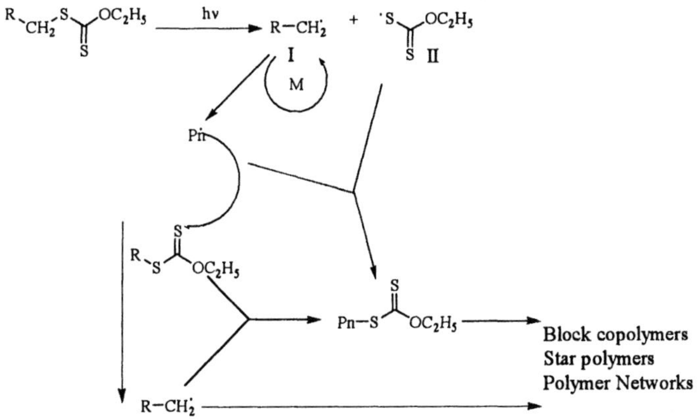

hν
I
M
II
Block copolymers
Star polymers
Polymer Networks

Scheme 1. Xanthate mediated controlled radical polymerization

2. Experimental

2.1. Materials

Isophorone diisocyanate (IPDI) was purchased from Hüls. Polyethyleneglycol(6) monoacrylate (PEA6), polypropyleneglycol monoacrylate (PPA6), polypropyleneglycol monomethacrylate (PPM5S) and 2-hydroxyethyl acrylate were kindly provided by Laporte Performance Chemicals. Hexanediol diacrylate (HDDA) was purchased from BASF. 2,2-bis(methylol) propionic acid (Bis-MPA) and ditrimethylolpropane (DITMP) were purchased from Perstorp AB. Dibutyltin dilaurate (DBTDL) catalyst and Potassium ethyl xanthogenate were obtained from Merck, and 1-hydroxy-cyclohexyl-phenyl-ketone (Irgacure 184) photoinitiator was kindly provided from Ciba. All chemicals were used as received.

2.2 General

^{1}H NMR and ^{13}C NMR spectra were recorded in *d*-chloroform and dimethyl sulfoxide-d_6 on a Varian GEMINI 200 spectrometer operating at 200 MHz. The FT-IR spectra of the samples were recorded with a Bomem MB-102 spectrometer. Differential scanning calorimetry (DSC) measurements were carried out on a Perkin Elmer Pyris 6 DSC analyser.

The complex dynamic viscosity (η^*) of oligomers were measured with a Rheometrics mechanical spectrometer RMS-605 operating in rate sweep mode, using a cone and plate geometry. Dynamic mechanical measurements of cured oligomers were performed on Rheometrics RMS-605 in the temperature sweep mode. The experiments were carried out on the rectangular bars (1 x 12.5 x 63 mm) between 25 and 150 °C at frequency of 1 Hz. The samples for dynamic mechanical analysis (DMA) were cured in a mould consisting of two transparent polyester films separated by 1 mm thick spacer. The distance of the sample to the focal point of UV-lamp was 10 cm, and the sample was exposed to the UV-light for 5 minutes on both sides to ensure uniform curing of the sample.

Hyperbranched urethane-acrylates were diluted with HDDA (20 wt.%). The coating films were drawn on metal plates (40 ± 5 μm) and cured using 2" metal halide lamp (UVPS, 80 W/cm) at conveyor speed of 10 m/min. Hardness of coatings was determined by Persoz pendulum. The flexibility of the coatings was determined by measuring the Erichsen indentation.

Synthesized polymers were denoted using the following abbreviations: $H(X)_a(PEA)_b$, where **a** and **b** are the number of xanthate and acrylate groups per molecule, respectively. Residual OH groups are not denoted in the abbreviations and their numbers can be calculated from the following expression N(OH) = 16 - b - c. For example, $H(X)_8(PEA)_6$ describes a HBP-UA theoretically modified with 8 xanthate and 6 acrylate groups based on PEA6.

2.3 Synthesis

Synthesis of hyperbranched polyol

Hyperbranched polyester-polyol with ditrimethylolpropane as a core and ditrimethylolpropionic acid as a branching element, was prepared by a procedure described in the literature (11), which includes a one pot polycondenstion reactions with p-toluenesulfonic acid as catalyst on 140 °C. A 4-necked reaction vessel equipped with a stirrer, a nitrogen inlet, a thermometer and a water-trap was charged with ditrimethylolpropane (62.5 g, 0.25 mole). The temperature was raised to 140°C and ditrimethylolpropionic acid (402 g, 3 mole) and p-TSA (2.32 g, 0.0135 mole) were added. The course of the reaction was followed by acid value titration. The reaction was continued until an acid value of 8 mg KOH/g was obtained.

Esterification of hyperbranched polyols with α-haloacids

Hydroxyl groups of hyperbranched polyol were partially modified (50 % of OH groups) with α-haloacids (2-bromo propionic acid and monochloracetic acid). In a 4-necked reaction flask equipped with a stirrer, a N_2 inlet, a cooler and a water-trap were charged 100.0 g (0.974 mole OH) of the hyperbranched polyol (HBP) and 74.54 g (0.487 mole) 2-bromopropionic acid. The temperature was raised to 175°C over 2 hours and the reaction was continued at this temperature. The course of the reaction was followed by acid value titration. The reaction was continued until an acid value of 12.5 mg KOH/g was obtained.

Synthesis of hyperbranched polyesters with xanthate moieties

A 250 mL three-necked flask was charged with a solution of the 2-bromopropionic acid modified hyperbranched polymer (10.00 g) in ethyl acetate (25 mL). Potassium *O*-ethyl xanthate (4.70 g, 0.029 mole) was added in portions over a period of 1h. The reaction mixture was stirred at room temperature overnight. The obtained mixture was filtered off, the residual salt was extracted and washed with water, the organic phase was dried over anhydrous $MgSO_4$ and solvent was evaporated under reduced pressure. Yield was 7.85 g (70.2 %).

The same procedure was repeated with HBP modified with monochloracetic acid. Yield was 72.6 %.

Synthesis of hyperbranched urethane acrylate with xanthate moieties

Hyperbranched urethane acrylates was synthesized in a three-step procedure. In the first step, halogen groups from hyperbranched polyester (with already partially modified hydroxyl group with 2-bromopropionic acid or monochloracetic acid) were substituted with xanthate groups, as described in the previous preparation. The NCO-bearing adduct was synthesized in a separate flask, equipped with a mechanical stirrer, a thermometer, and a cooler by the reaction of equimolar amounts of IPDI and hydroxyalkyl (meta)acrylate at 35 to 40°C with DBTDL. Hydroxyalkyl (meta)acrylate was added over a 90-minute period while maintaining the temperature below 35°C. The reaction temperature was then allowed to increase from 35°C to 40°C within 30 minutes, and the reaction mixture was stirred at 40°C for 90 minutes. These two reaction mixtures were combined and the reaction was continued at 75°C for about 8 hours. The reaction was followed by FT-IR and stopped when the NCO-peak (2267 cm^{-1}) had disappeared. The obtained mixture was filtered off, the residual salt was extracted and washed with water, dried over anhydrous $MgSO_4$ and the solvent was evaporated under reduced pressure.

For the reasons of comparison the similar procedure was repeated with unmodified hyperbranched polyester and in that way hyperbranched acrylate without xanthate groups was obtained.

All other urethane acrylates were synthesized in the same procedure with different hydroxyl alkyl (meta)acryltes listed in Table 1 .

Table 1. The synthesized hyperbranched urethane acrylates

Urethane Acrylate	Chemical structure of hydroxyl acrylate	Yield (%)
$H(PEA)_6$	HO(CH₂CH₂O)₆C(O)CH=CH₂	-
$H(X)_8(HEA)_6$	HOCH₂CH₂OC(O)CH=CH₂	75.1
$H(X)_8(PEA)_6$	HO(CH₂CH₂O)₆C(O)CH=CH₂	83.9
$H(X)_8(PPA)_6$	HO(CH(CH₃)CH₂O)₆C(O)CH=CH₂	77.4
$H(X)_8(PPM)_6$	HO(CH(CH₃)CH₂O)₅C(O)C(CH₃)=CH₂	80.6

3. Results and discussion

Partial modification of hyperbranched polyols has great influence on their thermo-mechanical and chemical properties, as it can be seen from Table 2. Polyesters synthesized as in Sheme 2 were used as polyol core for the synthesis of hyperbranched urethane-acrylates.

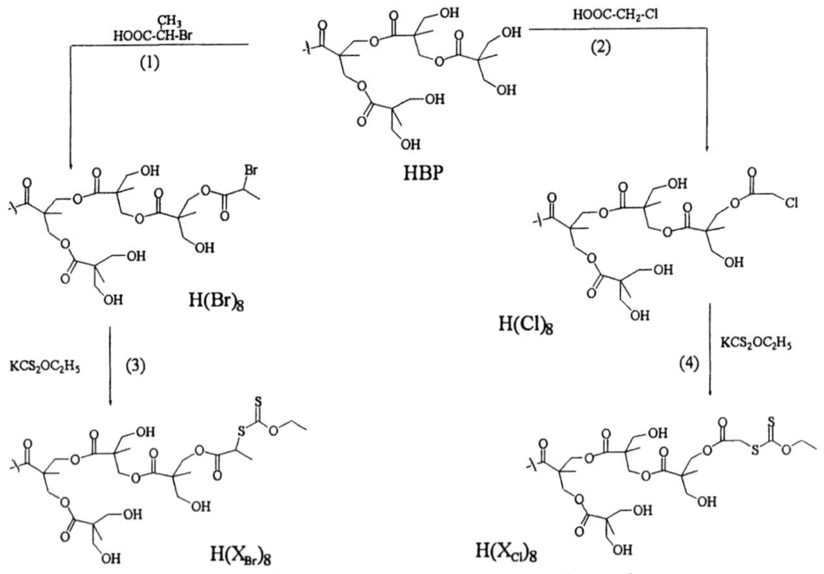

Scheme 2. Modification of hyperbranched polyester with xanthate groups

In addition, xanthate groups have lower polarity, as compared to hydroxyl groups, and this has influence in lowering the viscosity of hyperbranched polymers. As it can be seen from Figure 1 the modification of hydroxyl groups have great impact on thermal properties of hyperbranched polyesters. Contrary to the linear polymers, glass transition temperature for hyperbranched polymers has its origin in translational motions of macromolecules and therefore is more sensitive to the chemical nature of terminal units and less to molar mass (Table 2.) (12, 13).

Table 2. Properties of hyperbranched polyols

Sample	*η* (Pas) at 1Hz, 50 °C*	*Tg (°C)*
HBP	$8 \cdot 10^5$	19
$H(Br)_8$	218.1	-6
$H(X_{Br})_8$	98.5	0
$H(Cl)_8$	94.1	-13.8
$H(X_{Cl})_8$	44.2	-6

Urethane acrylate oligomers gave the UV curable materials with best mechanical and chemical properties, but because of high polarity lack from high viscosities. This was more pronounced with multifunctional oligomers where polar interactions were even stronger. In order to avoid this we used aloxylated (meta)acrylated monomers which gave the hyperbranched urethane acrylates with lower concentration of polar urethane groups, which in turn reduce the density of hydrogen bonding and viscosity of the oligomers (see Table 2) (14).

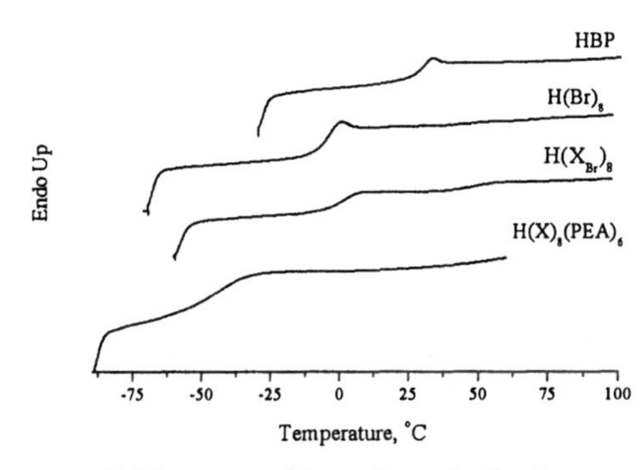

Figure 1. DSC curves of hyperbranched polymers

Hyperbranched urethane-acrylates were synthesized in a three-step process. In the first step, halogen groups from HBP (with already partially modified hydroxyl group with 2-bromopropionic acid or monochloracetic acid) were substituted into xanthate groups by reacting with potassium O-ethyl xanthate as shown in Scheme 2. The NCO bearing precursor was obtained from IPDI cycloaliphatic diisocyanate which possesses two NCO groups of the unequal reactivity (15). The secondary cycloaliphatic NCO group is up to 10 times more reactive than the primary aliphatic NCO group in urethane reaction (with DBTDL as the most selective catalyst). During the adduct synthesis there are some side reactions (reaction of primary NCO group; alophonate and biuret making reactions) which alter the stoichiometry and same unreacted IPDI remains in the reaction mixture. This unreacted IPDI can react with hyperbranched polyol giving high molecular weight products.

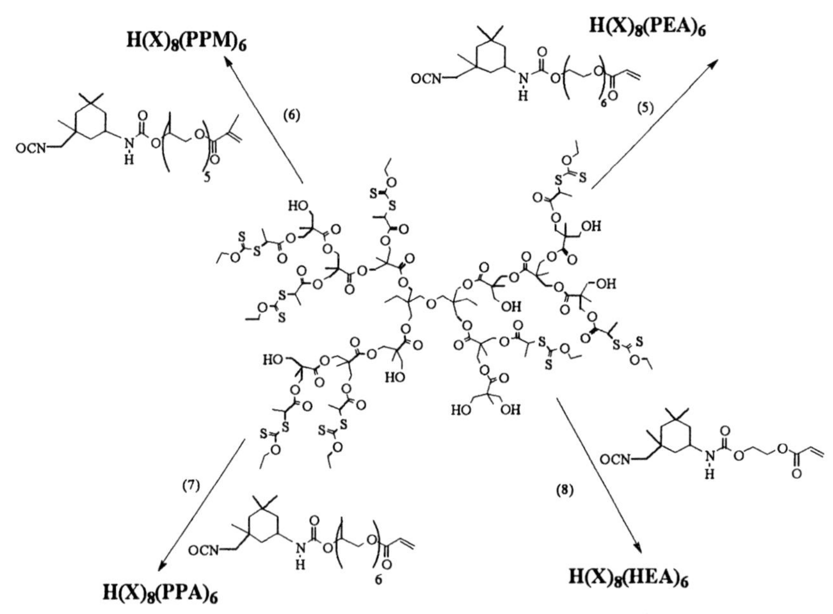

Schema 3. Synthetics routes for obtaining urethane-acrylates

This is also the reason for broadening the molecular weights distribution of the hyperbranched urethane acrylate compared to that of the starting polyols. It is well known that branched macromolecules have lower hydrodynamic volumes then their linear counterparts, and molecular weights obtained from GPC measurements are unrealistically low. From GPC measurements, only the distributions of molecular volumes (i.e., sizes) are reliable.

Table 3. Properties of hyperbranched urethane-acrylates

Urethane Acrylate	*η* (Pas) at 1Hz, 30 °C 20% HDDA*
$H(X)_8(PEA)_6$	23.45
$H(X)_8(PPA)_6$	11.74
$H(X)_8(PPM)_6$	6.70
$H(X)_8(HEA)_6$	119.80

The complex dynamic viscosity (η*) at 1 Hz of oligomers diluted with 20 wt.% HDDA are shown in Table 3. Frequency dependence of the viscosity is shown in Figure 2. It can be seen that the type of hydroxyalkyl (meta)acrylate has great influence on viscosity of the hyperbranched urethane acrylate polymers. This modification puts acrylate groups on the end of the polymers, thus reducing the polarity of the system. Although the urethane groups are very polar their influence is limited because they are now in the inner part of the molecule (Sheme 4.).

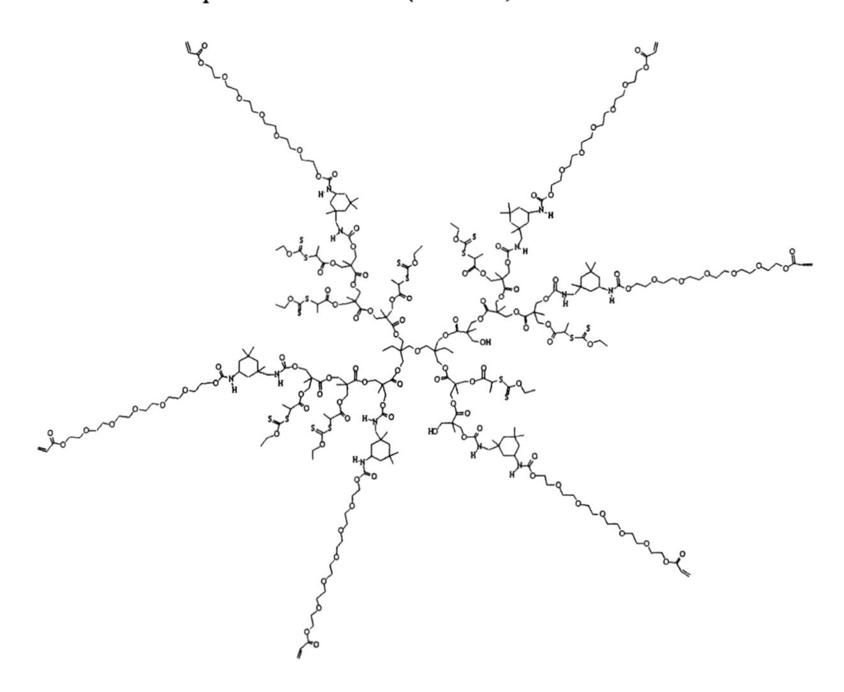

Scheme 4. Idealized structure of Hyperbranched urethane acrylate $H(X)_8(PEA)_6$

Urethane acrylate with short spacer between urethane and acrylate groups (2-hydroxy ethyl acrylate) had higher viscosities than those with polyethylenglycol or polypropyleneglycol spacer. At the same time, polymers with polypropyleneglycol spacer had the lowest viscosity due to higher mobility and lower polarity. It should be also mentioned that all HB-UA exhibit Newtonian behaviour (figure 2). This indicated that there were no intermolecular entanglements like in linear polymers. This was in good agreement with results obtained earlier for the more perfect dendrimers (16,17). Hence, it follows that hyperbranched urethane acrylate were small, compact (non-interpenetrating) "globules".

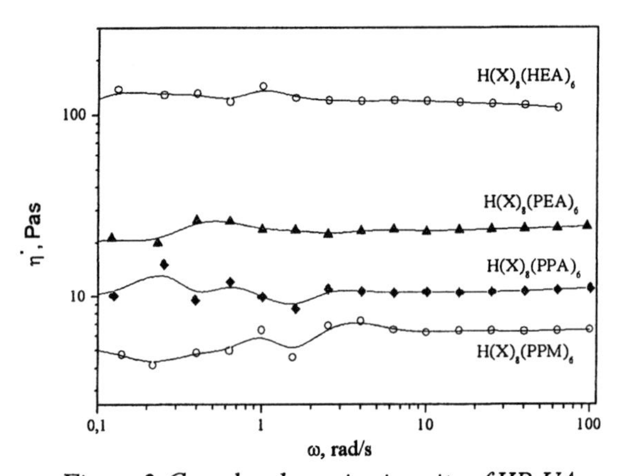

Figure 2. Complex dynamic viscosity of HB-UA

Dynamical mechanical analysis

Viscoelastic properties of UV-cured oligomers were characterized by DMA in a temperature-scanning mode. Once again, the properties of cured materials were largely influenced by type of modification of hyperbranched polyol. Figure 3 shows the tanδ curves of hyperbranched urethane acrylate diluted with 20% HDDA. Glass transition temperature, Tg, was determined as the temperature of the maximum on tanδ peak. HB-UA based on 2-HEA had highest Tg and broadest shape of tanδ curves. The broad tanδ curves indicate that obtained network was very inhomogeneous, i.e. there is a broad distribution of chain lengths between points of crosslink's. This is expected because the $H(X)_8(HEA)_6$ also had the broadest distribution of molecular volumes (by SEC). In addition, this polymer has largest concentration of acrylate and polar urethane groups compared with other urethane acrylates.

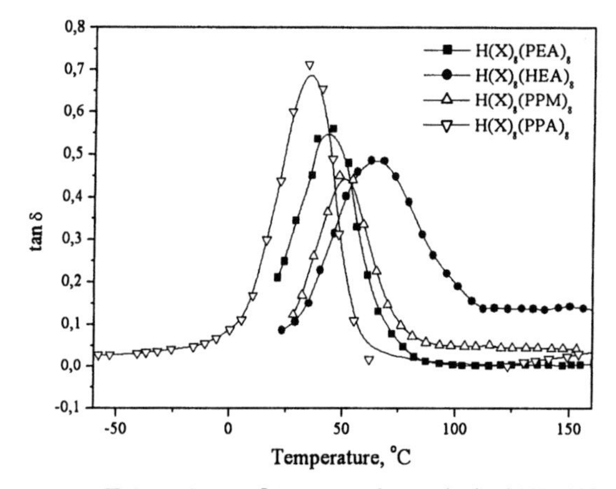

Figure 3. tanδ curves of crosslinked HB-UA

The introduction of polyalkyleneoxide chains, in other polymers, as flexible spacers between compact hyperbranched core and reactive groups, increased mobility of macromolecular segments in cured network and decrease the value of Tg. Therefore, hyperbranched polymers based on PEA6 and PPA6 had lower Tg's (34.1 and 44°C, respectively) compared with one based on 2HEA (75°C). Polypropyleneoxide chains, as already mentioned, are bulkier and less polar and therefore the crosslinked materials based on them will have lower Tg values. The height of tanδ curves of cured acrylate shows observable differences. The height of the tanδ peak reflects the mobility of the chain segments between points of crosslinks in the transition temperature region. Therefore, the heights of the tanδ curves decreased with decreasing lengths and mobility of alkyleneoxide chains in hyperbranched urethane acrylate, in the following order PPA6, PEA6, 2HEA.

The type of unsaturation (acrylate or methacrylate) also affected the properties of cured materials. Due to its stiffer structure, the methacrylate group gave material that generally exhibited higher Tg values, and in same time the lower height of tanδ peak. So, polymers based on PPM5S have higher Tg compared to polymer based on PEA6 (8°C higher) even with very similar concentration of functional groups and lover polarity of the polypropylene oxide chains. However, one should remember that the hyperbranched urethane acrylates consisted of approximately 60 or more weight percent of polyalkyleneoxide chains. Nevertheless, Tg values of 35°C and higher are unexpectedly high for this composition and this can be attributed to high level of functonalization of its hyperbranched\star structure.

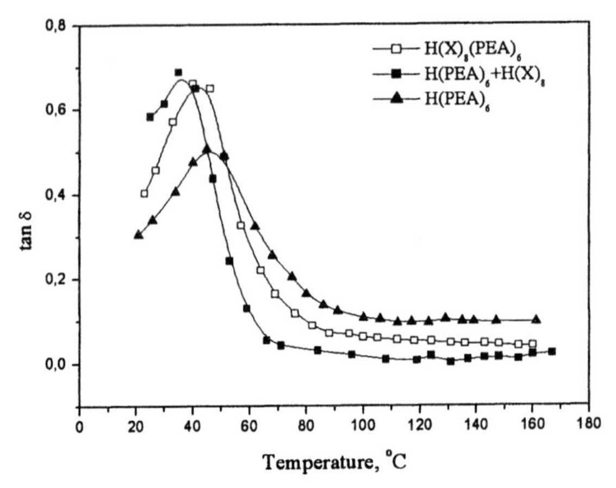

Figure 4. Comparasion of tanδ curves for different types of crosslinking and photoinitiation

The influence of xanthate groups on properties of crosslinked materials was twofold. On the one side, their presence decreased the concentration of urethane-(meta)acrylate groups and polarity of systems, thus decreasing Tg values. In addition, in modification of hyperbranched polyol with isocyanate adduct, the most exsposed hydroxyl groups had already been taken by xanthate groups so polar urethane groups were situated deeper in the hyperbranched polyester backbone. In that way the polarity of polymers were further decreased. On the other side, they were reactive groups as well, therefore increasing functionality (not in the same way as (meta)acrylate groups) and in the same time increasing Tg values. As it can be seen from Figure 4 the properties of cured materials were almost the same, so it can be concluded that the above mentioned effects cancelled each other. Hyperbranched polyester modified with xanthate groups can be used as macro photoinitiator without the attached (meta)acrylates groups. As expected, Tg value of the material cured in this manner was somewhat lower, because the average acrylate functionality of this system was also lower.

Hardness and flexibility

Properties of the UV curable coatings were determined by measuring coating properties such as pendulum hardness (Persoz hardness) and flexibility (Erichsen cupping). Coating formulations comprised of hyperbranched urethane acrylates and 20 wt% of reactive diluent (HDDA). Hardness represents ability of a film to resist surface abrasion and flexibility determined by the viscoelastic character of the polymer. Generally, UV curable oligomers give either soft, flexible substances or hard and brittle materials. Furthermore, lack of entanglements in hyperbranched polymers give them amorphous morphology and brittle film forming properties. However, these materials combine high crosslinking density with flexible segments between them and provide films with good hardness and flexibility (Table 3).

Table 3. Properties of UV cured coatings

Polymer	*Hardness, (Persoz, sec)*	*Erichsen Flexibility (mm)*	*Tg (DSC, °C)*	*Tg (DMA, °C)*
$H(PEA)_6$	120	8	22.5	46.0
$H(X)_8(PEA)_6$	92	8	28.0	44.0
$H(X)_8(HEA)_6$	150	8	37.0	65.1
$H(X)_8(PPA)_6$	40	8	13.6	34.9
$H(X)_8(PPM)_6$	110	8	41.1	50.5

Conclusion

The results of this study show versatility of hyperbranched urethane acrylates and synthetic ways to control their size and functionality during the synthesis. The thermal and mechanical properties of the hyperbranched urethane acrylates were investigated in order to determine the relationship between their chemical composition, structure and properties of crosslinked materials. Viscosity, reactivity and mechanical properties of crosslinked materials can be precisely controlled over a vide range of their values. In addition, novel type of living radical polymerization for obtaining highly crosslinked materials was investigated. The obtained materials show good compromise between hardness and flexibility and offer great possibilities in different applications.

References

1. A. Hult, M. Johanson, E. Malstörm, *Adv. Polym. Sci.*, **1999**, 143, 1.
2. Y. H. Kim, *J. Polym. Sci., Part A: Polym. Chem.*, **1998**, 36, 1685.
3. M. Johansson, T. Glauser, A. Jansson, A. Hult, E. Malstörm, H. Claesson, *Prog. Org. Coat.*, **2003**, 48, 194.
4. M. Johansson, A. Hult, *Journal of Coatings Technology*, **1995**, 67, 35.
5. Q. Wan, D. Rumpf, S. R. Schricker, A. Mariotti, B. M. Culbertson, *Biomacromolecules*, **2001**, 2(1), 217.

6. E. Dzunuzovic, S. Tasic, B. Bozic, D. Babic, B. Dunjic. *J. Serb. Chem. Soc.*, **2004**, 69, 441.
7. K. Matyjaszewwski, *Controlled Radical Polymerization*; American Chemical Society Ed.; Washington, DC, **1998**; Vol 685.
8. J. Chiefari, B. Y. K. Chong, F. Ercole, J. Krstina, J. Jeffrey, T. P. Le, R. T. Mayadunne, G. F. Meijs, C. L. Moad, G. Moad, E. Rizzardo, S. H. Thang, *Macromolecules* **1998**, 31, 5559.
9. D. Charmot, P. Corpart, H. Adam, S. Z. Zard, T. Biadatti, G. Bouhadir, *Macromol. Symp.* **2000**, 153, 23.
10. D. Taton, A. Z. Wilczewska, M. Destarac, *Macromol. Rapid. Comm.* **2001**, 22, 1497.
11. E. Malmström, M. Johansson, A. Hult, *Macromolecules,* **1995**, 28, 1698.
12. Y. H. Kim, R. Beckerbauer, *Macromolecules,* **1994**, 27, 1968.
13. Y. H. Kim, O. W. Webster, *Macromolecules,* **1992**, 25, 5562.
14. B. Dunjic, S. Tasic, B. Bozic, *Europ. Coat. J.*, **2004,** No. 6, 44.
15. R. Lomölder, F. Plogmann, P. Speier, *Journal of Coatings Technology,* **1997**, 69, 51.
16. S. Uppuluri, S.E. Keinath, D.A. Tomalia, P.R. Dvornic, *Macromolecules*, **1998**, 31, 4498.
17. S. Uppuluri, F.A. Morrison, , P.R. Dvornic, *Macromolecules*, **2000**, 33, 2551.

Chapter 16

Charge Percolation Mechanism of Ziegler–Natta Polymerization: Part II: Importance of Support Nanoparticles

Branka Pilic[1], Dragoslav Stoiljkovic[1,*], Ivana Bakocevic[1], Slobodan Jovanovic[2], Davor Panic[3], and Ljiljana Korugic-Karasz[4]

[1]Faculty of Technology, University of Novi Sad, Bul. Cara Lazara 1, 21000 Novi Sad, Serbia
[2]Faculty of Technology and Metallurgy, University of Belgrade, 11000 Belgrade, Karnegijeva 4, Serbia
[3]Technical Faculty, University of Novi Sad, 21000 Novi Sad, Trg Dositeja Obradovica 3, Serbia
[4]Department of Polymer Science and Engineering, University of Massachusetts, Amherst, MA 01003

In the heterogeneous Ziegler-Natta (ZN) polymerization, solid catalyst precursor $TiCl_3$ alone or solid supports ($MgCl_2$, Al_2O_3, SiO_2, graphite etc.), impregnated by liquid $TiCl_4$, are fragmented by ball-milling and during polymerization to the particles of nano-dimensions. Transition metal (Mt) ions are immobilized and distributed on these particles. The theory of active centre ensembles and charge percolation mechanism (CPM) of ZN polymerization have been applied to predict the effects of Mt distribution on catalysts productivities. **CPM - Part I: Fundamentals* has been presented in (*22*).

A half century has passed since Karl Ziegler and Giulio Natta discovered that olefins can polymerize in the presence of transition metal compounds as the precursors of active centres. It was discovered by K. Ziegler in 1953 that high density polyethylene was easily made at low pressures with a binary mixtures of metal alkyls and transition metals salts, such as $AlEt_3$ and $TiCl_4$. G. Natta in 1954 demonstrated that the same catalysts, and to a greater extent, catalysts containing lower valent transition metal chloride salts such as $TiCl_3$, could form isotactic polymers from α-olefins (*1*). Almost at the same time it was discovered that transition metal oxides could be used as the active centres precursors for olefin polymerization. Since that time a tremendous amount of research work has brought to the development of several generations of catalysts, including supported ones.

According to the broad definition, the Ziegler-Natta catalyst is a mixture of a base metal alkyl of the group I to III metals and a transition metal salt of groups IV to VIII metals (*1*). In addition to them, some Mt oxides (e.g. chromium, vanadium, molybdenum) can serve as active centre precursors, too. The most important from the scientific and commercial point are the systems: 1) Classical Ziegler-Natta complexes, i.e. $TiCl_3/AlEtCl_2$; 2) CW complexes, i.e. $TiCl_4/AlEt_3$; 3) Mt oxides based on Cr, Mo, V, e.g. Cr_2O_3.

Transition metal (Mt) precursors are activated by metal alkyls: an alkyl group and a vacant orbital are formed on Mt. It is assumed that monomer molecules activate Mt oxides. In addition to the alkylation and vacant orbital formation, a significant amount of Mt is reduced to lower oxidation states. As a result of preparation and activation, Mt exists in several oxidation states, i.e. Ti^{+4}, Ti^{+3} and Ti^{+2} in the case of Ti or from Cr^{+2} to Cr^{+6} in the case of chromium oxide.

Support Nano-Particles

In $TiCl_3$ based systems less than 1 % of the Ti is active. The remaining solid $TiCl_3$ acts as a support with the surface area up to 40 m^2/g (*1*). The β form of $TiCl_3$ has a linear structure, while α, γ and δ forms have a layer structure. The α-$TiCl_3$ modification has crystals resembling hexagonal shaped platelets.

The morphology of these titanium chloride particles can vary considerably by the method of preparation, e.g. the choice of the reducing metal alkyl or metal and the conditions of reducing during their synthesis.

$TiCl_3$ prepared by reduction of $TiCl_4$ with $AlEt_2Cl$ consists of small α, β and δ crystallites. Their coherent scattering regions (c.s.r.) in the direction 001, found from XRD data, are in the range of 2 to 12 nm (*2*). Those primary crystallites are densely packed to the bigger particles (K) that have the size of 15-200 nm. The particles K are agglomerated to the particles L that have the size of 100-500 nm. The particles L are agglomerated to the bigger particles (1-3 μm), that are further agglomerated to the almost spherical granules with the radius in range 10-30 μm.

α-$TiCl_3$ prepared by the electrolysis of $TiCl_4$ with metal Ti has irregular shape particles (90 %) and polyhedron particles (10 %) with various sizes. The α-$TiCl_3$ is described as staking of Ti and Cl atomic planes in the *c*-axes. The staking sequence is -Cl-Ti-Cl-Cl-Ti-Cl-Cl-Ti-Cl-. The Cl-Cl bond is weaker than the Ti-Cl bond, and there are surfaces of terraces (100-1500 nm) terminated with Cl planes (*3*).

When subject to a mild shear force, such as agitation during polymerization, the optically visible $TiCl_3$ granules are disintegrated into smaller primary crystalline particles, i.e. they defoliate easily producing small flat leaflets that are described as flat polygons whose diameters vary from 30 to 100 nm (*1*). The disintegration is presumed to take place by cleavage along the loosely held chlorine-chlorine layers.

CW complexes are prepared by supporting liquid TiCl4 on crystalline $MgCl_2$ and afterwards reducing by $AlEt_3$. The key of the success of $MgCl_2$ as a support is the crystal structure, which is very similar to that of $TiCl_3$ in both interatomic distances and crystal forms (*4*). The details on the structures of α, β and δ forms are described elsewhere (*4*). In all cases the $MgCl_2$ crystal consists of layers of Mg ions sandwiched between two layers of Cl ions.

The size of $MgCl_2$ particles varies from several tenths to several hundreds of nanometers (*5*) and decreases during milling to a greater extent in the direction perpendicular to the (001) plane than to the (110) plane (*6,7*). The shear forces during the milling cause the Cl-Mg-Cl layers to slide over each other, producing hexagonal $MgCl_2$ primary crystallites of only a few layers of a thickness of 3-4 nm and diameter of 20-40 nm (*8*). The surface area of milled $MgCl_2$ is up to several hundreds m^2/g.

Highly dispersed graphite (200-300 m^2/g), obtained by grinding in a vibratory mill, has been used as the support for $TiCl_4$, too (*9,10*). The size crystallite is 6-10.5 nm and the defect content is 0.8-1.6 %. It is well known that graphite has regular hexagonal layer lattices that can easily defoliate under the shearing force.

SiO_2 and Al_2O_3 are used as supports for chromium oxides. The support particle size is normally in the range of 30 to 150 μm (*11*). It is well known, however, that SiO_2 and Al_2O_3 have irregular hexagonal layer lattices. The surface area of those supports can be up to several hundreds m^2/g (*12,13,14*).

In all cases, the support particles, that have very high surface area, are physically defoliated and fragmented to leaflets and platelets, or can be represented by a great number of separate micro crystal blocks and small coherent regions jointed to one another. The possibility of a support to be fragmented into small flat nano particles is of great importance for polymerization.

Natta (*15*) noticed the importance of the layer lattice of the supports: "The production of highly isotactic polymers was found to depend on the presence of both a catalyst and of a highly crystalline substrate **(having a layer lattice).** Catalytic complexes containing transition metals and metal-organic groups, which alone do not polymerize the α-olefins, are not stereospecific catalysts when adsorbed on the surface of amorphous substrates, but become highly stereospecific when adsorbed on highly crystalline supports. Furthermore, there are some insoluble crystalline compounds **(having a layer lattice)** that alone, or even in the presence of aluminum alkyls, cause virtually no polymerization; but when added to soluble complexes (which again are inactive alone) yield stereospecific polymerization systems. **The structure of the crystalline support is of great importance.**"

Distribution of Transition Metal on Support

In Ziegler-Natta catalysts it is very important to know the distribution of Mt compounds on the support surface, because it is directly related to the catalysts performance (*16*). These Mt compounds are distributed over the leaflets, platelets and micro blocks of the support.

The active centres are found in the primary particles of $TiCl_3$. Two active centers are said to be present per primary particle (*1*). The titanium distribution on the surface of Ziegler-Natta catalysts has been observed by scanning Auger electron microscopy (*16*). All particles of the $TiCl_3$ showed the similar tendency, suggesting that Ti had a homogeneous surface distribution.

In the case of $TiCl_4$ supported on $MgCl_2$, a considerable reduction of Ti^{+4} by $AlEt_3$ takes place, not only to Ti^{+3} but to Ti^{+2} as well. $MgCl_2$ has a crystal structure very similar to that of violet $TiCl_3$. This indicates the possibility of an epitactical coordination of $TiCl_4$ units (or $TiCl_3$ units after reduction) on the lateral coordinatively unsaturated faces of $MgCl_2$ crystals. For both monomeric $TiCl_4$ and $TiCl_3$, coordination on the (110) face is favored relative to coordination on the (100) face (*17*).

It was found by scanning Auger electron microscopy that various states of titanium clusters existed, suggesting that $TiCl_4$ can not be supported evenly on the surface of $MgCl_2$ (*16*). It was shown that $TiCl_4$ reacts with the defects of $MgCl_2$ structure. The concentration of these defects is associated with the size of the coherent scattering range determined from X-ray data (*7*).

It is reported that the quantity of adsorbed $TiCl_4$ on Al_2O_3 is in the range of 1 to 3 Ti atoms/nm^2 (*14*).

In spite of more than 40 years of research effort, many aspects concerning the physico-chemical state of Cr species on silica remain controversial (*18*). Usually, a highly dispersed state of surface-stabilized chromate species can be achieved through many redispersion cycles of sublimation volatilization, spreading, deposition and stabilization of bulk CrO_3. The supported Cr species may be present on silica in a mixture of different valence (+6, +5, +4, +3, +2), coordination and distribution states, depending on the preparation conditions. When the content of CrO_3 exceeds the saturation coverage under a given condition, the excess of CrO_3 decomposes into aggregated Cr_2O_3 on the support surface during calcination. The Cr content on industrial Phillips CrOx/SiO2 catalysts is ca. 1 wt.% (i.e. 0.4 Cr/nm^2) (*18*). The average distance between Cr ions is in the range from 0.65 nm to 16 nm for Cr contents of 6 wt.% to 0.01 wt.%, respectively (*13*).

Characterization of chromia/alumina catalysts by X-ray photoelectron spectroscopy and proton induced X-ray emission has shown that chromia is well dispersed not only on the surface but also in the bulk (*19*).

$AlEt_2Cl$ and $TiCl_4$ were adsorbed irreversibly on graphite at the rate 0.8-2 molecules/nm^2, i.e. the surface of the support is almost completely covered by adsorbates (*9,10*).

The empirical data are related to the distribution of Mt on entire support surface. According to our knowledge, however, there are no data on the number of Mt atoms on the individual surface regions such as leaflets, platelets and crystal blocks.

Theory of Active Centre Ensembles

N. I. Kobozev developed a theory of active centre ensembles in 1939. (*20*). Panchenkov and Lebedev (*21*) presented a short outline of that theory. If atoms of a metal are applied to the surface of an ideal crystal, thermal motion will result in their spreading over the entire surface. When a crystal has a block structure, the atoms of a metal getting onto definite portions of the crystal surface need excess energy for surmounting the geometrical (and, consequently, energy) barriers and for motion over the entire surface. Thus, the surface of an adsorbent is divided into closed sections from a geometrical and energy viewpoint, in which at a given temperature non-activated motion of the atoms of the applied metal takes place. N. I. Kobozev called these section *regions of free migration* or *migration regions*. Therefore, the applied metal atoms arrange themselves on the surface of a support in the form of isolated aggregates or *ensembles* consisting of a certain number of atoms localized by the regions of free migration, as shown schematically in Figure 1.

In the simplest case it was found possible to derive a law of the distribution of the active centres on the support surface by assuming that the atoms of the metal being applied get onto a portion of the surface independently, i.e. the presence of one or more atoms of the metal in a given region does not change the probability of the following atom getting into the same region. This assumption is quite justified physically when the number of atoms of the applied substance is not too great relative to the dimensions of regions. It was shown by Kobozev that distribution of atoms over the surface of a support with a block structure obeys Poisson's law: $P_n = \nu^n e^{-\nu}/n!$. Here P_n denotes the probability of formation of *n*-atomic ensembles if the average number of atoms in one region is ν. *n* is an integer, i.e. *n*=1, 2, 3... The value of ν can be presented as $\nu = N{\times}s/S$. Here *N* denotes the total number of metal atoms applied onto a support that has the specific surface area *S* (m^2/g) fragmented into regions having the surface area *s*. The number of migration regions n_r is equal to $n_r = S/s$.

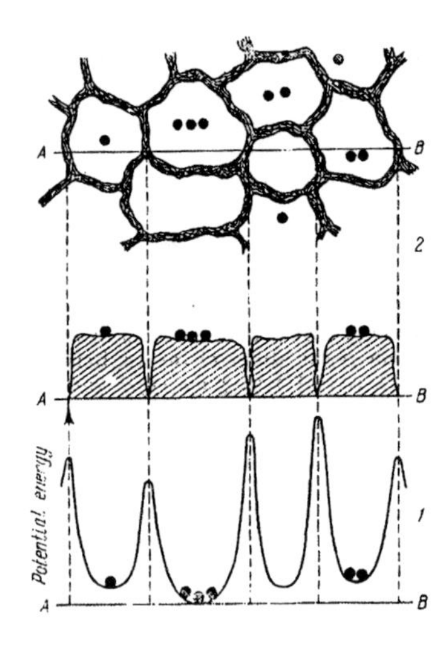

Figure 1. Distribution of Mt atoms by support regions (20,21)

Some examples of calculated Poisson's distribution of metal atoms onto support regions are given in Table I. For average number of metal atoms in one region $\nu = 2$, the probability that some region contains exactly $n = 2$ atoms is $Pn = 0.27$. For a high number of regions n_r, it means that 27% of all regions contain exactly $n = 2$ metal atoms. The other regions contain the higher ($n > 2$) or lower number ($n = 0$ or $n = 1$) of metal atoms.

According to the theory of active ensembles, for a given catalytic process, the active centre is an ensemble consisting of a definite number of metal atoms. For example, if an active centre consists of $n = 2$ metal atoms, than only those support regions that contain two metal atoms participate in the catalytic-process. The other regions, having more or less than two metal atoms, are not active.

Table I. Distribution of Metal Atoms onto Support Regions

n	Pn *for*		
	$\nu = 2$	$\nu = 3$	$\nu = 4$
1	0.27	0.15	0.07
2	0.27	0.22	0.15
3	0.18	0.22	0.20
4	0.09	0.17	0.20
5	0.04	0.10	0.16
6	0.01	0.05	0.10

Charge Percolation Mechanism of Ziegler-Natta Polymerization

Recently a new charge percolation mechanism (CPM) of olefin polymerization by supported transition metal (Mt) complexes has been presented (*22*). It is known that different oxidation states of Mt are obtained by activation, i.e. $Mt^{+(n-1)}$, $Mt^{+(n)}$ to $Mt^{+(n+1)}$, producing irregular charge distribution over the support surface. The tendency to equalize the oxidation states by a charge transfer from $Mt^{+(n-1)}$ (donor, D) to $Mt^{+(n+1)}$ (acceptor, A) cannot be fulfilled since they are immobilized and highly separated on the support. But, monomer molecules are adsorbed on the support producing the clusters with stacked π-bonds making a π-bond bridge between A and D (Figure 2). Once a bridge is formed (percolation moment), a charge transfer occurs. A and D equalize their oxidation states simultaneously with the polymerization of monomer. Polymer chain is desorbed from the support making the surface free for the subsequent monomer adsorption. The whole process is repeated by oxidation-reduction of another A-D ensembles immobilized on the same support region.

According to CPM, an active centre ensemble consists of two Mt ions, an acceptor (A) and an donor (D). These acceptors and donors are randomly distributed over leaflets and platelets of support that are used in ZN polymerization. Each leaflet and platelet corresponds to an individual region of Kobozev theory. Only those regions are active in polymerization that contain at least one AD ensemble. Different distributions of acceptors and donors onto the regions of support are achieved depending on the surface concentration of Mt ions, the relative ratio of acceptors and donors and size of surface regions. For example, if the average number of Mt ions per one region is $n_{Mt} = 2$, the regions that contain no Mt ion ($n_{Mt} = 0$) and only one Mt ion ($n_{Mt} = 1$) are not active. Also, the regions having $n_{Mt} = 2$ are not active if the both ions are acceptors or the both are donors. The regions having $n_{Mt} = 3$ Mt ions are active if there is an AD ensemble. Hence, the regions with three A or three D are not active.

So, the theory of active centers ensembles (developed by Kobozev in 1939) should be adopted in order to apply to the CPM of ZN polymerization.

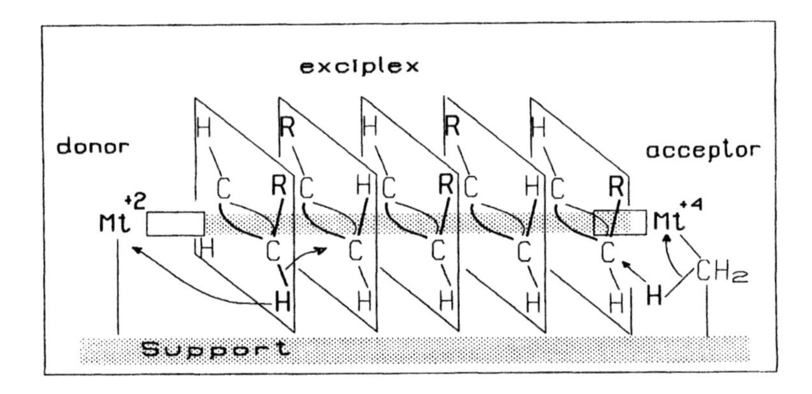

$$\begin{array}{l} \quad\;\; R \qquad\;\; R \qquad\;\; R \qquad\;\; R \qquad\;\; R \\ CH_2{=}C\text{-}CH_2\text{-}CH\text{-}CH_2\text{-}CH\text{-}CH_2\text{-}CH\text{-}CH_2\text{-}CH_2 \end{array}$$

$[Mt]^{+3}H$ $\qquad$ $H_2C\text{-}Mt^{+3}$

Figure 2. Monomer selforganization and charge percolation mechanism of Ziegler-Natta polymerization (upper) and polymer detachment (below)

Computer Simulation of ZN Polymerization by CPM

A Monte-Carlo procedure is used to simulate the ZN polymerization of olefins by CPM. The input parameters for simulation are: the type of support lattice, the fragmentation of support surface into the regions of definite sizes, the number of adsorption sites and the number of active centres on the support. The program algorithm enables one to simulate:
- The random immobilization of active centres on the support before the monomer addition is started (SAM-sequence, i.e. support + active centre formation + monomer adsorption) or
- The initial random adsorption of monomer before the active centres are formed followed by continuous monomer adsorption (SMA-sequence, i.e. support + monomer adsorption + active centres formation).

The one-dimensional (1D) and two-dimensional (2D) percolation processes have been analyzed. In 1D percolation the linear clusters of adsorbed monomer are formed along the monomer monolayer at the support. In 2D percolation the monomer clusters are formed in several directions in the monomer monolayer at the support (for hexagonal, square and triangular percolation lattices).

The simulation is performed by admittance one by one of monomer molecules to a randomly chosen adsorption site. A monomer molecule can be adsorbed at the site if it is free. If it is not, another monomer molecule is admitted to another randomly chosen adsorption site. So, the monomer monolayer is gradually formed. After adsorption of each monomer molecule, it is checked whether the bridge between A and D is completed or not. If it is completed, A and D are deactivated, and a polymer chain is formed. The polymer chain is desorbed and all adsorption sites, previously occupied by just polymerized monomer cluster (the percolation path), become free. A next admittance of a monomer molecule is performed, etc. The simulation can continue until the acceptors and/or donors are completely consumed.

The results of simulation are reported as polymer yield, rate of polymerization, number and mass average molecular mass and molecular mass distribution of polymer. Only the effects of the surface concentration of active centres (A and D) and fragmentation of supports on catalyst productivity are presented in this article. The catalysts productivity is presented as transition metal productivity (P_{Mt}) and support productivity (P_S). The P_{Mt} and P_S represent the quantity of polymerized monomer per one mole of the initial quantity of activated Mt and per unit of support, respectively.

Three series of simulations were performed that correspond to the real experimental conditions in ZN polymerization:

1. Series - Fragmentation of support is variable: the total number of adsorption sites and total numbers of acceptors and donors were kept constant

(n_S = const.; n_A = n_D = const., respectively) but the number of regions (n_r) was varied using different number of adsorption site in the regions.

2. Series - Specific surface area of support is variable: the number of adsorption sites at the support was varied by the number of regions (n_r), but the number of adsorption sites of one region (n_l) and the number of acceptors and donors were kept constant.

3. Series - Concentrations of acceptors and donors are variable: the numbers and the sizes of regions were kept constant, but the number of acceptors and donors were varied.

These three series were performed for 1D and 2D percolation (triangular, square and hexagonal lattices) and for SAM- and SMA-sequences of components addition to the polymerizing system.

A great number of simulations were performed and very good agreement with the more than one hundred points of experimental data was obtained. Only a small part of these results is presented in this article.

The results of simulation according to CPM showed that P_{Mt} initially increases, approaches a maximum value ($P_{Mt,max}$) and then decreases with the increase of surface concentration of Mt (Figure 3). P_S increases with the increase of surface concentration of Mt.

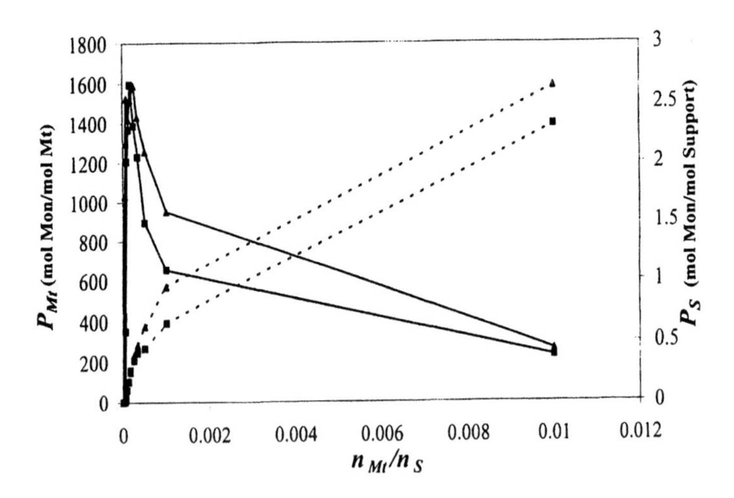

Figure 3. Effect of surface concentration of active centres n_{Mt}/n_S on P_{Mt} (solid lines) and P_S (dotted lines) for SAM-sequence (bottom lines) and SMA-sequence (upper lines) (prediction by 1D simulation; series 2; n_r=3 to 720; n_l=24000; n_A=n_D=360)

At very low surface concentration of activated Mt (n_{Mt}/n_S), the most of regions have no one active AD ensemble. Only a small number of regions posses them. Hence, the both P_{Mt} and P_S have low values.

With an increase of n_{Mt}/n_S more regions are populated by AD ensembles. Hence, both P_{Mt} and P_S are increased. In the case of SAM-sequence, a sharp maximum $P_{Mt,\ max}$ is obtained at $(n_{Mt}/n_r)_{max} \cong 3$ and $(n_{Mt}/n_r)_{max} \cong 5$ for 1D and 2D (for any type of lattice) percolation processes, respectively (Figure 4). The value of $P_{Mt,max}$ depends on the size of region, i.e. the number of adsorption sites (n_l) in it: higher n_l - higher $P_{Mt,max}$.

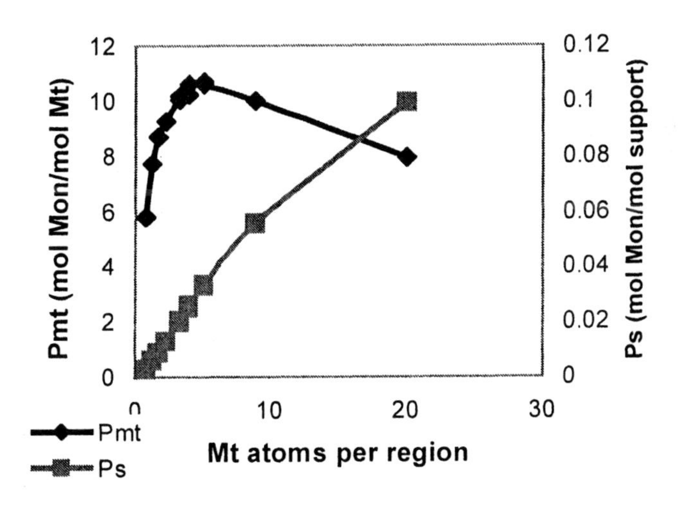

Figure 4. Effect of number of Mt atoms per region on P_{Mt} and P_S (Simulation conditions: 2D percolation, 2nd series, SAM-sequence, hexagonal lattice, n_{Mt}=8000, n_l=1600, n_r=400-10000)

In the case of $(n_{Mt}/n_r)>(n_{Mt}/n_r)_{max}$, the number of AD ensembles exceeds the number of regions. Hence, any region is many times active in the polymerization depending on the number of AD ensembles in it. So, the higher n_{Mt}/n_r - the higher P_S. But, the high n_{Mt}/n_r value means a short distance between acceptors and donors and a small number of monomer molecules in the clusters between them. Hence, each deactivation of AD ensemble includes a small number of polymerized monomer molecules. Because of that, P_{Mt} decreases sharply.

Similar results are found in the case of the SMA-sequence, but the corresponding values for P_{Mt}, P_S and n_{Mt}/n_r depend not only on the size of the region, but also on the relative ratio of Mt activation rate (i.e. AD ensemble formation) and monomer adsorption rate.

There are many published experimental data that agree with the predicted trends of productivities. One example is presented in Figure 5 for propene polymerization by $TiCl_4$ supported by $MgCl_2$ and activated by $AlEt_3$ (*23*).

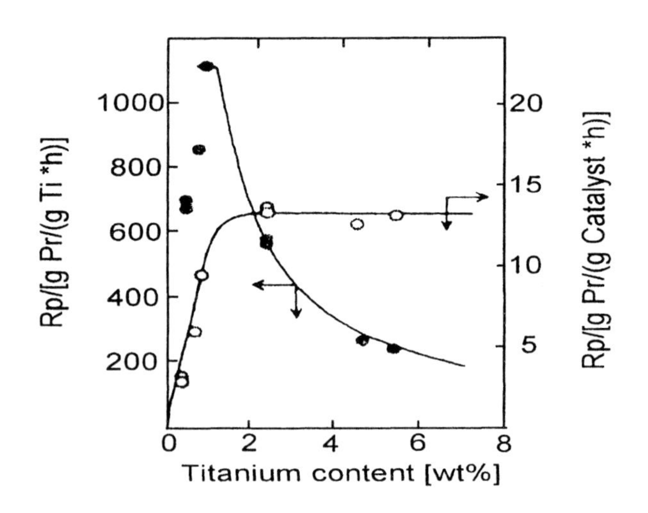

Figure 6. Productivity of $TiCl_4/MgCl_2/AlEt_3$ in propene polymerization [Experimental data (23)](With the permission of Wiley-VCH)

Another interesting example is P_{Mt} for propene polymerization by $TiCl_4/AlEt_2Cl$ on graphite with different specific surface areas, i.e. 61.5 and 6.5 m^2/g (Figure 7, upper) (*9,10*). According to CPM, the same values of $P_{Mt,max}$ indicate that both graphites have the same sizes of regions (proposal 1), but different numbers of regions (proposal 2). In that case, the ratio of values of Ti contents at maximum of productivities (0.7% and 0.08%) should be equal to the ratio of specific areas of graphites (61 and 6.5). Indeed, both ratios have the values ≅ 9. The computer simulation, based on the above mentioned proposals, confirmed that the qualitatively same shapes of the curves are obtained (Figure 7, bottom) as the experimental ones. (The different units in the upper and bottom graphs due to Monte-Carlo simulation procedure that manipulates by arbitrarily time units and adsorption sites of the support.)

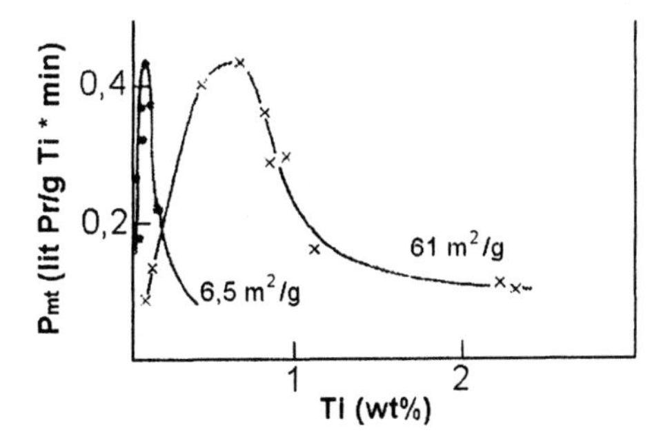

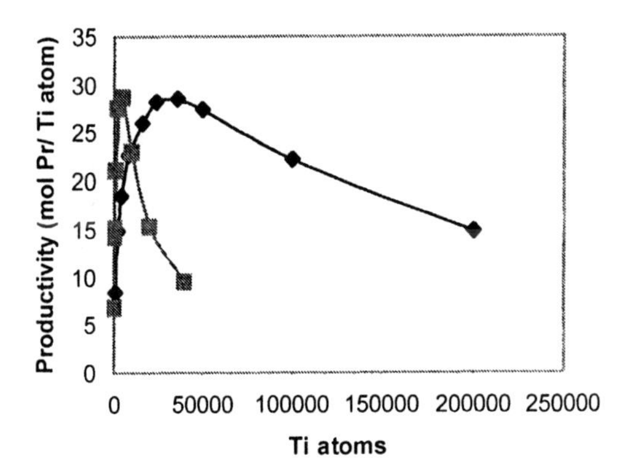

Figure 7. Propene polymerization by $TiCl_4/AlEt_2Cl$ on graphite with specific surface areas, i.e. 61.5 and 6.5 m^2/g: Top - experimental data (9,10)- (With the permission of Academizdatcentre "Nauka");Bottom - computer simulation

Conclusion

The charge percolation mechanism of Ziegler-Natta polymerization combined with an adopted Kobozev theory of active centers ensembles can successfully predict and explain experimental results on catalyst productivity. Furthermore, the same concepts have also been applied very successfully to explain and predict the molecular mass and molecular mass distribution of polymer produced by Ziegler-Natta complexes (*22*).

References

1. Boor, J. *Ziegler-Natta Catalysts and Polymerizations;* Academic Press: New York, 1979.
2. Zakharov, V.A. *Kinetika i kataliz* **1981,** *22,* 480.
3. Higuchi, T.; Liu, B.; Nakatani, H.; Otsuka, N.; Terano, M. *Appl. Surf. Sci.* **2003,** *214*, 272.
4. Moore, E. P. *The Rebirth of Polypropylene: Supported Catalysts;* Hanser Publishers: Munich, 1997; p 36
5. Kiss, E.; Lomic, G.; Skrbic, B.; Stoiljkovic, D.; Dingova E. *Acta Per. Technol.* **1997**, *28*, 123.
6. Keszler, B; Bodor, G.; Simon, A. *Polymer* **1980,** *21*, 1037.
7. *Catalytic Polymerization of Olefins;* Keii, T.; Soga, K., Eds.; Kodansha-Elsevier: Tokyo-Amsterdam, 1986; p 71.
8. *Handbook of Polyolefins;* Marcel Dekker Inc.: New York, 2000; p. 1.
9. Nedorezova, P. M.; Galashin, N. M.; Tsvetkova, V. I., Sukhova, T. A.; Saratovskikh, S. L.; Babkina, O. N.; Dyachkovskii, F. S. *Eur. Pol. J.* **1996,** *32*, 1161.
10. Nedorezova, P. M.; Tsvetkova, V. I.; Kolbanev, I. V.; Dyachkovskii, F. S. *Vysokomol. Soed.* **1989,** *A31*, 2657.
11. Hsieh, H. L. *Cat. Rev. - Sci. Eng.* **1984,** *26*, 631.
12. *History of Polyolefins;* Seymour, R. B.; Cheng, T., Eds.; D. Riedel Publ. Co.: Dordreht, 1986, p 271.
13. *Applied Industrial Catalysis;* Leach, B. E., Ed.; Academic Press: New York, 1983; Vol. I, Ch. 6.
14. Czaja, K. *Polymer*, **2000,** *41*, 3937.
15. Natta, G. *J. Polym. Sci.* **1959**, *34*, 21.
16. Hasebe, K.; Mori, H.; Tomitori, M.; Keii, T.; Terano, M. *J. Mol. Cat. A: Chemical* **1997,** *115*, 259.
17. Monaco, G.; Toto, M.; Guerra, G.; Corradini, P.; Cavallo, L. *Macromolecules* **2000,** *33*, 8953.
18. Liu, B.; Terano, M. *J. Mol. Cat. A: Chemical* **2001,** *172*, 227.
19. Rahman, A.; Mohamed, M. H.; Ahmed, M.; Aitani, A. M. *Appl. Catalysis A: General* **1995,** *121*, 203.
20. Kobozev, N. I. *Zhourn. Fiz. Khim.* **1939**, *13*, 1.
21. Panchenkov, G. M.; Lebedev, V. P. *Chemical Kinetics and Catalysis;* Mir; Moscow, 1976.
22. Stoiljkovic, D.; Pilic, B.; Jovanovic, S.; Panic, D. in *Current Achievements on Heterogeneous Olefin Polymerization Catalysts*; Terano, M., Ed.; Sankeisha. Co.: Nagoya, 2004; p 135.
23. Keii, T.; Suzuki, E.; Tamura, M.; Murata, M,; Doi, Y. *Makromol. Chem.* **1982**, *183*, 2285.

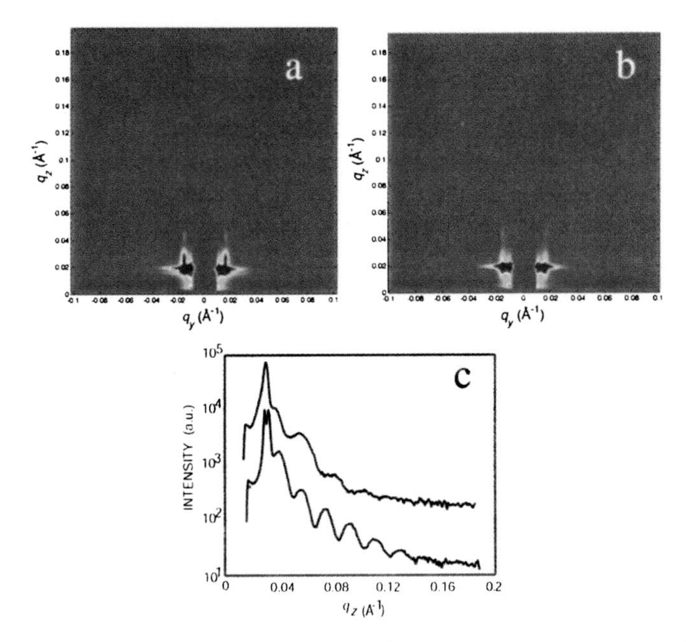

Plate 12.2 Grazing incidence small angle x-ray scattering pattern of LSM71k diblock copolymer thin films prepared on a substrate modified with random coplymer after being dipped into (a) acetic acid and (b) cyclohexane. (c) The qz scans qy=0.0178 Å-1 of both GISAXS patterns.

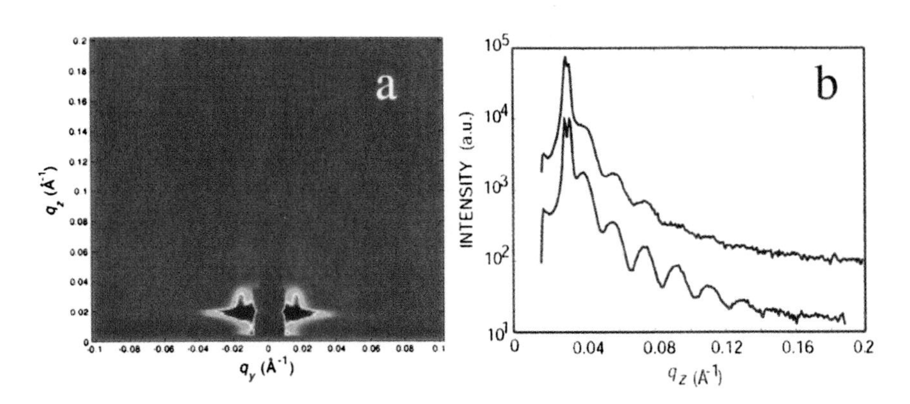

Plate 12.3(a) Grazing incidence small angle x-ray scattering pattern of an LSM71k diblock copolymer thin film prepared on a passivated substrate after being dipped into acetic acid. (b) The qz scans qy=0.0178 Å-1 of the GISAXS patterns of Plate 12.2a and 12.3a.

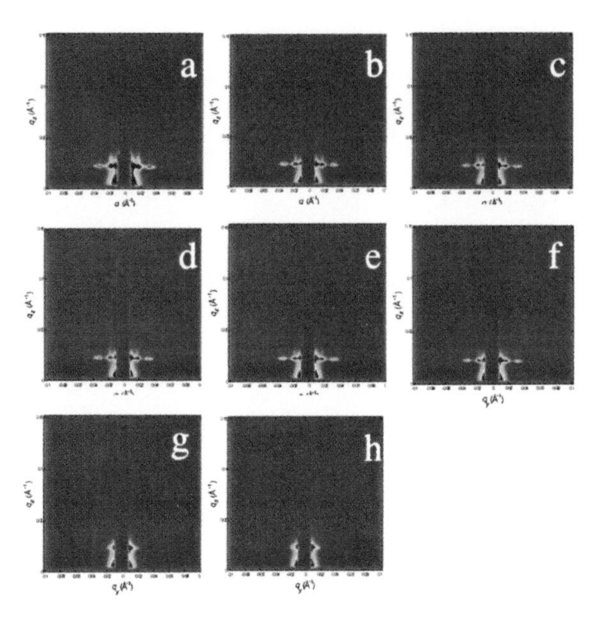

Plate 12.4 Grazing incidence small angle x-ray scattering patterns of the re-annealing process of a CSM69k diblock copolymer thin film swollen with acetic acid. (a) 60°C, (b) 70°C, (c) 80°C, (d) 85°C, (e) 90°C, (f) 100°C. The film was annealed at each temperature for 6 minutes. (g) was taken after annealing the sample at 100°C for 15 minutes.

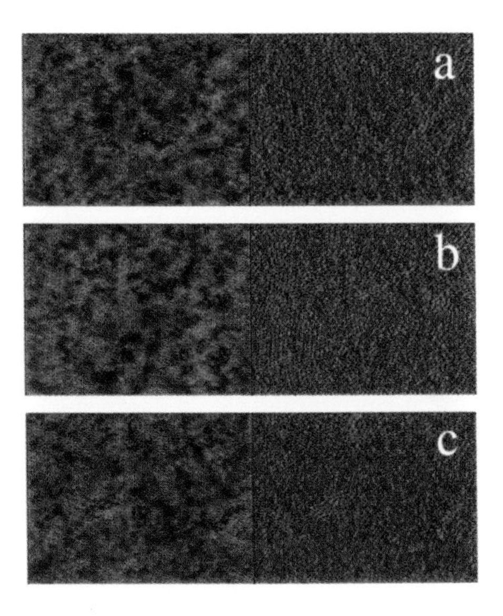

Plate 12.6 AFM images of CSM69k diblock copolymer thin films after being (a) swollen with acetic acid at room temperature and annealed at (b) 80°C, (c) 90°C, (d) 100°C and (e) 110°C for an hour. The height scale is 15 nm and the phase scale is 15° for all the images.

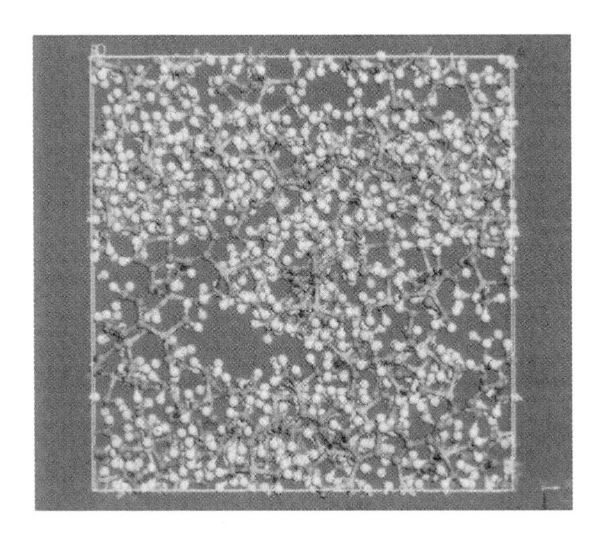

Plate 14.1. A typical packing model for PTMSP. Large cavities are observed.

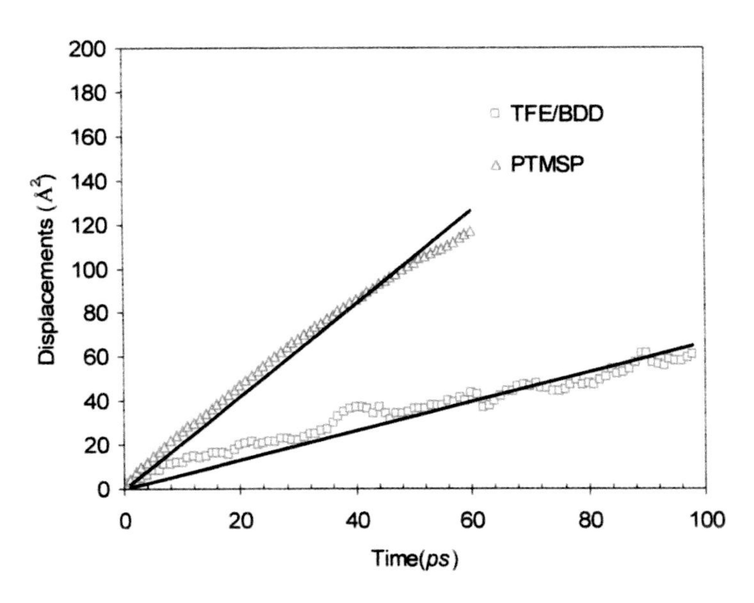

Plate 14.4. Mean-square displacement of CO_2 in PTMSP and TFE/BDD87 at 298K during NVT molecular dynamics. Lines represent linear fit of data.

Chapter 17

Preparation of Molecularly Imprinted Cross-Linked Copolymers by Thermal Degradation of Poly(methacryl-*N,N'*-diisopropylurea-*co*-ethylene glycol dimethacrylate)

Radivoje Vuković, Ana Erceg Kuzmić, Grozdana Bogdanić, and Dragutin Fleš

Research and Development Sector, INA-Industrija nafte d.d., Lovinčićeva bb, P.O. Box 555, 10002 Zagreb, Croatia

Dedicated to the 70th birthday of Professor Frank E. Karasz

Crosslinked porous copolymers of methacryl-isopropylamide with ethylene glycol dimethacrylate, poly(MA-iPrA-co-EDMA), were prepared by thermal degradation of crosslinked copolymers of methacryl-diisopropylurea with ethylene glycol dimethacrylate, poly(MA-DiPrU-co-EDMA), at temperature of 180-250^{0}C in vacuum or under TGA conditions in nitrogen. Nonporous copolymers which represent the model compound of porous copolymers of the same structure were also prepared. Both, porous and nonporous model copolymers have almost the same thermal stability as shown by TGA measurements. At the same time the T_g of porous and nonporous model copolymers are different and indicate that nonporous copolymers have higher structural ordering.

Introduction

Synthesis of copolymers which contain porous cavities of defined geometry and distribution has attracted a large attention during the last few decades. Of special importance is that cavities in the polymer matrix contain definite recognition sides which have the affinity to the selected analytes. These polymers are mostly applied as enantioselective or structurally selective carriers in the chromatographic separations, as mimics of enzymes, or as carriers of drugs or catalyst in organic reactions or in biomedical applications. Of special interest is the use of porous copolymers as sensors in the field in which polymers are used as substitutes for biological materials (*1-5*).

Molecularly imprinted polymers, which contain specific recognition groups, can be prepared by several well-described procedures. Two basically different methods are mostly used: the first method involves the formation of covalent adduct between print molecule and the functional monomers (*6*), while the second approach involves the use of non-covalent interaction between print molecules and functional monomers (*7*).

One problem in both covalent and non-covalent molecular imprint procedure is the separation of imprint molecules from the template after the copolymerization. Although the amount of templates, which after the extraction contain imprint molecules is small, and amounts to less than 1%, it usually causes difficulties in the analytical application of imprinted molecules (*8*).

A new method, which enables the preparation of crosslinked molecularly imprinted polymers, was recently developed in our Laboratories. The method is based on the thermal degradation of crosslinked copolymers of acryl-and methacryl-disubstituted urea copolymerized with EDMA (*9-11*).

The first porous copolymers obtained by thermal degradation of crosslinked copolymers of disubstituted urea were prepared from the poly(acryl-dicyclohexylurea-co-EDMA), (*9,10*). Thermal degradation was performed by heating the crosslinked copolymer at temperature of 180-450^{0}C. The decomposition proceeds by a two-step mechanism under the separation of volatile cyclohexylisocyanate ($C_6H_{11}NCO$), at temperature of 180-250^{0}C, and formation of cavities in the crosslinked matrix. The amount of volatile fraction closely correlates to the amount of terminal cyclohexylamine in urea: -CO-NHC_6H_{11}.

Following the same experimental procedure we have recently prepared porous copolymers by thermal decomposition of methacryl-dicyclohexylurea (MA-DCU) with EDMA under the separation of $C_6H_{11}NCO$ (*11*).

In continuation of our work on the thermal degradation of crosslinked copolymers which contain disubstituted urea derivatives, we prepared the porous copolymers of methacryl-isopropylamide (MA-iPrA) with EDMA, under the separation of isopropylisocyanate (iPrNCO).

Experimental and Discussion of Results

Synthesis of N-Methacryl-N,N'-diisopropylurea (MA-DiPrU)

The title compound was prepared by condensation of 6.3 g (0.05 mol) of diisopropylcarbodiimide in 10 mL of tetrahydrofuran with 4.3 g (0.05 mol) of MAA and 0.2 g hydroquinone in 10 mL of THF at room temperature. The crystalline diisopropylurea was filtered off, the mother liquor evaporated to dryness and treated with 30 mL of petroleum ether (b.p. 40-60^{0}C) yielding 3.4 g (32.1%) of MA-DiPrU melting at 77-78^{0}C.

Analysis of MA-DiPrU: Calculated for $C_{11}H_{20}N_2O_2$ (212) (%): C, 62.26; H, 9.43; N, 13.21; Found: C, 62.18; H, 9.43; N, 13.06.

The monomer structure was determined by NMR spectroscopy measurements.

Copolymerization of MA-DiPrU with EDMA at Molar Ratio of 0.5 to 0.5 in the Feed

To a mixture of 0.636 g (0.003 mol) of MA-DiPrU and 0.594 g (0.003 mol) of EDMA was added 6 mL of butanone which contains 2% (0.025 g) of Bz_2O_2 as initiator, and the reaction mixture was heated for 48 hours at 70^{0}C in a stream of nitrogen. After cooling, the copolymer was washed with butanone, followed by methanol yielding 0.832 g (67.7%) of the crosslinked copolymer. Under the described conditions of copolymerization, the ratio of comonomers in the crosslinked copolymer is 0.26 molar ratio of MA-DiPrU and the molar ratio of EDMA is 0.74.

Calculated amount of nitrogen in copolymer is 3.66% while the experimental value of nitrogen in copolymer has a value of 3.44%.

Thermogram of crosslinked copolymer of MA-DiPrU-co-EDMA at a monomer-to-monomer ratio of 0.5 to 0.5 in the feed is shown in Figure 1. It is evident that the copolymer decomposes by a two-step mechanism losing 11.5% of weight at 250°C.

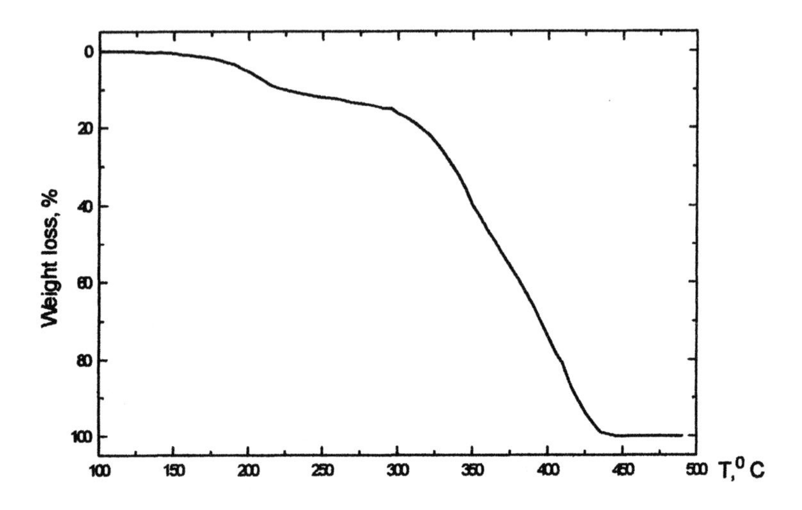

Figure 1. Thermogram of crosslinked copolymer of MA-DiPrU with EDMA at molar ratio of 0.26 to 0.74 in copolymer (0.5 to 0.5 in the feed).

Preparation of Porous Crosslinked Copolymer of Methacryl-isopropylamide (MA-iPrA) with EDMA by Thermal Degradation of Poly(MA-DiPrU-co-EDMA)

Porous copolymers MA-iPrA with EDMA were prepared by thermal degradation of poly(MA-DiPrU-co-EDMA) in vacuum or in thermogravimetric analyzer in nitrogen.

In vacuum experiment, a sample of 0.702 g of crosslinked copolymer poly(MA-DiPrU-co-EDMA) prepared at monomer ratios of 0.5 to 0.5 in the feed, was heated for 60 min at 200-210°C in vacuum of 0.26 kPa, yielding 0.615 g (87.60%) of solid residue and 0.087 g (12.40%) of volatile fraction identified as iPrNCO.

Elemental analysis of solid residue (%): C, 59.83; H, 7.55; N, 1.97.

Based on nitrogen the copolymer contains 0.18 molar fraction of MA-iPrA.

Thermal degradation of the same copolymer was also performed in the stream of nitrogen at heating rate of 10°C/min in TGA instrument: A sample of

2.206 mg of copolymer was heated for 5 min at 250^{0}C yielding after cooling 1.940 mg (87.94%) of solid porous copolymer and 0.266 mg (12.06%) of volatile fraction. Volatile fraction was calculated as the difference to 100% of starting copolymer sample.

Elemental analysis of solid residue (%): C, 59.16; H, 7.35; N, 2.16.

Based on nitrogen the copolymer contains 0.19 molar fraction of MA-iPrA.

It is of interest to note that the amount of volatile fraction obtained by thermal degradation of MA-DiPrU-co-EDMA in vacuum or by the discontinuous measurements in thermogravimetric analyzer is practically the same: 12.40% and 12.06% respectively. Closely related is also the loss of weight by TGA measurement at 250^{0}C shown in Figure 1, which amounts 11.50%.

In order to compare the properties of porous crosslinked copolymers with those of nonporous copolymer of the same composition, we prepared the nonporous model compound: poly(MA-iPrA-co-EDMA) by copolymerization of MA-iPrA with EDMA.

Synthesis of Methacyl-isopropylamide (MA-iPrA)

To a solution of 5.9 g (0.08 mol) of isopropylamine in 30 mL of ether was added 15.6 g (0.1 mol) of MA-anhydride and the reaction mixture was left overnight at room temperature. The ether solution was washed with aq. NaOH solution, followed by water, dried with Na_2SO_4 and ether was evaporated to dryness. The residue was treated with petrolether yielding 2.09 g (20.0%) of MA-iPrA; m.p. 89-90^{0}C. The structure of the product was proved by elemental analysis and by NMR spectroscopy.

Analysis of MA-iPrA: Calculated for $C_7H_{13}NO$ (127) (%): N, 11.02; Found: N, 10.59.

Copolymerization of MA-iPrA with EDMA at Molar Ratio of 0.5 to 0.5 in the Feed (Nonporous Model Compound)

Comonomers mixture of 0.127 g (0.001 mol) of MA-iPrA with 0.198 g (0.001 mol) of EDMA in 2 mL of butanone with 2% Bz_2O_2 was heated for 48 hours at 70^{0}C under a stream of nitrogen. Crude polymer was washed with butanone, yielding 0.295 g (90.67%) of crosslinked polymers.

Elemental analysis (%): C, 59.85; H, 8.12; N, 3.67.

Based on the nitrogen content, the copolymer contains of 0.33 molar ratio of MA-iPrA and 0.67 ratio of EDMA.

Thermograms of porous copolymers of MA-iPrA with EDMA obtained as a solid residue after the removal of iPrNCO from poly(MA-iPrU-co-EDMA) prepared at molar ratio of comonomer of 0.5 to 0.5 in the feed, and nonporous model compounds prepared at the same molar ratio of 0.5 to 0.5 in the feed of MA-iPrA with EDMA are shown in Figure 2. Thermograms in Figure 2 indicate that both copolymers have similar thermal stability, thus indicating that the porous structure has no significant influence to the thermal stability of crosslinked copolymers.

In order to prove the difference between crosslinked copolymers prepared by thermal degradation of MA-DiPrU-co-EDMA and model copolymers MA-iPrA-co-EDMA, we compared T_g's of the mentioned crosslinked copolymers. It is shown that the T_g of copolymer obtained after the removal of iPrNCO is 207^0C, while the T_g of model compound exhibits the value of 190^0C and a transition at 267^0C, thus indicating that model compound has higher structural ordering. Preliminary results obtained by comparison of x-ray diffractograms of both porous and nonporous copolymers, also indicates the higher structural ordering of model copolymers. It is of interest to note that in the preliminary experiments, it is shown that the density of crosslinked poly(MA-DiPrU-co-EDMA) is higher than that of porous crosslinked poly(MA-iPrA-co-EDMA).

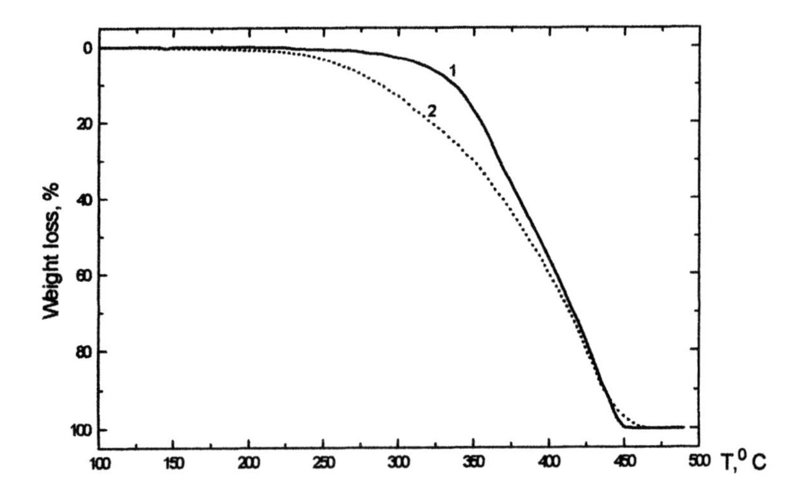

Figure 2. Thermograms of porous residue obtained after removal of iPrNCO from the copolymer of MA-DiPrU with EDMA (1) and model copolymer of MA-iPrA with EDMA (2). Both copolymers are prepared at molar ratios of 0.5 to 0.5 of comonomers in the feed.

The possible mechanism of the thermal degradation of poly(MA-DiPrU-co-EDMA) is shown in Scheme 1.

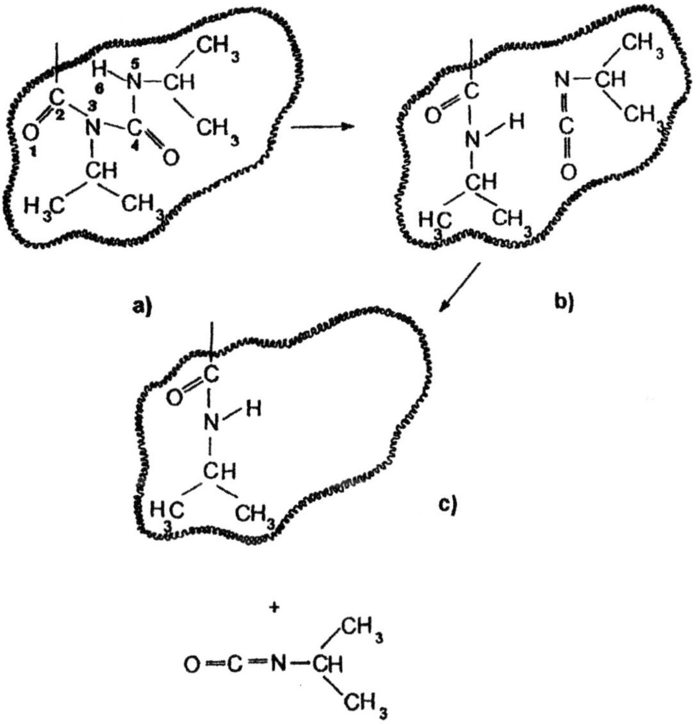

a)

b)

c)

$O{=}C{=}N{-}CH(CH_3)_2$

Scheme 1. Mechanism of thermal degradation of crosslinked copolymers of MA-DiPrU with EDMA.

By applying the coiled structure of the terminal chain in repeating unit of the MA-DiPrU trapped in the crosslinked copolymer matrix, the numbering of atoms can be represented by the Rule of Six proposed by Newman (*12*) as indicated by the Scheme 1a. As represented by the structure of the terminal unit in the Scheme 1a, the hydrogen atom **6** is close to the nitrogen atom **3** and at the temperature of about 200^{0}C hydrogen **6** reacts with iPrN= **3** thus causing the separation of iPrNCO as shown in the Scheme 1b. In Experimental part of this article is shown that iPrNCO which has a boiling point of 74^{0}C can be easily removed in vacuum at a temperature of 200-210^{0}C, or by heating the crosslinked copolymer at 250^{0}C in nitrogen.

After the removal of volatile iPrNCO, the nanopores, which contain the recognition groups -CO-NHCH(CH_3)$_2$ are formed, thus indicating that the obtained porous copolymers described in this article could be of interest in various practical applications.

It is further of interest to note that since various N,N'-disubstituted ureas can be easily prepared, the procedure in the article can be considered as a general method for the preparation of porous crosslinked copolymers which contain the desired recognition groups in pores.

Conclusion

Copolymer of MA-DiPrU with EDMA at the monomer-to-monomer molar ratio of 0.5 to 0.5 in the feed was prepared. The yield of crosslinked copolymer, which contains 0.26 to 0.74 molar ratios of comonomers, was 67.7%. In the thermogravimetric analysis it is shown that the prepared crosslinked copolymer decomposes by two-step mechanism under almost quantitative evaporation of isopropylisocyanate (iPrNCO) at a temperature of 250^{0}C. The residue obtained after the removal of iPrNCO is thermally stable, and decomposes by a one-step mechanism between 280 and 450^{0}C. After the removal of volatile iPrNCo, the crosslinked porous copolymer which contains the recognition groups -CO-NHCH(CH_3)$_2$ are formed.

In view of the possibility to modify the copolymer composition and geometry of N,N'-disubstituted urea, we consider that these copolymers may be of interest as host specific for desired molecules for various applications, especially in the thin layer chromatography or as carriers of various functional analytes.

Acknowledgement

The Ministry of Science and Technology of Croatia supported this work.

References

1. Mosbach, K. Molecular Imprinting. *Trends Biochem. Sci.* **1994,** *19,* 9-14.
2. Shea, J. K. Molecular Imprinting of Synthetic Network Polymers: The De Novo Synthesis of Macromolecular Binding and Catalytic Sites. *Trends Polym. Sci.* **1994,** *2(5),* 166-173.
3. Kempe, M.; Mosbach, K. Molecular Imprinting Used for Chiral Separations. *J. Chromatogr.* **1995,** *A694,* 3-13.

4. Wulff, G. Molecular Imprinting in Cross-Linked Materials with Aid of Molecular Templates-A Way Toward Artificial Antibodies. *Angew. Chem. Int. Ed. Engl.* **1995**, *34*, 1812-1832.
5. Muldoon, T. M.; Stanker, H. L. Plastic Antibodies: Molecularly-Imprinted Polymers. *Chem. Ind.* **1996**, *24*, 204-207.
6. Wulff, G.; Biffis, A. Molecular Imprinting with Covalent or Stoichiometric Non-covalent Interactions. In: Molecularly Imprinted Polymers: Man-Made Mimics of Antibodies and Their Applications in Analytical Chemistry. Ed.; Sellergren, B. Elsevier: Amsterdam. **2001**., pp. 71-111.
7. Sellergren, B. The Noncovalent Approach to Molecular Imprinting. In: Molecularly Imprinted Polymers: Man-Made Mimics of Antibodies and Their Applications in Analytical Chemistry. Ed.; Sellergren, B. Elsevier: Amsterdam. **2001**., pp. 113-184.
8. Sellergren, B.; Shea, J. K. Influence of Polymer Morphology on the Ability of Imprinted Network Polymers to Resolve Enantiomers. *J. Chromatogr.* **1993**, *635*, 31-49.
9. Erceg Kuzmić, A.; Vuković, R; Bogdanić, G.; Fleš, D. Separation of Cyclohexylisocyanate from the Crosslinked Copolymers of N-Acryl-Dicyclohexylurea with Ethylene Glycol Dimethacrylate or Divinyl Benzene. *J.Macromol.Sci.-Pure Appl.Chem.* **2003**, *A40(1)*, 81-85.
10. Erceg Kuzmić, A.; Vuković, R.; Bogdanić, G.; Fleš, D. Synthesis of Nanoporous Crosslinked Poly(acryl-N-cyclohexylamide-co-ethylene glycol dimethacrylate) by Thermal Degradation of Poly(acryl-N,N'-dicyclohexylurea-co-ethylene glycol dimethacrylate). *J. Macromol. Sci.-Pure Appl. Chem.* **2003**, *40*(8), 747-754.
11. Erceg Kuzmić, A.; Vuković, R.; Bogdanić, G.; Fleš, D. Preparation of Nanoporous Crosslinked Poly(Methacryl-N-cyclohexylamide-co-ethylene glycol dimethacrylate) by Thermal Degradation of Poly(Methacryl-N,N'-dicyclohexylurea-co-ethylene glycol dimethacrylate). *J. Macromol. Sci.-Pure Appl. Chem.* **2004**, *A41*(8) (in Press.).
12. Newman S. Melvin. Some Observations Concerning Steric Factors. *J. Am. Chem. Soc.* **1950**, *72*, 4783-4785.

Chapter 18

Confined Macromolecules in Polymer Materials and Processes

Peter Cifra*[*] and Tomas Bleha

Polymer Institute, Slovak Academy of Sciences, Dúbravská cesta 9, 842 36 Bratislava, Slovakia
[*]Corresponding author: cifra@savba.sk

In polymeric systems such as polymer-particle colloid dispersions, composites and nano-composites, clay-intercalates with polymers the geometrical confinement of macromolecules becomes one of crucial features influencing final properties. Confinement is a driving force also in characterization of macromolecules by liquid chromatography (LC). Macromolecular confinement near geometrical obstacle gives rise to a depletion of polymer concentration in the vicinity of wall and to the depletion force. We address the interplay of the confinement, the solvent quality, polymer adsorption, polymer concentration and its consequences for LC and colloid particle-polymer phase behavior. Particularly intriguing is the case of compensation of macromolecular exclusion and adsorption by the confining walls which marks the characteristic inversion of behavior in these fields.

Introduction

The thermodynamic, conformational and dynamic properties of macromolecular chains are strongly affected by geometrical confinement. This phenomenon is relevant to numerous applications of polymer systems but poses also scientific challenges (*1*). Polymer chains when confined to a small space exhibit properties distinctly different from those in the unconfined space. Confinement operates also in polymer materials of complex morphology with abundant interfaces which can be approximated by solid walls. There is an important difference between surface/interface and bulk properties of macromolecular chains. Confined flexible polymer chains experience a loss in configuration entropy and this loss can be outweighed by adsorption energy of macromolecules attracted to the surface. The resulting complex behavior poses difficulties to an analytical treatment as well as to experimental studies (*2*). In this situation molecular simulations become a useful tool (*3-6*). Simulation can establish a full equilibrium (in contrast to experimental studies often characterized by an incomplete equilibrium and presence of history in adsorptive systems). It can provide local structural information on confined chains and it is possible to deal with perfectly flat surfaces and disregard the surface heterogeneity, surface steps or pre-adsorbed components. Of course, it is possible to study the individual effects either separately or to include the full complexity at later stages.

In this paper we will analyze two examples of confined macromolecular systems which represent general features of phenomena involved in similar systems. First, it is the partitioning of chains into slits, which is underlying the macromolecular characterization and separation by size exclusion chromatography (SEC) and, second, the depletion effects in colloid-polymer dispersions stability. This analysis is accompanied by analysis of concentration profiles including depletion profiles, adsorption profiles and compensation (adsorption *vs* steric exclusion) profile. The confinement was so far presented mainly through entropic effect (a loss in number of conformations). The effect of confinement, however, varies in the presence of interactions (adsorption, variable solvent quality), chain stiffness, polymer concentration and geometry of confinement or chain architecture. It is the interaction of all these factors which should be well understood.

The partitioning of polymer into porous medium is characterized by the partition coefficient K which is the pore(I)-to-bulk(E) concentration ratio at equilibrium, $K=\phi_I/\phi_E$. In original theoretical treatment (*7*) the pure steric exclusion of ideal chains in infinitely dilute solutions was addressed. The analytical approach based on the analogy between diffusion motion of a particle and conformation of a freely-jointed polymer chain provided the relations between the partition coefficient K and the coil-to-pore size ratio $\lambda=2R_g/D$, where R_g is the radius of gyration of a polymer coil and D is the characteristic dimension of a pore. According to theory, the coefficient K depends on the confinement ratio λ only and not on each of these two parameters separately. K is related to the free energy of confinement, $\Delta A = -kT \ln K$, which represents a penalty for the transfer of a molecule from the bulk solution to the solution in pore. Theoretical models of partitioning of flexible macromolecules were recently reviewed (*8*). The purely steric partitioning of freely-jointed and excluded volume chains into pores with repulsive chain-walls interaction was examined by Monte Carlo (MC) simulations (*3-6,9*). It was found that partitioning rules for ideal chains according to Casassa (*7*) have to be modified in the case of excluded volume chains to account for solvent quality and for solute concentration.

The confining porous media in many cases involve pore walls which are attractive to polymer. Polymer partitioning in such cases proceeds via a combined steric exclusion/adsorption mechanism. In order to explain this process the diffusion-equation theory (*7*) of partitioning of ideal chains was extended (*10*) to include a short-range segment-wall attraction potential ε. Partitioning of macromolecules in adsorbing pores is examined by theory (*10-12*) and by simulations (*13-14*) especially in order to clarify the separation mechanism in LC. The combined effect of purely steric partitioning and wall attraction is particularly noticeable in critical chromatography (*15-16*) operating close to the critical adsorption strength ε_c. Macromolecular solutes adsorb from solution onto a substrate if attractive potential ε exceeds the critical value ε_c, (the adsorption/depletion threshold (*1*)). At critical condition the entropy loss due to chain confinement into the pore should be compensated by the attractive energy gain and the partition coefficient, K=1, may become independent of the chain length N of solute macromolecules (the compensation point). This compensation is used in complex separations, for instance of mixtures including block copolymer. The idea is to find a compensation point for one of the components(blocks) in the mixture. All chain lengths of this component are then moving freely, K=1, between the bulk solution and porous medium, while at the same time the other components are separated according to the chain length. In this way one of the components in LC separation is effectively masked.

Similar confinement situation is characteristic for mixtures of colloid particles and polymers, which are of great practical importance in number of industrial, environmental and medical applications. In general, interaction between polymers and particles can dramatically affect the stability of particle suspensions. The forces between two particles in a solution depend mainly on whether the polymer is attracted (adsorbed) to or repelled (depleted) from the particle surface. An attachment of polymers to the surface, by grafting or by strong adsorption, is a common method of stabilizing large particles in a solution. On the other hand addition of non-adsorbing macromolecules to a colloidal suspension can result in polymer-mediated "depletion attraction" between particle pairs.

The depletion forces in polymer-colloid systems are of a long-standing interest in last few decades. Investigations treated mostly the "colloid limit" where the particle radius R_c greatly exceeds the coil size expressed by the radius of gyration R_g. The depletion mechanism in the colloid limit was first described by Asakura and Oosawa (*17*). Depletion attraction often leads to demixing into the polymer- and colloid-enriched phases at sufficient polymer concentration. Theoretical and experimental studies on depletion-induced phase separation in colloid-polymer mixtures have recently been reviewed (*18*). In the colloid limit $R_c >> R_g$ a colloidal particle resembles a hard planar wall on a relevant length scale. Thus depletion interaction is frequently treated in theory and simulations by a model of a slit formed by two parallel repulsive walls immersed in a polymer solution.

Effect of depletion interaction was so far studied almost exclusively in athermal systems of colloidal particles modeled by hard (repulsive) walls (*18, 19*). These entropy-driven phase transitions originate solely from the excluded volume interactions. It is of major interest to explore how depletion forces are affected by introduction of a weak attraction potential between polymer and colloidal particles. Inclusion of the energy contribution to the depletion force may assist in portrayal of phase behavior of more realistic colloid-polymer mixtures than are hard particle systems.

Much of the characteristic behavior of confined systems is furnished in concentration profiles at the confining walls. Below we present a general analysis of concentration profiles for the depletion and for the adsorption mode as well as for compensation profile. This is followed by systems with interplay of confinement, polymer-wall and polymer-solvent interactions, variable polymer concentration, together with an illustration of consequences for complex LC separations and colloidal stability. We examine depletion effect and depletion interaction in a weakly attractive slit, particularly in the region around the critical point of adsorption. The method used in Monte Carlo simulation of confined systems has been described in previous paper (*20*). The simulation of confined polymer chains is performed on a cubic lattice of a lattice unit a with the nearest neighbor non-bonded interactions of segments ε_{ii} (in kT units) to account for solvent quality and a similar interaction ε of polymer segments with the wall to account for the chain attraction to the wall.

Concentration profiles and compensation effect

Depletion profiles. Concentration profiles $\phi(x)$ shown in Figure 1 depict the situation in the slit represented by hard walls, $\varepsilon = 0$. Depletion of polymer chains at each wall is seen in all profiles. The concentration profiles are shown in the slit of D/a =12 for chain length N = 100 and for the athermal solutions (good solvent) at concentrations ϕ_1 = 0.303, 0.195, 0.120, and 0.032 and for the theta solution at ϕ_1 = 0.286, 0.183, 0.112, and 0.039 in volume fractions. Radius of gyration at infinite dilution and the first overlap concentration for these chains are R_{g0} = 6.45 (5.34) and ϕ^*=0.120 (0.207), for good and theta solvent respectively (*9*). There is an uneven filling of the space seen in the slit, especially at low concentration. The arrow indicates filling of the space at the walls by increasing concentration. Furthermore, there is another tangible phenomenon that shows a difference between athermal and theta solutions and can be ascribed to the unfavorable segment-solvent interactions. At lower concentrations the two profiles are similar, with a large depletion layer at the slit walls. As the concentration builds up, the athermal chains distribute segments uniformly and are able to fill the depletion layer readily. In the theta solutions, however, the chains do not fill the depletion layer even at high concentrations, but rather build up the concentration in the center region of the slit. Favorable interchain and intrachain interactions allow for this build-up. Alternatively, a lower osmotic pressure in the theta solution in comparison to athermal solution, due to missing second virial coefficient in the theta solvent, is unable to drive the polymer coils into the depletion layer.

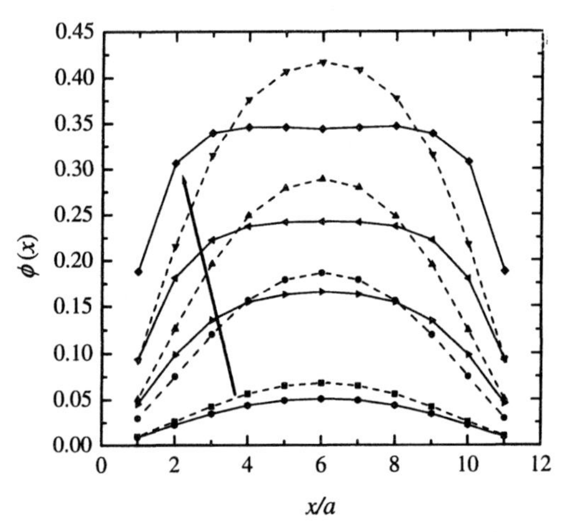

Figure 1 Depletion concentration profiles for good solvent (solid lines) and theta solvent (dashed lines). (Reproduced with permission from reference 9. Copyright 2000 Am. Inst. of Physics)

Broader profiles produced by much longer chains in wider slit allow for analysis of the law governing these depletion profiles. Instead of individual profiles as presented above we want to stress universality in these profiles. A reduced plot in terms of the reduced concentration $\phi(x)/\phi_{bulk}$ as a function of reduced distance x/ξ was constructed (*21*) from simulations of semidilute solutions of athermal chains conducted for N=2000 (R_g=37.7) in the slit of D/a =30, Figure 2. The average density ϕ_{AV} within the slit, reduced by ϕ^*=0.013 was between 4.7 and 8.6 for N=2000. These concentrations are sufficiently high to allow to estimate the concentration correlation length in the bulk solution by $\xi = R_{g0}(\phi_{bulk}/\phi^*)^{-\nu/(3\nu-1)}$, where $\nu/(3\nu-1)$=0.766. There was a well-defined concentration plateau ϕ_{bulk} in the density profile in the center section of the slit in each of the simulation results; ξ/a for these concentrations is expected to be less than 9.25 for N = 2000. It means that the effective confinement was weak at these concentrations. At lower concentrations, a plateau in the density profile could not be seen. All the data compiled from different concentrations fall on a single master curve with the correct scaling exponent 1.69 as predicted by scaling theory, $\phi(x)/\phi_{bulk} = (x/\xi)^{1/\nu} = (x/\xi)^{1.69}$. (For dilute solution this profile scales as ~ $(x/R_g)^{1.69}$.) However, a small shift in distance x to the wall by 0.36a had to be introduced. This shift was ascribed to a lattice approximation used in

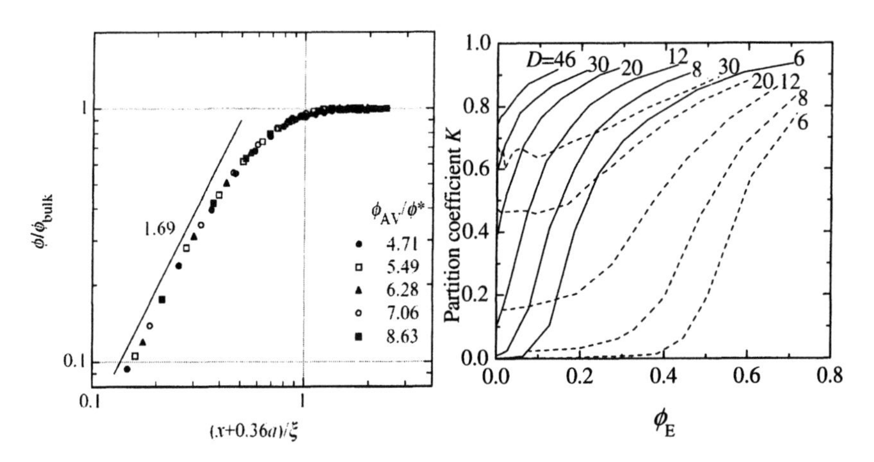

Figure 2 Reduced depletion profile for long chain and five concentrations (Reproduced from reference 21. Copyright 2004 ACS)

Figure 3 Weak-to-strong penetration transition for good solvent (solid curves) and theta solvent (dashed curves) and series of slit widths D.

the simulation (*21*) but a similar correction was introduced also in analogous off-lattice simulation study (*19*).

Increasing the concentration of the polymer in the bulk solution increases the partition coefficient K above its value in the dilute limit, Figure 3, and is consistent with the filling of depletion layer by increased concentration, Figure 1. The existence of the weak-to-strong penetration transition was confirmed by simulations (*4,9*) in slits with repulsive walls and semidilute athermal and theta solutions. The penetration into pores in semidilute solution is governed by the same scaling law as in dilute athermal solutions, only the nature of the entity entering the pore changes from the coil size to a size of concentration blob.

The change of solvent quality brings about a qualitative alteration of this behavior in the form of a delayed weak-to-strong penetration transition found on increasing concentration in the theta solvent, Figure 3. Repulsion between chains becomes here effective in pushing the chains into pores at higher concentrations than in good solvents (*9*) due to missing second virial coefficient. These concentration effects were also studied in solutions of polymer mixtures with the aim to develop an effective technique for separation of macromolecules at higher concentrations and variable solvent quality (*22,23*).

Adsorption profiles. Concentration profiles with polymer adsorption exhibit an enhancement of polymer concentration at the wall in contrast to the depletion profiles. These profiles, however, depend on the adsorption strength of walls towards the polymer chains. First, we concentrate on a strong adsorption of polymer at confining walls. Generally the adsorption profile of macromolecules is, according to de Gennes, divided into several regimes depending on the distance from the surface. From the three typical regimes (proximal, central and distal) for the case of strong adsorption and for semi-dilute concentration the central regime becomes important (*24*). This is because the proximal range reduces to the monomer size in the limit of strong adsorption and the distal range is effective only in dilute solutions at low adsorption. The universal concentration profile in the central regime is given by the following scaling law, $\phi(x)/\phi_{bulk} = (x/\xi)^{-(d-1/\nu)} = (x/\xi)^{-1.299}$, and for the case of strong adsorption and semidilute solution covers the whole range of profile at the wall. Here $d = 3$ and $\nu = 0.588$ was used. This scaling was recently verified by simulation (*26*) for the chain length N=1000 (ϕ^*=0.0208) and five concentrations in the range of ϕ_{bulk}/ϕ^* =1.15-8.41, for which the correlation length ξ/a ranges between 22.92 and 4.99 respectively (compare to R_{g0}/a=25.52). The chains were athermal representing the good solvent conditions and the adsorption was set to $\varepsilon = -1$. It is noteworthy that the profile at the wall reaches the correct theoretical slope and the curves for different concentrations are superimposed, *i.e.* form a master curve, Figure 4.

Increasing concentration interestingly washes out large differences in adsorption profiles seen between the weak and strong wall attraction (*25*) in dilute solution. This is seen in partitioning of polymer into attractive pores in

Figure 5 for chains of N=100 in good solvent. Differences in partitioning seen in the dilute solution for different adsorption strength ε decrease with concentration and the osmotic pressure induced drive of macromolecules into the pore starts to dominate. The theoretical curves were obtained using Flory (dashed curves) or Huggins (solid curves) relations for the concentration dependence of chemical potential of polymer in partitioning equilibrium between the free and confined solution, $\mu_E(\phi_E) = \mu_I(\phi_I) - kT\ln K_o$, where K_o is the partition coefficient in dilute solution.

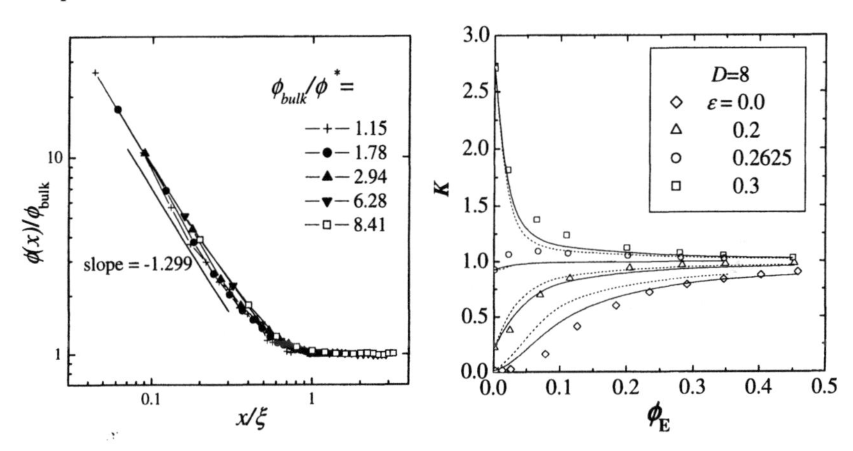

Figure 4 Reduced adsorption profiles for long chains and five concentrations. (Reproduced with permission from reference 26. Copyright 2002 MTS)

Figure 5 Variation of partitioning with concentration for attractive pores in good solvent. (Reproduced from reference 25. Copyright 2002 ACS)

Compensation of size exclusion and adsorption. When adsorption strength ε of the slit walls varies in the range between weak adsorption and strong adsorption, ε = -1, multitude of concentration profiles appears in which for a certain adsorption strength the characteristic depletion profile turns into the adsorption profile with the opposite curvature. From these many possible concentration profiles we single out three typical situations shown in Figure 6. For a long chain, N=1000, in good solvent at infinite dilution in the slit of width D/a=50 we show a depletion profile (still present for a weak adsorption, ε = -0.2), an adsorption profile for a stronger adsorption ε = -0.3 and a typical compensation profile for ε_c = -0.2793 with a flat horizontal concentration profile in the wide intra-pore zone except the close proximity to the walls here. The segment concentration is almost independent of the distance from the walls. For perfect compensation the ideally flat concentration profile implies the overall ideal behavior of coils which do not sense the walls as an obstacle. As a result there is a free penetration ($K = 1$) of coils into pores at the compensation point. This was shown in recent simulation study (*14*). However, a weak dependence of compensation point on the chain length of polymer was observed which represents a complication in LC separations. Recently a compensation point was estimated not from the compensation condition (K=1) but as a true transition point from the intersection of the partitioning curves (K *vs* ε) for different chain lengths (*27*). The value ε_c obtained by this procedure was slightly shifted with respect to the previous method but the weak chain length dependence in partitioning at this critical condition still remained.

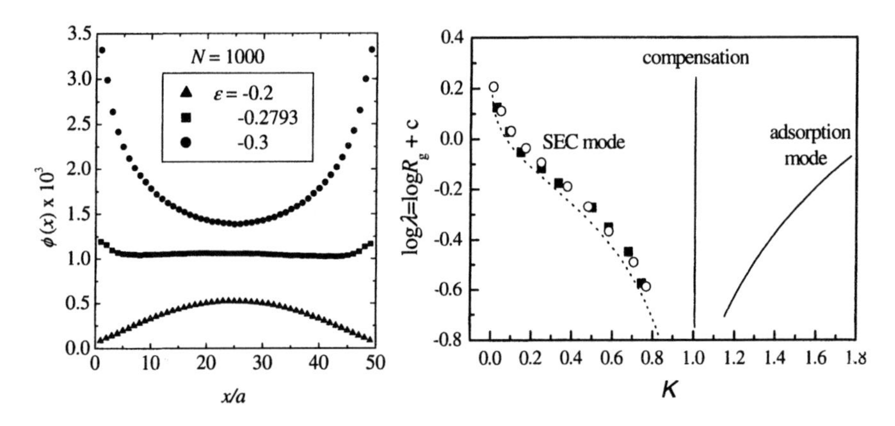

Figure 6 Three typical concentration profiles for indicated weak adsorption close to compensation point. (Reproduced from reference 20. Copyright 2004 ACS)

Figure 7 Separation modes in liquid chromatography of polymers

In Figure 7 we illustrate three separation modes used in LC, related to three concentration profiles shown in Figure 6, plotted in coordinates employed in this technique. Partitioning coefficient K in x-axis is proportional to an elution volume and y-axis to logarithm of molar mass. The standard SEC mode of separation is illustrated by the simulation results on chains (N =100) in good and theta solvent and by respective classical theoretical curve due to Casassa (*7*) for ideal chains. The adsorption and the compensation mode in LC characterization are shown schematically. The compensation point marking the transition between SEC mode and adsorption mode is sensitive to adjustment conditions as observed also in complex LC studies. This effect is intensively studied (*27, 15-16*) and is becoming a useful tool for the analysis of complex systems.

Depletion interaction in colloidal systems

In addition to already mentioned characteristics of confinement, namely the partition coefficient K and the free energy of confinement ΔA, the derivative of the free energy $d(\Delta A/kT)/dD = -f/kT$, termed the confinement force, is determined from the simulations of equilibrium of athermal chains in a repulsive slit (*6,19,28,29*). This force f contributes to the pressure p_I exerted by the confined molecules on the walls inside slit (*30, 31*). The slit walls also experience a pressure p_E from the surrounding bulk solution. The magnitude of the pressures p_I and p_E affects the stability of colloidal dispersions. The polymer molecules are excluded from the region near to particle surface (the depletion zone). As the particles approach, the depletion zone minimizes and the coils are pushed out of the gap between the surfaces. The negative value of net pressure $\Delta p = p_I - p_E$ between depletion zones and bulk solution brings about the depletion effect i.e. an effective attraction between colloidal particles. This situation is illustrated in Figure 8.

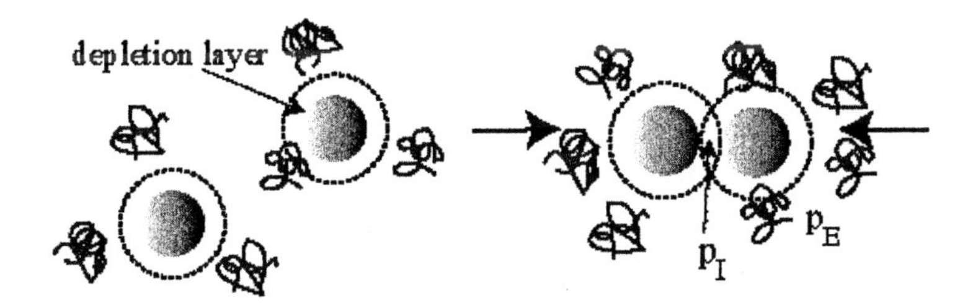

Figure 8 Depletion interaction between colloid particles at low polymer concentration

The data on the free energy of confinement $\Delta A/kT$ in the dilute solution limit are presented in Figures 9. The term ΔA represents the free energy of transfer of a macromolecule from a bulk solution of concentration ϕ_E to a solution of concentration ϕ_I in a slit. The variation of the free energy of confinement $\Delta A/kT$ with the slit width D normalized by the coil diameter $2R_g$ in the dilute solution limit $\phi \rightarrow 0$ is shown for chain lengths $N=100$. The curve for $\varepsilon=0$ corresponds to a purely steric exclusion in a repulsive slit. The confinement penalty $\Delta A/kT$ diminishes by an increase in the attraction strength ε. The free energy $\Delta A/kT$ is about zero at $\varepsilon=-0.2625$, close to the critical point of adsorption ε_c for this chain length, where the steric exclusion and adsorption effects in the partitioning fully compensate. Perturbation of chains by a presence of the slit walls seems thus fully screened at critical conditions. At the attraction strength higher than the critical one, a molecule in a slit becomes stabilized relative to the bulk phase ($\Delta A/kT<0$) and the stabilization increases with the degree of confinement.

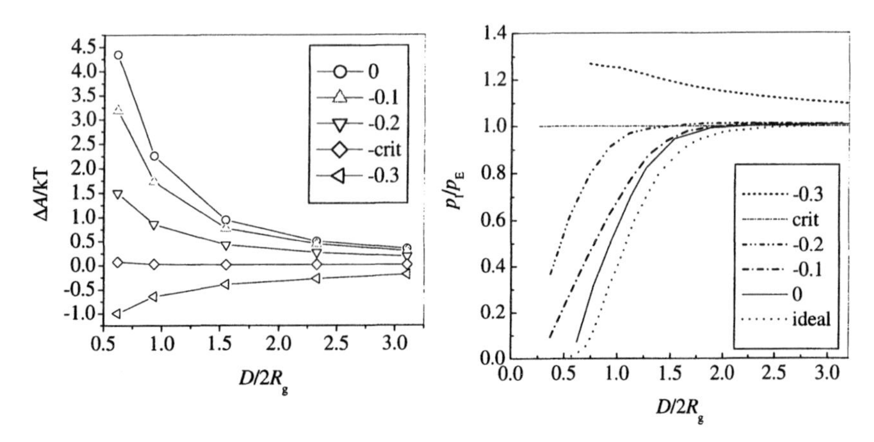

Figure 9 Free energy of confinement as a function of the reduced slit width for different wall attraction ε

Figure 10 Pressure ratio as a function of slit width for different adsorption strength ε. The dot line represents the ideal chain behavior (7). (Reproduced from reference 20. Copyright 2004 ACS)

The plot of free energy for purely steric exclusion of athermal chains in Figure 9 resembles the corresponding function for ideal chains (*7*). The Casassa free energy function of ideal freely-jointed chains confined in a repulsive slit predicts the proportionalities $\Delta A \sim D^{-1}$ in wide slits and $\Delta A \sim D^{-2}$ in narrow slits. The scaling theory (*8*) that applies only in the limit of narrow repulsive slits suggests the relation $\Delta A \sim D^{-5/3}$ for athermal chains.

The confinement force calculated from ΔA resembles closely the plot of free energy. The force term f/kT is positive at the weak attraction potential ε and decreases monotonically with increasing separation D. The repulsive nature of the force diminishes by increasing $-\varepsilon$ and the force vanishes at the critical point of adsorption. In supercritical region above ε_c the force is negative due to adsorption of macromolecules to a single wall or to both walls (particle bridging). In dilute polymer solutions where interchain interactions are negligible the pressure p_I is related to the confinement force f by following relation (*30,31*) $p_I = (kTn_I/V_I)(1 + Df/kT) = \pi^{id}_I(1 + Df/kT)$ in which n_I/V_I is the number density of a polymer chain within a slit and π^{id} is the osmotic pressure of ideal solution. In repulsive narrow slits $Df/kT >> 1$ and the force term is the leading term. In wide slits or in weakly attractive slits the contribution of the term Df/kT to the pressure p_I becomes insubstantial. The external pressure p_E on slit walls in dilute solutions is determined by the osmotic pressure term π^{id}_E.

The above relations lead to the ratio of intra-pore and bulk pressure at full equilibrium in dilute solutions (*32*) as $p_I/p_E = (\pi^{id}_I/\pi^{id}_E)[1+Df/kT]= K[1+Df/kT]= K+D(dK/dD)$. This equation was employed (*32*) to compute the ratio p_I/p_E as a function of the plate separation D under full equilibrium conditions by an analytical theory of solutions of ideal polymers in a repulsive slit. The distinct depletion effect was observed at small separations D. The effect was amplified with an increase in the chain length. Data presented in Figure 9 allow calculating depletion interaction in cases involving non-ideal (excluded volume) chains and weakly attractive slits. The resulting relative pressure p_I/p_E for chains of N=100 is plotted in Figure 10 as a function of $D/2R_g$ for various attraction strength ε. It is seen that in a repulsive slit (ε=0) the intra-slit pressure p_I drops significantly below the bulk pressure p_E at the wall separations less than about $3R_g$, due to expulsion of molecules from the slit. However, this manifestation of the depletion effect is slightly weaker in excluded-volume chains than in ideal chains. In contrast to force-distance profiles between particles covered by grafted or strongly adsorbed polymers, depletion attraction in systems of non-adsorbing polymers is much smaller and notoriously difficult to measure. Nonetheless, depletion attraction of the type shown in Figure 8 was observed in recent experiments by the AFM technique (*33*) and the surface force apparatus (*34*).

An increase in the potential ε substantially reduces the depletion effect, Figure 10; the depletion onset moves to the left below the critical point ε_c. By an increase in ε the mean intra-slit concentration ϕ_I approach the bulk concentration ϕ_E and the ratio of osmotic pressures (π^{id}_I/π^{id}_E) converges to unity. As a consequence, the force term Df/kT has to diminish and finally vanish on increase in ε in the subcritical region. At critical conditions p_I/p_E=1 in Figure 10; the net pressure Δp=0 and the depletion interaction becomes completely ineffective. Above ε_c, in the supercritical region, the adsorption layer is formed near the walls and the mean concentration of a polymer in a slit ϕ_I exceeds the bulk concentration ϕ_E.

The variable p_I/p_E in Figure 10 can be changed to the relative pressure difference $\Delta p/p_E$. An integration of the net pressure Δp gives the interaction potential per unit area between the plates *W(D)* (depletion potential) (*17,18*). The potential *W(D)* between two flat walls is related by Derjaguin approximation to the force between two spherical colloidal particles of radius R_c. A knowledge of the effective depletion potential *W(D)* is a prerequisite of a good understanding of the phase behavior of colloid-polymer mixtures (*18*). Here we note that at critical condition the depletion potential *W(D)* vanishes for all plate separations, including the region $D/2R_g<2$ since net pressure Δp=0 here. Experimental measurements of the depletion force of neutral polymers between surfaces or of the depletion-induced phase separation of colloid-polymer mixtures are commonly rationalized (*18*) by using the theoretical models appropriate for ideal chains and repulsive slits. In analogy, the present simulations offer alternative depletion characteristics in situations where polymer-particle attraction potential (or temperature) is used in experiments to regulate the depletion force and/or the phase behavior.

Depletion effect as described above operates in the dilute solution. It should be noticed, however, that depletion effect builds up with concentration until the coil overlap concentration and then starts to diminish. This can be seen from the confinement force which scales as $f/kT \sim \phi^{-5/4} D^{-8/3}$ for the semidilute concentration (*29*).

Conclusions

Confinement of macromolecular systems is traditionally viewed only through the entropy loss. Including interactions of polymer segments with confining walls and with solvent combined with the effect of concentration brings variability with characteristic complex behavior.

Depletion of macromolecules in concentration profiles near the walls diminishes with increasing bulk concentration and varies with solvent quality. Wall attraction in porous medium affects strongly the confinement. At critical attraction the steric exclusion by confinement is compensated by adsorption and this has consequences for instance for complex liquid chromatography separations or for the regulation of stability and phase behavior of polymer-colloid dispersions.

Increased polymer concentration in bulk solution brings a penetration of polymer even into the strong confinement. This penetration transition shifts to higher concentrations for the theta solvents relative to the good solvent and both the shift from the good to the theta solvent and the increased concentration was shown to be interesting for novel complex LC separations.

References

1. Fleer, G.J.; Cohen Stuart, M.A.; Scheutjens, J.M.H.M.; Cosgrove, T.; Vincent, B. *Polymers at Interfaces,* Chapman & Hall: London, 1993;
2. Frantz, P.J.; Granick, S. *Phys. Rev. Lett.* **1991**, *66*, 899-902.
3. Bleha, T.; Cifra, P.; Karasz, F.E. *Polymer* **1990**, *31*, 1321-1327
4. Wang, Y.; Teraoka, I. *Macromolecules* **1997**, *30*, 8473-8477.
5. Cifra, P.; Bleha, T. *Macromolecules* **2001**, *34*, 605-613.
6. Chen, Z.; Escobedo, F.A. *Macromolecules* **2001**, *34*, 8802-8810.
7. Casassa, E. F. *J. Polym. Sci. Polym. Lett. Ed.* **1967**, *5*, 773-&
8. Teraoka, I. *Progr. Polym. Sci.* **1996**, *21*, 89-149

9. Cifra, P.; Bleha, T.; Wang, Y.; Teraoka, I. *J. Chem. Phys.* **2000**, *113*, 8313-8318
10. Gorbunov, A.A.; Skvortsov, A. M. *Adv. Colloid Interface Sci.* **1995**, *62*, 31-108
11. Guttman, C.M.; DiMarzio, E.A.; Douglas, J.F. *Macromolecules* **1996**, *29*, 5723-5733
12. Kosmas, M.K.; Bokaris, E.P.; Georgaka, E.G. *Polymer* **1998**, *39*, 4973-4976
13. Davidson, M. G; Suter, U. W.; Deen W. M. *Macromolecules* **1987**, *20*, 1141-1146
14. Cifra, P.; Bleha, T. *Polymer* **2000**, *41*, 1003-1009
15. Berek, D.; *Progr. Polym. Sci.* **2000**, *25*, 873-908
16. Pasch, H.; Trathnigg, B. *HPLC of Polymers*, Springer-Verlag: Berlin, 1998
17. Asakura, S.; Oosawa, F. *J. Chem. Phys.* **1954**, *22*, 1255-&.
18. Tuinier, R.; Rieger, J.; de Kruif, C.G. *Adv. Colloid Interface Sci.* **2003**, *103*, 1-31.
19. Milchev, A.; Binder, K. *Eur. Phys. J. B* **1998**, *3*, 477- 484.
20. Bleha, T.; Cifra, P. *Langmuir*, **2004**, *20*, 764-770
21. Teraoka, I.; Cifra, P. Yongmei Wang *Macromolecules* **2001**, *34*, 7121-7126
22. Wang, Y.; Teraoka, I.; Cifra, P. *Macromolecules* **2001**, *34*, 127-133
23. Cifra, P.; Wang, Y.; Teraoka, I. *Macromolecules* **2002**, *35*, 1446-1450
24. Daoud M. C. R. *Acad. Sci. Paris, Series IV* **2000**, *1*, 1125-1133
25. Škrinárová, Z; Bleha, T.; Cifra P. *Macromolecules* **2002**, *35*, 8896-8905
26. Cifra P. *Macromol. Theory Simul.* **2003**, *12*, 270-275
27. Yingchuan Gong; Yongmei Wang *Macromolecules* **2002**, *35*, 7492-7498
28. De Joannis, J; Jimenez, J.; Rajagopalan, R.; Bitsanis, I. *Europhys. Lett.* **2000**, *51*, 41-47.
29. Bleha, T.; Cifra, P. *Polymer* **2003**, *44*, 3745-3752
30. Vrij, A. *Pure Appl. Chem.* **1976**, *48*, 471-&.
31. Clark, A.T.; Lal, M. *J. Chem. Soc. Faraday Trans. 2* **1981**, *77*, 981-996.
32. Ennis, J.; Jönsson, B. *J. Phys. Chem. B* **1999**, *103*, 2248-2255.
33. Piech, M.; Walz, J.Y. *J. Colloid Interface Sci.* **2002**, *253*, 117- 129.
34. Ruths, M.; Yoshizawa, H.; Fetters, L.J.; Israelachvili, J.N. *Macromolecules* **1996**, *29*, 7193-7203.

Chapter 19

Chain Conformational Statistics and Mechanical Properties of Elastomer Blends

Milenko Plavsic[1], Ivana Pajic-Lijakovic[1], and Paula Putanov[2]

[1]Faculty of Technology and Metallurgy, Belgrade, Serbia and Montenegro
[2]Serbian Academy of Sciences and Arts, Belgrade, Serbia and Montenegro

Design of mechanical properties, especially the relation between tensile and dynamic moduli in terms of chain conformational statistics is considered. Data on new elastomer blends with gradient properties, able for mixing on molecular level and partial networking with separation of phases, are analyzed in parallel to classical commodity materials. Especial attention is paid to synergetic effects of the increase of blend modulus, relative to moduli obtained by linear "rule of mixtures" for component polymer moduli, described by Kleiner-Karasz-MacKnight equation, and new models based on self-similar scaling of elastomer network dynamics. New data on model blends exposing synergetic effects with the change of conformational statistics are presented as well.

Polymer blends achieved among commodity materials in last forty years importance, which alloys realized for many centuries of metal production experience. One of most attractive opportunities that blends afford is obtaining desirable properties that may not be available in the component homopolymers, in quantitative or qualitative sense. For example, an increase of elastic modulus of some plastics as well as elastomer blends, over the modulus of each component, formulated quantitatively by Kleiner- Karasz- Mac Knight equation, is often of high practical importance (*1-3*):

$$E_b = \varphi_1 E_1 + \varphi_2 E_2 + \varphi_1 \varphi_2 \beta_{12} \tag{1}$$

where E_1 , E_2 , φ_1 and φ_2 represent the moduli and composition of components respectively and β_{12} represents relation between 50/50 blend and components moduli. Fundamentally, it indicates some synergetic interactions between component polymers. It is derived considering tensile properties of a series of blends of truly miscible components: polystyrene (PS) and poly(2,6-dimethyl-1,4-phenylene oxide) (PPO), which provide the basis also for the family of engineering thermoplastics called Noryl®. Similar increase of tensile moduli exhibits a number of blends of noncompatible elastomers , without and with filler, e.g. chlorobutyl rubber (CIIR) with natural rubber (NR), silicone rubber (MQ) with NR, styrene-butadiene rubber (SBR) and styrene-acrylonitryle (NBR) (*3-6*). At the first glance, such synergetic improvement of mechanical properties seems to be effect of quite a general type for polymer blends. But, more detailed analyses face at once a number of difficulties. The change of plateau moduli G_N^0 with blend composition of PS/PPO estimated from dynamic measurements of Porter and Lomellini (*7,8*), exhibit just the opposite effect: lower values of blend moduli then the simple linear "rule of mixtures", described by equation: $G_{Nb}^0 = \varphi_1 G_{N1}^0 + \varphi_2 G_{N2}^{0,}$ predicts (see also Figure 1). Kleiner at al. correlated in elegant way eq 1 with a simple structural parametar as blend density gradient (*1*). A couple of years later Gressley and Edwards correlated G_N^0 in general, with number of chains per unit volume and chain length, considering entanglement effects (*9*). Recently, Edwards and Terentjev published rigorous analyses of real entangled networks at low-frequency regimes, but as Terentjev pointed, there is still much to be understood about mechanism of dynamic response of elastomer networks (*10,11*). On the other hand, a number of mechanical properties of elastomer blends can be correlated easily with conformational features of component molecules for behavior predicting and design of high performance compounds. Moreover, these relations, in some cases, could be further correlated to self-similar scaling of network properties.

In this contribution we try both, to review "the state of art" in fully cured network moduli scaling, and to correlate moduli with chain conformational statistics. We look in particular for explanation of and possibilities for synergetic effect in some new elastomer blends, and design of materials in general.

Theoretical Background

The basic knowledge providing research background for the issues of interest here, are polymer networking theories and rubber elasticity theories, connected with chain conformational statistics. The one, who gave fundamental contribution to the all three fields, with his theories, ideas and intuition was P. J. Flory. The first theory of polymer networking is formulated by Flory in 1941 (*12*). It provides formula for probability p_c of a network arising in polymer systems: $p_c = [1 / (f\text{-}1)] \sim (1 / N)$. It starts from reaction probabilities of functional groups of molecules building the network, which are simply proportional to their fractions in the reaction mixture, and takes into account neither the space architecture nor the steric hindrances in the course of network developing. Still, it has been very successful in prediction behavior of many systems, in particular vulcanization of rubbers. (In previous equation f is functionality of growing structures and N degree of polymerization). That encouraged development of new theories, to include network architecture, as e.g. cascade theory of Gordon and others (*13*). It connects networking kinetics with space organization, describing process by cascade of network branches, extending all over the system, what immediately makes connection to graph theory in mathematics. But, we know from vulcanization chemistry, that grow of network starts from accelerator-activator complex nuclei, spread all over the system. It extends in space and time, till percolation threshold i.e. an especial type of bond connectivity. It can be described as appearance of indefinitely polydisperse ensemble of randomly branched polymers at the gel point (GP). Further reaction progress makes the network denser. Such clustered structure can be described in elegant way, in terms of fractal geometry developed by Mandelbrot, by fractal dimension d_f (*14*). Both analogues, graphs and fractals, provide a background for generalization, bat have hidden danger of misinterpretation of network connectivity. For example, conformation of a polymer chain can be also described by fractal dimension. But, it should be considered as an effect resulting from topology of bond connections, short and long range interactions of the chain and environment influence as well, and not a geometrical feature. It is in fact multi -body interaction effect, what is very tough problem in physics. Flory was the first who succeeded to avoid some difficulties for polymer systems and offer solution with his excluded volume theory (*12*), what can be expressed also by power low:

$$N = AR_G^D \tag{2}$$

where R_G is characteristic length (radius of gyration) and N degree of polymerization of the chain. Exponent describing non-asymptotic scaling of a polymer chain with increase of the length due to excluded volume effect, is D=5/3. If the effect is screened, D=2 ("Flory's theorem") In terms of the theory of scaling of physical properties with structure relations, developed by P. J. de Geenes (*15*), exponent D is a global feature of general validity for polymer molecules (or generally speaking, for some class of systems) describing some essential features of chain structure, while prefactor A represents local properties characteristic for each polymer type. But, as we know, scaling exponents determined experimentally for some new scaling relations, vary significantly. It indicates interfering of some other phenomena that should be extracted from scaling relation, or a need of redefinition of class of systems, it is valid for.

Considering cluster connectivity, (what is of especial interest for our problem), should be pointed to spectral dimension d_s, defined first by Alexander and Orbach (*16*) It could be extracted from scaling relations for clusters of different topology e.g. different types of networks. By definition, scaling of N_t (the number of distinct sites on the lattice visited by random walk) with t, (number of steps, or time) is described by $d_s/2$ The average radius of walk can be easily connected to radius of gyration R_G in eq 2.

Returning to the network connectivity described by cascade theory, it should be pointed that, according to percolation theory developed by Stauffer and de Gennes,(*17*) scaling of cluster distribution in the vicinity of GP is false. Reason is simply, the false representation of connectivity of the cascade, not taking into account steric hindrances.

Modulus and Connectivity Levels

De Gennes is the first who recognized connectivity relationship in modulus scaling ("electrical analogy" (*15*)). On the other hand, connectivity is extensively discussed issue in rubber elasticity theories. In the first theory, based on contributions of Kuhn, Wall and Flory (*3,12*), proposed Flory incorporation in kT/Mc prefactor correction $(1 - 2Mc / M)$ (M is initial molecular weight), because dangling chains do not transfer stress. Later, extensive literature has been accumulated on semi- phenomenological model of entanglement connectivity, including theories of entanglement effects in blends as well (*3*). But, opposite to entanglement view, Flory in the mean field approach of constraint on junctions theory, incorporated cycle rank of the cluster as connectivity measure: $\Psi = n_p + 1 - n_c$. Here n_c and n_p are numbers of vertices and chains in a spanning tree. For a perfect phantom network (having no dangling chains and large n_c), $\Psi = n_p - n_c$ and $n_p = (f/2)N_c$,(where f is crosslink functionality). Thus, the prefactor in James and Gouth type modulus is just the cycle rang: $n_p(1-2/f)\ kT/2$, while for affine network; $E = n_p kT / 2$. Using "electrical analogy" can be shown (*3*) that the mean - to -end network segment separation $<R^2>$, affine deformed, but with the contribution of a fluctuation term independent of deformation, gives the same moduli relation. Dependence on $2 / f$ for phantom network turns up in the application of path integral method, as well. So, essential difference of real network (being with modulus between phantom and affine model) is additional degree of constrains on fluctuations, in the mean field approach. In can be understood simply, as decrease of connectivity with extension in dense system of high sequential mobility, (as can be seen from theories of Flory, Erman and Mark (*18*)).

Viscoelasticity aspect is introduced to percolation theory postulating distribution of relaxation times due to existence of fractal domains, with θ_0 and θ_z as lower and upper time limit, respectively. Indeed, first Winter and Chanbon and a number of authors later (*19-22*), showed that at the gel point, the distribution of relaxation times obeys a power law:

$$H(\theta)\ d\ln\theta \propto \theta^{-\Delta} d\ln\theta \tag{3}$$

The shear relaxation can be obtain by :

$$G(t) = \int_0^{\infty} [H(\theta)/\theta]\ e^{-t/\theta} d\theta \propto t^{-\Delta} \qquad (\theta_0 < t < \theta_z) \tag{4}$$

and by applying the Fourier transform we obtain

$$G(\omega) \propto \omega^{\Delta} \quad (1/\theta_z < \omega < 1/\theta_0) \tag{5}$$

It appears that two limits exist: $\omega = 1/\theta_z$ and $\omega_0 = 1/\theta_0$ and between these values G' and G'' vary according to a power law :$G'(\omega) \sim G''(\omega) \sim \omega^{\Delta}$ where $\omega^* < \omega < \omega_0$. The measurement frequencies are usually very low compared to ω_0 and only ω^* is taken into account. Than, three characteristic situations can be distinguished: (a) before GP the medium can be considered as a Newtonian liquid for which $G'(\omega) \sim \omega^2$ and $G''(\omega) \sim \omega$, (b) after GP the medium is equivalent to an elastic solid for which $G'(\omega) \sim$ const. and $G''(\omega) \sim \omega$, (c) at GP ω^* tends toward zero. Based on that, Winter and his colleagues at Amherst developed an elegant method for GP determination by measuring dynamic moduli of reacting system. It has been successfully tested by numerous authors on different systems (*20*). But, values for global exponent Δ varied in the whole range from 0 to 1, depending on molecular composition and crosslinking conditions. Mathukumar pointed (*23*) on screening effect on Δ and suggested that exponent Δ for fully screened excluded volume effect becomes: $\Delta = d(d + 2 - 2d_f) / 2(d + 2 - d_f)$. But, another problem arises, considering vulcanizates cured up to optimal mechanical properties, as "fully networked" systems. Statement, that fully cured networks have infinite relaxation time, contradicts ergodic theorem application. Also, from the statements of percolation theory follows, that moduli in the vicinity of GP scale by $G \sim | p\text{-}p_c |^z$. It is hard to imagine, that fractals clusters embedded to network of increasing density, immediately loose their fractality, although one accepts that network impose some strain to them In the case of ideal gel with "phantom network", the elastic modulus is inversely proportional to chain degree of polymerization i.e. $G \sim kT/N$. The energy kT is energy stored per correlation volume. Than if, $\xi \sim | p\text{-}p_c |^{-\nu}$ is the correlation length of fractal cluster, each of effective chains of phantom network included in a correlation volume ξ^d constitutes a "cluster cell" of fractal dimension d_f. Thus N is proportional to ξ^d and elastic modulus of elastic cell will scale as $G_\infty \sim (kT/\ \xi^d)$. Alexander and Orbach (*16*) concluded from various simulations, that the spectral dimension is close to 4/3 in the case of percolation with a space dimension d between 1 and 6. Using that value for spectral dimension, and previous formula Martin at al. calculated value Δ =0.6 for a network pretty far from GP. Indeed, Adolf and Martin obtained experimentally for fully cured poli(dimetilsiloxane) networks Δ=0.55 (*24*). It raises a number of issues, we have to consider in next sections more in detail. But, before considering some of them, let us turn back for a moment to the beginning of theoretical section, with a couple of comments on the well known and intriguing question: how the first and most simple networking theory has been so successful, compared with some vulcanization experiments, in spite of all difficulties quoted above? Let as describe relative fluctuations of a percolated gel structure by

$$\left[\frac{\Delta n_{gel}}{n_{gel}}\right]^2 \sim \frac{p - p_c}{\xi^d} \tag{6}$$

The correlation length for vulcanization is certainly proportional to the typical scaling length of conformational chain statistics i.e. the Kuhn end-to-end distance of a "phantom chain": $N\, l^{1/2}$. Normalizing relation for correlation length relative to p_c and using Flory's expression for p_c, ,one obtains:

$$\left[\frac{\Delta n_{gel}}{n_{gel}}\right]^2 \sim N^{-2}\left[\frac{1}{p - p_c}\right]^{3-d/2} \tag{7}$$

For dense systems as rubber melts scaling dimension d incorporates upper limit of spectral dimension d_s due to high topological connectivity, what is 6, as we have seen above. It follows that second part of the right side of eq 7 vanish. So, it follows, that fluctuations in vulcanizate network, that Flory neglected in the mean field approach, really are low, and the approximation provides good results. But, for co-vulcanizing blends, can be shown that fluctuations are significant, and can not be neglected (*25*). It follows that issue of scaling of moduli for fully cured networks considered above, will be even more pronounced for some blends. We will come back to that, later.

Experimental Data on Dynamic Moduli Scaling

There are very few data in the literature on scaling of moduli with deformation frequency for fully cured elastomer blend materials, as well as for "incipient" blend networks, close to GP, especially not for scaling of moduli with (independently determined) extent of vulcanization reaction. It is a bit surprising because dynamical measurements are the main characterization method in rubber laboratories today. In addition should be noted attractive idea of Dusek and his coworkers on "total design" of network materials (from basic chemical reactions up to moduli prediction) with significant success developed by Prague group for some networks of other type (*26*). But, vulcanization kinetics is rather complicated and the extent of reaction analyses is usually based on indirect data from rheology and swelling thermodynamics, instead vice verse. But, data on scaling of “fully cured” one-polymer elastomers, are here also of interest for comparison of moduli scaling sensitivity to network structure We try to review in short all that kinds of data in that and next section, but to present results of our calculations and correlations based on some of them, together with some our new results in section: Results and Discussion

Nelson at al. (*27*) investigated blends of polybutadiene (BR) and ethylene-propylene-ethylidienenorbornene rubber (EPDM), in 25/75, 50/50 and 75/25 weight ratios, cured by electron beam to different extent of networking. Both, dynamic and permanent tensile moduli are determined, and correlated with morphology, but only temperature sweep of 50/50 blend is presented explicitly. We found, from available data in the paper, excess of blend moduli relative to "rule of mixtures" for both, permanent-tensile and dynamic measurements. Unfortunately, appropriate data for check of frequency scaling are not available, in addition to reserves due to the method of vulcanization applied.

Winter and collaborators (*20*) presented detailed analyses of poly(ε-caprolactone)diol (PCL) network blended with high molecular weight poly(styrene-co-acrylonitrile) (SAN) as a inert component. PCL network exposed elastomer properties and SAN was significantly above entanglement molecular weight. The other phase in PCL/SAN blends is glassy and uncured so, results can be used here for comparison of completely (maximally) cured PCL and “fully cured” blend to see influence of another phase on scaling. The authors did not consider that aspect in the paper, mainly oriented to GP point determination. We found from G'' sweep of cured pure PCL the slope Δ=1. But, in the presence of another phase (roughly 80/20 ratio) is approximately Δ=0.4. The results are not in accordance with Martin's theory numerical predictions, but are in accordance with trends of scaling change with blending, predicted in previous section..

Martin and Adolf (*24*) prepared PDMS gels with different stoichiometric ratio of silane to vinyl group *r*. For *r*= 0.45 (just enough crosslinker to reach the GP) was Δ=0.65±0.065. But, with *r*=0.52 at GP was Δ=0.56±0.04 and with *r*=0.80 at GP was Δ=0.51±0.04 and for the last two fully cured networks they obtained $\Delta \cong 0.55$, as we quoted in the previous section.

Recently, Terentjev and collaborators (*11*) investigated under oscillatory shear, at very small strains, pure PDMS, fully cured but in a special way, to produce two networks of very different topology: the first (assigned SCIE) was with lot of steric hindrances, randomly linked and the other (UIE) with fixed-length strands, uniformly cross-linked at the ends. In the rubber region, storage moduli G' of both materials were found independent of frequency and increasing linearly with temperature, but with a substantial negative offset, in consistent with straightforward enthalpy effects. The loss moduli G'', followed time-temperature superposition, decreasing with increasing temperature and with power-low dependence on frequency in the range 10^{-4} to 10^{4} Hz. Master curve for G' of SCIE can be parted in to three regions with corresponding middle points at 10^{-3}, 10^{0}, 10^{2} Hz and slopes Δ_I=1, Δ_{II}=1/2, Δ_{III}=2/3, respectively. For UIE network there is only one region in the range from 10^{-5} to 10^{-3} with the slope $\Delta = 2/3$.

Rempp at al.(*22*) investigated self similar behavior of polyurethane networks made from poly(ethylene oxide) (PEO) and pluriizocyanates as crosslinkers at GP but under different reaction conditions: (a) for stoichiometric conditions, (b) for an excess isocyanate, (c) for excess of hydroxyl ended POE, (d) for stoichiometric conditions but in addition of zero functional PEO to produce blend. Oscillatory shear experiments have been performed on these reaction media and the variation of G' and G'' versus ω has been monitored through the GP. The following results were obtained (a) under stoichiometric conditions at GP moduli become congruent, regardless of the frequence applied, with Δ = 1/2, (b) in the case of excess isocyanate the plots $\log G'$ and $\log G''$ versus ω follow the same general pattern as observed under stoishiometric conditions. The moduli move closed as the reaction proceeds and they are congruent, with Δ = 1/2 at GP.(c) In the case of excess polymer with OH-functions the behavior is quite different. The stage at which both moduli, obey power laws with the same exponent Δ over entire frequency spectrum is for exponent very close to $\Delta = 2/3$.

Some interesting analogies can be made between these results and previously described behavior of fully cured siloxane networks of SCIE and UIE, starting from bead and spring models of Rose and Zimm All quoted experiments exhibit scaling exponent corresponding to dynamics described by Rose (with $\Delta = 1/2$) and Zimm (with $\Delta = 2/3$). The Zimm model differs from the Rose treatment in its incorporation of intramolecular hydrodynamics interactions. But here change from Rose to Zimm-like behavior in experiments of Rempp appears with change of reaction conditions (excess of PEO) and for SCIE in Terentjev's results, with the shift of frequency range. For siloxane networks, obviously irregular, clustered SCIE network exposes three regions of self-similar behavior. The first region (for $\omega \rightarrow 0$) can be understood as response of the network as a whole, like continuum solid of high inertia. Than, the second region could correspond to response of polydisperse clusters, embedded to network as a loose Rose-like string of beads. And, following that logic, the third region is the response of the networks inside clusters with higher density of interactions and higher connectivity. It results in hydrodynamic intramolecular effects, like Zimm model. But, why than the same, incipient, polydisperse, loose claster at GP in Rempp's experiments switch from Rose-like to Zimm-like behavior? It is only in the case with excess of polymer component in the mixture but not isocyanates. It could be understood in terms of Flory's theorem or screening density of interactions by high molecular solvent. It changes scaling exponent D in eq 2 and clusters become more compact and stiff. In favor of this explanation (and contra to hypothetical reaction reasons) are Rempps experiments described as (d) group: blending with non-reacting polymer of the same kind produce the same effect of shifting to Zimm model. Moreover, blending in the case (d) gives the same effects (of the shift to Zimm-like behavior) with the progress of the reaction to higher degree of networking -as Rempp *at al.* noted. It is in agreement with our hypothesis, in previous section, on more clustered, self-similar fully cured networks structure, due to blending.

Experimental Data on Contribution of Phases

It is widely accepted that blend viscoelastic properties are dominated by two major factors: the nature and density of connections between components and types of mechanical coupling between phases. There is a huge number of papers and several reviews considering contribution of phases to blend (and composite) moduli, by mechanical coupling (*2-4,6,28*). But, models of Takayanagi's type based on phenomenological approach do not predict synergetic effects and provide no insight to molecular structure at all. Similarly is with variational methods, but self-consistent schemes are more promising (*3*). Here, of especial interest are mechanical models in "reverse mode" leading to extraction of actual properties of phases in a blend. They realize it, accounting for morphology composition and viscoelastic characteristics of pure components, and those of blended systems. But here again, as in previous section there is very few papers considering reverse mode approach. Bauer is the first who applied it to elastomer blends in terms of Gaussian moduli $G_g = \sigma / (\lambda - \lambda^{-2})$, where σ is stress and λ is relative strain. Bauer extended Takayanagi's phenomenological approach, incorporating some molecular characteristics, as change of network density of phases, due to blending(*4*). He developed the method to estimate change of the density by extrapolation phase modulus from a series of model compounds of different network density and check it by comparison with results from Takayanagi coupling of phase moduli. Bauer at al. successfully applied the procedure on NR/CIIR model blends (*4*). Our group obtained also good results for NR/MQ, and SBR/MQ blends, in that way (*5,6,29,30*). But, synergetic effects are in Bauer's method implicitly connected with change of network density by interphase diffusion of curatives. Another approach of Marle at al. to reverse mode, is based on phase interlayer contributions to moduli, also with good fit for some plastics blends (*28*). Interlayer approach is not valid to elastomer blends but, also curative redistribution has a number of contra arguments. Looking for possible other synergetic effects, we made two series of NR and SBR gum vulcanizates and NR/SBR blend from masterbatches in 70/30 weight ratio, with same curative system (sulfur 2.5, zinc oxide 2.5, Santocure CBS 1.5 pph). An interesting approach to phase influence synergetic effects can be obtained also from experiments of Kader *at al.*(*31-33*). They blended acrylate - (Nipol® AR51) (ACM) and fluorocarbon rubbers (Viton® A200) (FKM) obtaining both, monophase and polyphase blends at appropriate conditions. Analyses of our experiments (see Figure 2) and our reconsideration of ACM/FKM blends in view of synergetic effects of all that results are presented in the next section.

Results and Discussion

In Figure 1 are presented data from Kleiner at al., Porter at al. and Lamellini papers on PS/PPO blend moduli (*1,7,8*). As can be seen, the increase (not of density but) gradient of material density with blending (which Kleiner at al. in elegant way correlated with increase of tensile moduli), can not be correlated with the change of blend moduli, (at least not) in general. The effect is opposite for plateau moduli, G_{Nb}^{0}. It follows that equation of state- and lattice fluid theories can not be used for explanation of different trends in moduli gradient. But, approach from a density of interaction change, due to different chain conformational statistics, can give a straight explanation, as we will see.

The good compatibility in PS/PPO and moduli increase can be attributed (at least in part), to additional interactions of component dipoles in blends of that type. But, according to theories of Tsenoglou and Wu (*34,35*) in "athermal case" of polymer blending, components will mix with unperturbed entanglement probability i.e. thermodynamic interactions are not able to modify the entanglement probability. It follows that lower values of G_N^0 than athermal ones, mean a decrease of entanglement probability by blending. So, one can come to more general conclusion that polar attractions render more difficult real entanglement of different polymers in blends. Here, it is in complete agreement with the previous results. Lower average entanglement in the blend, compared to

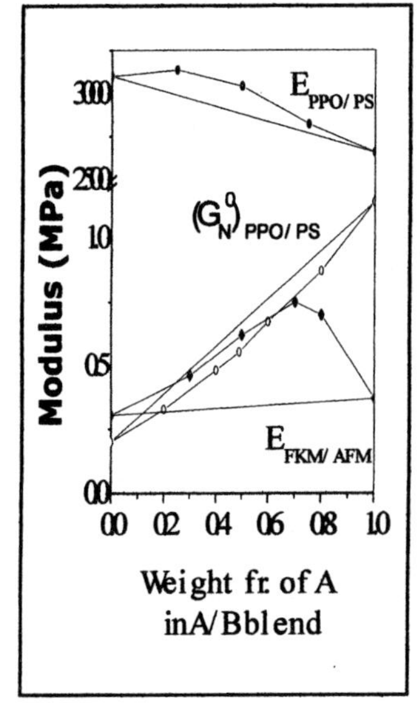

Figure 1. Tenzile(E)- and plato (G^0_N)- moduli for PPO/PS and FKM/ACM.

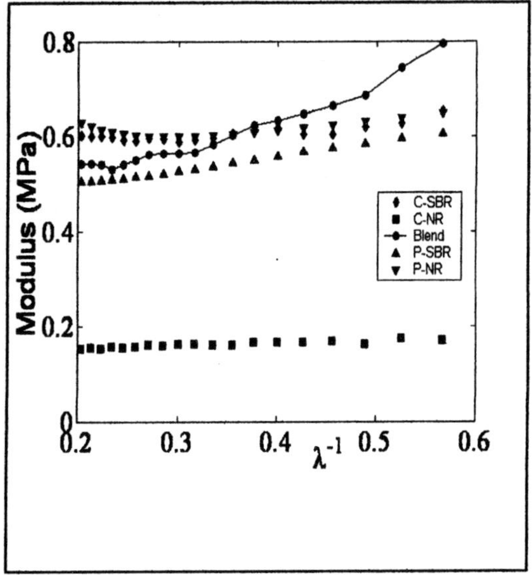

Figure 2. Mooney-Rivlin plots for NR/ SBR blend (B), components (C-) and phases (P-).

single polymers, make blend moduli in oscillatory mode lower. But, it helps to easier orientation and better interactions between polar groups (and segments) in permanent tension mode. That increase density of interactions (and probably material density during tension), and total connectivity. So, tensile modulus. is higher. Conformational characteristics of PS/PPO system are specific, due to large, flat, rigid aromatic groups, able for high density of interactions in appropriate position, but with bulky and space hindering effects in other conformations (*36-40*). If the chains were more flexible the moduli increase will appear in oscillatory mode, as well. It is just the case with ACM/FKM blends. It can be seen from Figure 1 for tensile moduli and curve for T=20 ^{0}C in Figure 3a.The curves for ACM/FKM blends we made on the base of experimental data published by Kader *at al.* (*31-33*), on poly(ethyl acrilate) and fluoro rubber blends already described in previous section. Similarly, to PS/PPO show ACM/FKM blends synergetic mechanical behavior, due to polar interactions and mixability on molecular level, even in the rubbery region. As we expect, more flexible chains of elastomer blend enable better contacts between polar groups and the increase of blend moduli relative to component moduli, both in permanent tensile and oscillatory mode, as can be seen from Figures 1 and 3a. But, significant increase of conformational dynamics would decrease contact density and total connectivity, resulting in decrease of moduli. It exactly happens that way, according to results presented in Figure 3a, with increase of temperature. Here should be noted some difference in conformational rigidity sources in PS and ACM systems. Conformational properties of PS are under influence of high rigidity of aromatic groups but, ACM expose high intramolecular polar attractions of ester groups, oriented to anti-parallel position of dipole vectors (*36-39*). The polar interactions with other polymer in a blend decrease intramolecular attractions and make the chain even more flexible than in one-component material. It follows that one should be careful with above quoted generalization, that polar attractions disturb real entanglements (*8*).

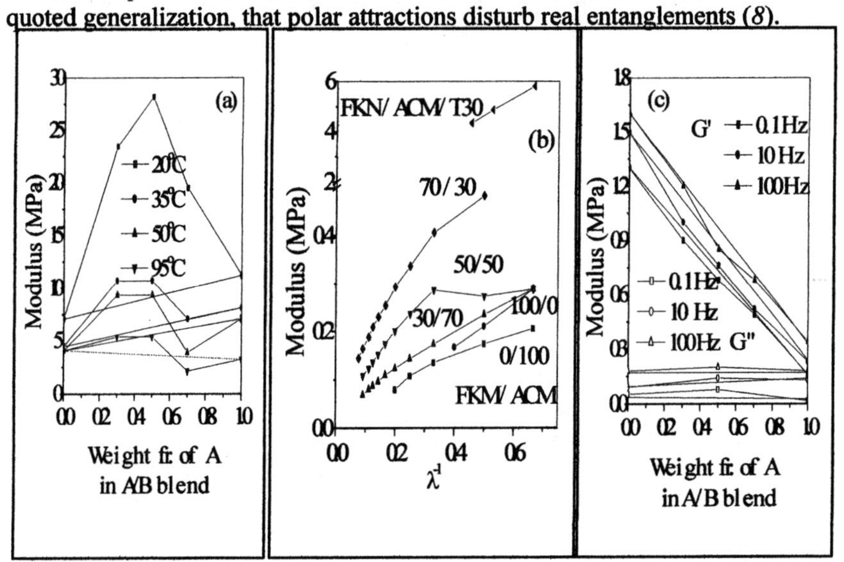

Figure 3. FKM/ACM blend properties versus composition (w/w): (a) G^0_N for temperature T, (b) Mooney-Rivlin plots, (c) G' and G" for frequency – ω.

Conformational partition function of the main chain and pendant flexible groups, can not be factorized, what all that mutual changes indicate. But, it opens numerous possibilities (*3,25*). The opposite effect, (of moduli decrease with permanent tension), is presented with our results for SBR and NR, in the form of Mooney-Rivlin curves, in Figure 2. Both effects, the increase of moduli with extension for some systems, and the decrease for others, are based on change of chain conformational statistics. In the first case, orientation provides better contacts between polar groups and high connectivity, but in the second higher conformational fluctuations result in lower connectivity. These effects are not in contradiction, as can be seen by comparison of results for ACM/FKM in Figures 1 and 3b. Moduli at the same level of deformation are higher for all blends relative to component polymers, (due to polar interactions), but decrease for each of these blends, with increase of deformation. The moduli measured, have been result of both effects. The Gaussian moduli of NR/SBR blend also decrease with extension as can be seen in Figure 2. It is in accordance with increase of fluctuations, as previously described. But, now we face properties of another main group of elastomer blends: polyphase system. There is no direct influence of one type polymer on conformational statistics of the others. Still, indirect influence is of high importance, as we will see. In Figure 2 also blend moduli excess can be seen, but only for low deformation. It can be ascribed to redistribution of compound ingredients among the phases leading to higher degree of networking as described in the previous section. But, it could not be the reason here, because:(a) blend phases are generated directly from masterbatches, (b) concentration differences of additives between phases are low, (c) there is significant resistance to diffusion on the phase boundary, (d) a significant change of network density will shift complete curve in the same direction. Still we applied Bauer method and obtained hypothetic contribution of phases to blend modulus, also given in Figure 2. As expected, the method gives unrealistic increase of NR-phase moduli and much more parallel shift of curves than observed. What could be than reasons of the intriguing behavior at low deformations? It is well known that conformational statistics change in phase boundary layers. The entropy force due to ordering at the boundary will produce some tension in the network, especially for fine dispersed phases, in well prepared compounds, as the NR/SBR blend. It will make network clusters inside drops or layers of another phase and even in the matrix, stiffer. It seems to be kind of collective mode response (in terms of de Gennes) of network clusters, trying to resist to changes at low deformations. At higher deformations, the increase of fluctuations will compensate that resistance. Indeed, the same effect at low deformations can be seen from data of Nelson *at al.* (*27*), already described. The soft EPDM phase makes much higher resistance to BR phase (which is there glassy and inert) at lower deformations. Finally, considering dynamical properties of polyphase elastomer blends, we correlated data for frequency sweep and change of composition for ACM/FKM/T30 blends. They have the same polymer component (as above for monophase system) but, in addition of a constant amount of trimethylolpropane triacrylate (T30) as third component, which causes separation of phases (*32*). Results presented in Figure 3c indicated decrease of storage moduli and increase of loss moduli with blending. (as expected).In analogue to NR/SBR, in Figure3b FKM/ACM/T30 (50/50/T30), two-phase-blend shows increase of Gaussian moduli, at low deformations. But, due to a number of possible influences of T30, that analogue should be taken as a crude estimate, and we will not go to detail analyses.

Conclusions

We considered in this contribution change of elastomer blend moduli, especially synergetic effects, from several main aspects: with structure, modes of deformation, and levels of deformation. Three types of structure are considered: with change of constitution, change of type of networking and change from mono-phase to poly-phases systems. Two modes of deformation are considered: permanent -tension and oscillatory deformations with appropriate constitutive parameters. The light -motif associating all aspects above, was comparison of connectivity responsible for deviations of blend moduli from linear combination of component moduli. Three causes are found of special interest for synergetic effects considered: (a) influence of polar interactions and real entanglements on moduli, (b) dominance of fluctuations of junctions at higher deformations, (c) tendency of compact network structures in polyphase systems to resist deformation, and its response with collective mode, what could be connected to self-similar cluster behavior. Although, the relations seem to be rather complex, chain conformational statistics provided theoretical background for analyses and prediction of mechanical behavior in all cases considered.

References

1. Kleiner,L.W.;Karasz,F.E.; MacKnihgt, W.J.*Polym. Eng. Sci.* **1979,** *19,* 519.
2. Mangaraj, D. *Rubbberr Chem. Technol.* **2002,** *75,* 365.
3. Plavsic, M. B. *Polymer Matherials Science and Engineering;* Naucna Knj: Belgrade, SCG, 1996. (and references therein)
4. Bauer, R. F.; Crossland, A. H.; *Rubber Chem. Technol.* **1988,** *61,* 585.
5. Popovic, R. S.; Plavsic, M. B. *Kautsch. Gummi, Kunstst.* **1991,** *44,* 336.
6. Popovic, R. S.; Plavsic, M. B. *Kautsch. Gummi, Kunstst.* **1996,** *49,* 826.
7. Prest, W. M.; Porter, R. *J. Polym. Sci., Part A-2* **1972,** *10,* 1639.
8. Lomellini, P. *Macromol. Theory Simul.* **1994,** *3,* 567.
9. Graessley, W. W.; Edwards, S. F. *Polymer,* **1981,** *22,* 1329.

10. Edwards, S. F.; Takano, H., Terentjev, E. M. *J. Chem. Phys.* **2000,** *113,* 5531.
11. Squires, A.; Tajbokhsh, A.; Terentjev, E. *Macromolecules* **2004,** *37,* 1652.
12. Flory, P. J. *Principles of Polymer Chemistry;* Cornell Un,.Ithaca, NY, 1953.
13. Vilgis, T. A.; In *Comprehensive Polymer Science*; Allen G.; Bevington, J. C. Eds.; Pergamon Press, New York, USA, 1989; Vol. 2, pp 227-279.
14. Mandelbrot, B. B. *The Fractal Geometry of Nature*, Freeman and Co., San Francisco, CA, 1982.
15. De Gennes, P. G. *Scaling Concepts in Polymer Physics;* Cornell: Unv. Ithaca, NY, 1979.
16. Alexander, S.; Orbach, R. *J. Phys. (Paris) Lett.* **1982,** *43,* 625.
17. Stauffer, D.; Coniglio, A.; Adam, M. *Adv. Plym. Sci.* **1982**, *44*, 103.
18. Mark, J. E. *J. Phys. Chem. B* **2003,** *107*, 903.
19. Chambon, F.; Winter, H. H. *Polym. Bull.* **1985,** *13,* 499.
20. Izuka, A.; Winter, H. H.; Hashimoto, T. *Macromolecules* **1997,** *30,* 6158.
21. Chambon, F.; Petrovic, Z. S.; MacKnight, W. J.; Winter, H. H. *Macromolecules* **1986**, *19*, 2146.
22. Muller, R.; Gerard, E.; Dagand, P.; Rempp, P.;Gnanou, Y. *Macromolecules* **1991,** *24,* 1321.
23. Muthukumar, M., *Macromolecules* **1989**, *22*, 4656.
24. Adolf, D.; Martin, J. *Macromolecules* **1991,** *24,* 6721.
25. Plavsic, M. B. *to be published*
26. Dusek, K. *Adv. Polym. Sci.* **1986,** *78,* 1.
27. Nelson, C. J.; Avgeropoulos, G.; Weissert, F. C.; Bohm, G. G. A. *Angew. Makromol. Chem.* **1977,** *60/61,* 49.
28. Colombini, D.; Merle, G.; Alberola, N. *Macromolecules* **2001,** *34,* 5916.
29. Popovic, R. S; Plavsic, M. *Kautschuk-Gummi-Kunststoffe* **1997**, *50*, 860.
30. Plavsic, M. B.; Pajic-LIjakovic, I.; Cubric, B.; Popovic, R. S.; Bugarski, B.; Cvetkovic, M; Lazic, N. *Material Science Forum* **2004**, *453/454*, 485.
31. Kader, M. A.; Bhowmick, A. K. *Rubber Chem.Technol.* **2000,** *73,* 889.
32. Kader, M. A.; Bhowmick, A. K. *Rubber Chem.Technol.* **2001,** *74,* 662.
33. Kader, M. A.; Bhowmick, A. K. *Polym. Eng. Sci.* **2003,** *43,* 975
34. Wu S., *J. Polym. Sci. Phys. Ed.* **1987**, *25*, 557, 2511.
35. Tsenoglou, C. *J. Polym. Sci., Part B: Polym. Phys.* **1988**, *26*, 2329.
36. Plavsic, M. *Int. Symp IUPAC-Macro 29.*Bucharest, Sep. 5-9 **1983,** *4,* 71.
37. Saiz, E.; Hummel, J.; Flory, P. J.; Plavsic, M. *J. Phys. Chem.* **1981,** *85,* 3211.
38. Yarimagaev, Y.; Plavsic, M.; Flory, P. J. *Polym. Prep.* **1983,** *24,* 233
39. Plavsic, M. B. *Croat. Chem. Acta* **1987,** *60,* 129.
40. Plavsic, M. B. *Material Science Forum* **1996,** *214*, 123.

New Techniques for Polymer Characterization

Chapter 20

Polymer Dynamics and Broadband Dielectric Spectroscopy

Graham Williams

Department of Chemistry, University of Wales at Swansea, Singleton Park, Swansea SA2 8PP, United Kingdom

Broadband dielectric spectroscopy (BDS) provides a direct means of studying the molecular dynamics of polymers in the amorphous, crystalline and liquid crystalline states, in blends and other composites. Recent experimental and theoretical findings from BDS concerning the structural (α) relaxation process in amorphous polymers, polymer blends and partially-crystalline polymers are examined here in some detail.

Introduction

Broadband Dielectric Spectroscopy (BDS), with its exceptional frequency range 10^{-6} to 10^{+12} Hz, has been used for the past sixty years or so to study the motions of dipolar groups within polymer materials (*1*). In the early years of such studies BDS remained a tedious and difficult technique in all bands of its vast frequency range, while other methods, such as multi-nuclear NMR, ESR, dynamic light scattering and transient fluorescence depolarisation, rose to prominence through their development and use of modern techniques. However, in recent years modern dielectric instrumentation using on-line data-processing became available, enabling measurements of the complex dielectric permittivity $\varepsilon(\omega)$ of polymers to be made quickly and accurately in the range 10^{-4} to 10^{+10} Hz over wide ranges of temperature and applied pressure, even for materials of low dipole strength (e.g. polyisoprenes) and small (sub-micron) film thickness. Also real-time BDS studies of polymerizing systems were achieved. These practical developments, together with advances in dielectric theory, have allowed BDS to take its place alongside those other techniques used for the study of chain dynamics in simple and complex polymer materials.

The dielectric and dynamic mechanical behaviour of amorphous polymers is well-documented (*1-4*). Early studies, and their interpretations using phenomenological and molecular theories of polymer dynamics, were described by McCrum, Read and Williams (*1*) and covered many materials including acrylate and methacrylate polymers, hydrocarbon polymers, polyvinyl esters, polyvinyl halides, polyesters and polycarbonates. Updated accounts of dielectric properties by Williams (*2*) and Runt and Fitzgerald (*3*) were followed by the recent research text of Kremer and Schönhals (*4*). For amorphous polymers, the primary (α) process, associated with the glass transition, and the (faster) secondary (β) process, due to limited motions of dipolar chains or side groups, are generally observed, where the relaxation strength $\Delta\varepsilon_\alpha$ for the α-process usually exceeds that for the β-process, a notable exception being syndiotactic polymethyl methacrylate (*2a*).

As is well-known (*1-4*) the dielectric α-process in most amorphous polymers exhibits broad, asymmetric loss peaks in the f-domain that are well-approximated by the Havriliak-Negami (HN) function with two stretch parameters or the Kohlrausch-Williams-Watts (KWW) function with one stretch parameter β_{KWW}. The β-process is even broader, but symmetrical, and can be fitted using Cole-Cole or Fuoss-Kirkwood functions. The temperature dependences of the average relaxation times $<\tau(T)>$ for α and β processes are given, approximately, by the well-known Vogel-Fulcher-Tammann-Hesse (VFTH) and Arrhenius equations, respectively. The dielectric properties of low molar mass (LMM) glass-forming liquids (*5-7*) exhibit the same pattern of behaviour as that for amorphous polymers. Also α, β, αβ relaxation data for amorphous polymers (*1-4,7-9*) and glass-forming liquids (5-8,10*)*, obtained from NMR relaxation, dynamic mechanical relaxation, quasi-elastic light-scattering, time-resolved fluorescence and optical relaxation experiments, parallel those obtained by BDS. The need for an appeal to Occam's razor to rationalise the generalities of structural relaxation behaviour in molecular glass-forming materials was recognised by M. Goldstein in the early 1970's.

Consequently, one of the great challenges in recent years in condensed matter physics, polymer physics and materials science has been to obtain an understanding of the ubiquitous relaxation behaviour of amorphous materials. So, what efforts have been made in this direction? Space does not allow a full account here. A reader is referred to the recent major review of this subject by Angell, Ngai, McKenna, McMillan and Martin (*7*), which contains over 500 references, mostly recent, together with special issues (*8*) of *J. Non Crystalline Solids* resulting from International Meetings on *Relaxations in Complex Systems* organised by Ngai and associates. These document relaxation phenomena in a wide range of glass-forming liquids and amorphous solids, including polymers, and define and discuss a range of problems that impede our full understanding of the observed behaviour.

In view of the large number of papers appearing on this subject, it is essential to assess the understanding gained in certain key areas. This paper focuses on two topics which are presently of much interest, discussion and debate. (i) Broadening of the α-relaxation in amorphous polymers, polymer blends and partially-crystalline polymers, (ii) The combined effects of temperature and applied pressure on the α-process in amorphous polymers. Due to limitations of space most details of the experimental results and mathematical formulations will be omitted; key references only will be given, where further information, in abundance, will be found.

Broadening of the α Relaxation

Amorphous Polymers

The α-relaxation in amorphous polymers (*1-4,7-9*), such as as polycarbonates, and in LMM glass-forming molecular liquids (*5-8,10*), such as o-terphenyl, as observed using different relaxation, scattering and spectroscopic techniques exhibits 'stretched exponential' (SE) behaviour. This has received much attention from theorists in recent years. The HN or KWW functions used to fit dielectric data in the f- and t-domains are phenomenological, so give no information on the molecular origins of SE behaviour. Such is obtained through comparisons of experimental data with the predictions of model theories and computer simulations of chain motions in bulk polymers. Prominent among the many approaches to explain SE behaviour are the 'mode-mode' coupling theory (MCT) originated by Sjögren and Götze (see e.g. (*11*)) and the Ngai Coupling Model (see e.g. (*12*)). In MCT an unknown correlator C(ω) for relaxation in the frequency domain, which is the Fourier transform of the correlator C(t) in the t-domain, is related to an unknown second order memory function $M_2(\omega)$ via a Mori continued fraction. Choosing $M_2(t)$ to be a particular function of C(t) closes the mathematical problem, allowing the determination of both C(ω) and $M_2(\omega)$ (and hence C(t) and $M_2(t)$) using the truncated continued fraction. However, such a theory is phenomenological since the results depend on the choice made for the relation between $M_2(t)$ and C(t). Their 'F_{12} model' (*11*) produces C(t) as a KWW function over most of the structural relaxation range. This memory function approach has been criticized by Williams and Fournier (*13*) since the KWW function yields unphysical results for the first order memory function $M_1(\omega)$ at high frequencies. In the Ngai approach (*12*) C(t) for the α-process is assumed to be exponential at very short times then to change to a KWW function at longer times. The KWW function (or $dC_{KWW}(t)/dt$)) is assumed at the starting point of the theory. No specific model for the long-time relaxation is contained in this proposition, so this approach does not establish mechanistic origins of KWW behaviour for the α-process and, in contrast to molecular theories, cannot differentiate between the relaxation functions obtained from different relaxation, spectroscopic and scattering studies.

Turning now to model theories for chain motions, the concepts of 'dynamic heterogeneity' and 'energy landscapes' have been used increasingly in recent years to rationalise KWW-type behaviour for the α-process in glass-forming materials. Using multinuclear NMR relaxation, Schmidt-Rohr and Spiess (*14*) established that a sub-ensemble of relaxors in bulk polyvinyl acetate (PVAc) above its T_g appeared to be dynamically-heterogeneous, whilst this ensemble retained the thermodynamic stationary property of being ergodic in time. Thus the SE function of KWW-type arises in this case from a distribution of transient spatially-heterogeneous regions, each with their own relaxation behaviour. Further evidence for dynamic heterogeneity and 'energy landscapes' in amorphous polymers and LMM glass-forming liquids were given by (i) Böhmer et al (*15*), (ii) Sillescu et al (16) (see also Ch.14,15,17 in (*4*) on NMR relaxation, solvation dynamics and pulsed and non-resonant hole-burning BDS in polymers) and (iii) Ediger et al. (*6*) using optical relaxation techniques .

The energy landscape approach to structural relaxation in amorphous materials assumes that relaxors move by thermally-activated transitions between local free-energy states, governed by a master equation of motion with detailed balancing of species. Such a model is insufficient to predict dielectric, NMR or other molecular relaxation behaviour (see e.g. Williams in (*4*), Ch.1). It is necessary to 'decorate' a relaxor 'i' with vectors $[\mathbf{r}_i, \mathbf{u}_i]$ where **r** and **u** refer to centre of mass position and angular orientation, respectively, so the trajectory of its motions in the energy landscape are tracked in time, allowing time correlation functions (TCFs) appropriate to particular experimental observations to be determined. By specifying the geometric coupling of $\mathbf{u}_i$ with transitions in the energy landscape, Diezemann, Sillescu, Hinze and Böhmer (*16c*) determined first and second rank two-time orientational TCFs $<P_{1,2}(\cos\theta(t))>$ and higher order (four-time) orientational TCFs for models of reorientational motions ranging from small to large angle jumps in energy landscapes having chosen gaussian densities-of-states (DOS). They found (*16c*) that KWW-type relaxation was obtained for these TCFs and that the ratio $R = <\tau_1(\text{dielectric})>/<\tau_2(\text{NMR})>$ of the average relaxation times for models with small, intermediate and large angle jumps was close to unity in all cases if the DOS was very broad. They observed that if $R \approx 1$ is obtained experimentally for a polymer material, it says nothing about the geometry of the reorientational motions in the energy-landscape if the DOS is broad (*16c*). Thus comparative data on P_1 and P_2 TCFs for chain motions are insufficient to define the orientation process. Further information is required, e.g. from (i) higher time-order TCFs, as in the works of Spiess, Schmidt-Rohr, Böhmer et al (*14,15*), and (ii) the relaxation functions from deep optical bleaching, solvation dynamics and pulsed and non-resonant hole-burning BDS experiments (see e.g. Ediger (*6*), (*15*) and Ch.14,15,17 in (*4*)).

Kivelson et al (*17*) developed a model of the thermally-activated dynamics of 'frustration-limited' domains in LMM glass-forming liquids that predicts KWW behaviour for structural relaxation. Also Wolynes et al (*18*) constructed a 'mosaic picture' of a supercooled liquid that is both dynamically-heterogeneous and ergodic. Relaxation in the different regions are thermally activated, with configurational entropy fluctuations being the main driver for the width of the relaxation process. Dynamic heterogeneity (DH) is common to both models (*17,18*) leading to distributions of relaxation times and hence KWW-behaviour for the α-process. These models for LMM glass-forming liquids translate readily to apply to α-relaxations in amorphous polymers.

In landmark experiments, Israeloff and coworkers (*19*) used scanning-probe microscopy techniques to observe the dielectric α-process in mesoscopic regions (~50 nm) of amorphous PVAc a few degrees above its T_g. The dielectric relaxation of a small region exhibited fluctuations due *both* to intrinsic heterogeneities at smaller length scales *and* the cooperative dynamics of the chains. The heterogeneous, evolving dynamics summed to give KWW behaviour for the overall α-process, which was the same as that for a bulk sample. This provides further support for the results of Spiess, Böhmer, Ediger, Richert and their coworkers (*15*), which demonstrated that DH is a source of KWW-type behaviour for the α-process in amorphous polymers and LMM glass-forming liquids (see also Ch.14,15,17 in (*4*)). The concepts of 'energy-landscapes', 'frustation limited domains' and 'mosaic structures' are all consistent with DH since a relaxor moves at different rates depending on its location in an energy landscape, or to which frustration-limited domain or mosaic it belongs in an ensemble at t = 0. This raises the questions – what information is contained in orientational two-time TCFs and how may further information on a relaxation process be obtained?

Consider the two-time *ensemble-averaged* TCF (eq 1) and the *time-averaged* TCF (eq 2) for a dynamical variable A of a relaxor e.g. for dielectric relaxation, it is the dipole moment vector (for details see e.g. (*20*)).

$$C_{ens}(t) = \iint A(p,q;\tau)A(p,q;t+\tau)f(p,q)dpdq = \langle A(\tau)A(t+\tau)\rangle \qquad (1)$$

$$C_{time}(t) = \lim_{T \to \infty} \left[(1/T) \int_0^T A(\tau)A(t+\tau)d\tau \right] \qquad (2)$$

(p,q) are the conjugate momentum and position of the relaxor, f(p,q) is its phase-space equilibrium distribution function (*20*). $C_{ens}(t)$ = $C_{time}(t)$ for an ergodic system such as an isotropic liquid or a bulk amorphous polymer above its T_g.

$\langle P_1(\cos(\theta(t))\rangle$ data from BDS and $\langle P_2(\cos\theta(t))\rangle$ data from NMR, depolarized light-scattering, fluorescence depolarisation, optical relaxation experiments are determined as $C_{ens}(t)$. However eq 2 involves the real-time trajectory of A(t), which contains more information than the two-time TCFs in eq 2 and 3. 'Molecular Dynamics' (MD) and 'Monte Carlo' (MC) simulations of polymer ensembles reveal the time-series A(t) for different dynamical variables. While it is unlikely that BDS can be adapted to study the motions of single dipole groups within polymer chains, Vanden Bout et al (*21*) succeeded in making time-series measurements of the optical anisotropy of single Rhodamine-6G dye molecules dissolved in polymethyl acrylate just above its T_g by a optical polarization hole-burning method. In this case $C_{time}(t)$ = $\langle P_2(\cos\theta(t))\rangle$ was determined using eq 2 and was found to be of KWW-form with $\beta_{KWW} = 0.73$. So SE behaviour was observed for a single dye-molecule tumbling in an amorphous polymer. Using the methods of Vanden Bout it should be possible to study the orientational trajectories of single fluorescent dye-moieties in a polymer chain, as an extension of the studies by Monnerie et al (see (*2d*) pp 383-413) of transient fluorescence depolarisation of fluorophore-labelled polymers in solution. Schmidt-Rohr, Spiess et al (*14,15*) take a *sub-ensemble* of the distribution f(p,q) and determine how the component relaxors project in time, giving more information than that from the $\langle P_{1,2}(\cos\theta(t))\rangle$. Israeloff et al (*19*) study a region sufficiently small to exhibit fluctuations due to intrinsic heterogeneities, again giving more information than that from these TCFs. Thus BDS data probe the orientational motions of chain dipoles in terms of $\langle P_1(\cos\theta(t))\rangle$ (*1-4*) but in order to obtain more information about the α-process, data are required for small ensembles, as in (*19*), or via optical bleaching, solvation dynamics and BDS hole burning experiments ((*6*) and Ch.14,15,17 in (*4*)).

MD and MC simulations of polymer chain dynamics, as developed by Roe, Binder, Pakula and others, yield TCFs for the anisotropic motions of chain units that bring together dielectric and related relaxation properties of bulk polymers. This will not be considered here except to say that the time-series A(t) for different molecular dynamical variables contained in these simulations need to be examined in order to define further, and hence obtain physical insight into, the nature of dynamic heterogeneities in glass-forming materials.

Polymer Blends

A large literature has accumulated in the past 25 years for the dielectric behaviour of polymer blends, associated with MacKnight and Karasz, Fischer, Floudas, Colby, Adachi, Alegria and Colmonero, Runt et al and other groups; for reviews see Runt in (3) and Floudas and Schönhals in (4)). BDS studies of inhomogeneous blends reveal the α-processes of each component. Apparently homogeneous polymer blends have a single T_g and a broadened α-loss peak since each component experiences a range of local environments.

BDS studies include PSt/polymethylphenylsiloxane, PSt/polyphenylene oxide, PSt/polychlorostyrenes, polyvinylethylether/styrene-co-p-hydroxystyrene, PSt/polyvinylmethylether, polyisoprene/polyvinylethylene; PSt is polystyrene. While β_{KWW} values for the dielectric α-process in homopolymers lie in the range 0.4-0.7, additional broadening is observed and provides evidence for local 'concentration fluctuations' in such blends. As two examples, we refer to the recent works of Miura et al (*22*) and Leroy et al (*23*), which contain numerous references to the earlier BDS studies of homogeneous blends. Miura et al (*22*) studied P St/poly-o-chlorostyrene blends and observed a marked broadening of the α-loss peak with decreasing temperature, indicating that the localized concentration fluctuations were on the scale of the segmental motions, as their earlier studies had revealed. The distribution of the effective T_g values of the polar component in the blend was determined as a function of temperature. Leroy et al (*23*) consider whether the dynamic heterogeneity of the segmental dynamics in such blends is due to (i) chain connectivity effects or (ii) thermal concentration fluctuations. The 'self concentration' approach of Lodge and McLeish (*24*) had introduced a distribution of local chain concentrations involving a effective Kuhn length for a chain in a blend, which i s a ssociated more with (i) than (ii). Leroy et al (*23*) found their BDS data for PS/PVME and PS/PoCS blends could be fitted with a single parameter $<(\delta\Phi_{eff})^2>$, which is the variance of the effective concentration fluctuations that contains *both* self-concentration *and* thermal concentration effects. Thus B DS p rovides e ssential new information on spatial heterogeneity in polymer blends through analyses of the broadening of the overall α-process. It evident that BDS can be used in the future as a quantitative tool to determine the micro-to-nano structures of homogeneous and inhomogeneous polymer blends.

Partially Crystalline Polymers

For many crystalline polymers, e.g. linear polyesters, polycarbonates and polyamides, while the material is 100% spherulitic, as viewed in an optical microscope, it is typically less than 50% crystalline. The amorphous regions are contained within the spherulites and strong interactions exist between these regions and the lamellar crystals. BDS provides a direct means of studying the motions of these constrained amorphous regions inside the spherulites. Preliminary real-time studies by Williams and Tidy (see (*2a*)) of the BDS behaviour of polyethylene terephthalate (PET) showed that during isothermal crystallization the normal α-process for the amorphous phase was replaced by a slower, broader α' process which was due to motions of the constrained amorphous phase. Since that time Ezquerra, Balta-Calleja and their associates made extensive real-time dielectric studies, and simultaneous X-ray studies, of polymers during isothermal crystallization and have quantified how the dynamics of the amorphous regions change during the process (*25*). Their studies of PET (*25a*), PEEK (*25b*), polyaryl ketones (*25c*), ethylene/vinyl acetate copolymers (*25d*) and mixed aromatic polyesters (*25e*) show that real-time BDS provide a powerful means of monitoring the isothermal crystallization of bulk polymers and of characterizing the dynamics of the constrained amorphous regions within spherulites as crystallization proceeds; information not available from techniques such as optical and electron microscopies that give information only on polymer morphology and the average local structure of crystalline regions.

Effects of Temperature and Applied Pressure

The BDS studies in the early 1960's by O'Reilly (*26*) for PVAc and by Williams (*2a,27*) for polymethyl acrylate (PMA), polyethyl acrylate (PEA), amorphous polypropylene oxide (PPrO), polynonyl methacrylate and polyvinyl chloride established the effects of pressure on the frequency-temperature locations of the α-process of amorphous polymers. These and further studies by Sasabe, Saito et al (*28*) for alkyl methacrylate polymers and polyvinyl chloride were reviewed in 1979 by Williams (*2a*). Few studies of pressure effects on the dielectric properties of amorphous polymers were reported in the following 20 years, mainly due to instrumentation difficulties, which led Floudas to remark (see Ch.4 in (*4*)) that pressure became the 'forgotten' variable in BDS studies of polymers. However, the advent of modern high pressure dielectric assemblies led to several studies in the past five years by the groups of Roland, Rolla and Floudas and others. In his review Floudas (Ch.3 in (*4*)) lists BDS studies of PET, polyalkylmethacrylates, polyethylenes, polysiloxanes, PPG/salt complexes, PSt/PVME blends, Nafion, polyisoprene, polyvinylethylene and polypropylene glycols (PPG). The large/*small* effects of pressure on dielectric α/β processes are thus well-documented (*2a*, Ch.4 in (*4*)), as are the effects of pressure on (i) the splitting of the αβ-process in polyalkyl methacrylates, first observed by Williams (*27e*), (ii) polymer crystallization in polyalkyl methacrylates (*29*) and (iii) the dielectrically-active Rouse motions of chains in polyisoprenes (*30*).

We focus here on the debate concerning the relative effects of temperature and volume on the α-process in amorphous polymers. In the original studies of PVAc by O'Reilly (*26*) and PMA, PPrO and PEA by Williams (*27*) the effect of pressure on $<\tau_\alpha (T,P)>$ was large with $(\partial \log<\tau>/\partial P)_T$ being in the range 1-3 kb^{-1}. Using values of the thermal pressure coefficient $(\partial P/\partial T)_V$ Williams (*27b-27d*) determined the ratio of the constant volume and constant pressure apparent activation energies $\chi = Q_V(T,V)/Q_P(T,P)$ for PMA, PEA and PPrO and found $\chi > 0.6$ in all cases. Hoffman, Williams and Passaglia (*31*) pointed out that this presented serious difficulties for free volume theories - which predict $\chi \leq 0$! As a result they (*31*) favoured thermal activation of chain units over local energy barriers as the mechanism for the α-process. Free-volume theories are still being widely-used to describe α and T_g processes in amorphous polymers despite this refutation (*31*), which has been repeated several times (*2a-d,27d*)),.

Recent studies on the effects of pressure on the dielectric α-process were made by (i) Roland et al (*32*) for polystyrene, polyvinylmethylether, propylene glycol oligomers and 1,2-polybutadiene, (ii) Roland et al (*33*) for the LMM liquids phenylphthalein-dimethylether, diglycidylether of bisphenol A, 1,1'-di(4-methoxy-5-methylphenyl)cyclohexane,1,1'-bis(p-methoxyphenyl)cyclohexane. In related works Kivelson et al (*34*) studied (T,P) effects on the α-process for triphenyl phosphite and glycerol.

Casalini and Roland (*32b*) say that the relative importance of thermal energy and volume on the average relaxation time $<\tau_\alpha>$ is intensely debated, with contrary viewpoints being expressed; Williams (*2,27*) and Kivelson (*34*) favouring thermal energy as the dominant effect and Roland and coworkers saying that V effects can become as, or even more, important than temperature in some cases, referring to their results for the α-process in LMM glass-forming liquids (*33*). It is important to resolve the question raised by Roland et al (*32,33*) concerning the relative effects of temperature (or thermal energy) and volume on $<\tau_\alpha>$.

O'Reilly (*26*) and Williams (*2,27*) demonstrated that volume effects are important for $<\tau_\alpha>$ in amorphous polymers. The fact that $\chi > 0.6$ in these materials suggested the process is *thermally activated* and simple free volume theories, which have no thermal energy or dynamics in their concept, do not apply to the α-process or to T_g (*2,27,31*). These results are consistent with energy landscape models, which may contain as a sub-sets the 'frustration-limited domains' or 'mosaics' of Kivelson and Wolynes. We consider further the case for thermal activation of the α-process.

Eq 3-11 relate ratios of derivatives of $<\tau(T,P,V)>$ to $\chi = Q_V(T,V)/Q_P(T,P) = (\partial \ln\tau/\partial T)_V / (\partial \ln\tau/\partial T)_P$ (*35*)

$$(\partial \ln\tau/\partial V)_P / \partial \ln\tau/\partial V)_T = (1 - \chi)^{-1} \tag{3}$$

$$(\partial \ln\tau/\partial T)_V/(\partial \ln\tau/\partial T)_P = \chi \quad (4)$$
$$(\partial \ln\tau/\partial P)_V/(\partial \ln\tau/\partial P)_T = -\chi(1 - \chi)^{-1} \quad (5)$$
$$(\partial \ln\tau/\partial V)_T/(\partial \ln\tau/\partial T)_V = (\chi^{-1} - 1)/(\partial V/\partial T)_P \quad (6)$$
$$(\partial \ln\tau/\partial P)_T/(\partial \ln\tau/\partial T)_P = (\chi - 1)/(\partial P/\partial V)_T \quad (7)$$
$$(\partial \ln\tau/\partial P)_V/(\partial \ln\tau/\partial V)_P = -\chi/(\partial P/\partial V)_T \quad (8)$$
$$(\partial T/\partial V)_\tau(\partial V/\partial T)_P = (1 - \chi^{-1}) \quad (9)$$
$$(\partial T/\partial P)_\tau(\partial P/\partial T)_V = (1 - \chi) \quad (10)$$
$$(\partial V/\partial P)_\tau(\partial P/\partial V)_T = \chi \quad (11)$$

Eq 10 was derived by Williams (*27a*) while eq 9 is given by Casalini and Roland (*32b*). Thus an understanding of the various plots of $\log\tau(X)$ vs. X at constant Y, where the (X,Y) are permutations of (T,P,V), may be gained through the origins of χ.

As explained above, simple free volume theories are unable to rationalise the experimental result $\chi > 0.6$, so should be discarded. Williams (*27a*) extended the Eyring transition state theory to apply to a dielectric α-process and showed that

$$\chi = 1 - T(\partial P/\partial T)_V \, \Delta V^*/\Delta H^* \quad (12)$$

ΔV^* and ΔH^* are, respectively, the apparent activation volume and activation enthalpy for the thermally activated process (for definitions of standard states see (*27a*) and (*36*)). Thus the result $\chi \geq 0$ for the α-process in amorphous polymers may be explained in terms of a thermally activated process in a free energy barrier system, giving eq 12, as was made clear by Williams in 1965 (*27e*).

Alternatively we may adopt an Arrhenius relation for thermal activation involving a local energy barrier Q(V) = Q(T,P) determined by the interactions of the relaxor with its local environment (*31*). It follows that (*35*)

$$\chi = [1 - T(\partial \ln Q/\partial T)_P]^{-1} = [1 + T(\partial P/\partial T)_V(\partial \ln Q/\partial P)_T]^{-1} > 0 \quad (13)$$

Increase in T at constant V yields Q_V = Q(V)). Q decreases as T is raised at constant P while Q increases as P is raised at constant T.

Thus these models of thermal activation, giving eq 12 and 13, are able to (i) rationalise (T,P,V) effects on the α-process, (ii) give meaning to the result $\chi > 0$ and (iii) show how the paired relative variations of $\langle\tau_\alpha\rangle$ with (T,P,V) are related to χ (eq 3-11).

The energy landscape for an amorphous polymer will be complicated, so to further the concept of Q = Q(V) consider the model case of a rotator phase crystal pentachlorotoluene (PCT) in which the molecules undergo thermally-activated reorientations in a local barrier system of C_6 symmetry. Garrington and Williams (*37*) obtained BDS data for PCT as ε" vs. logf/Hz over ranges of (T, P) and found that $Q_P = 35$ kJ mol^{-1}, $\chi = 0.70$. Thus Q(T,P) increases when P is raised at constant T due to compression, and decreases when T is raised at constant P due to dilation, of the material. So $0 < \chi \leq 1$ for this thermally activated process in a rotator-phase crystal. The same result applies to energy landscapes, frustrated domains and mosaics in glass-forming materials, including amorphous polymers. The local barriers to reorientation are a function of local volume, and hence of (T, P). Any apparent controversies concerning the relative effects of thermal energy and volume (*32b*) are removed in this inclusive approach, which interprets variations of $<\tau(T,P,V)>$ through an understanding of the origins of χ in a thermally-activated process. Note also the paper by Struik (*38*) concerning the meaning of Q when the VFTH equation is applied near to T_g for glass forming materials.

Finally, we comment on the merging of the dielectric α and β processes to form the αβ process at high temperatures, first observed by Williams in 1966 (*27e*) in studies of polyethylmethacrylate (PEMA) under applied pressure. The 'Williams ansatz' (*2a*) gives the total dielectric relaxation function $\Phi_\alpha(t)$ arising from partial relaxation (β-process, $\phi_\beta(t)$) and total relaxation (α-process, $\phi_\alpha(t)$) of a chain dipole as follows

$$\Phi_\mu(t) = \phi_\alpha(t)[A_\alpha + (1 - A_\alpha)\phi_\beta(t)] \tag{14}$$

A_α is the average relaxation strength remaining after the β-process has partially-relaxed the dipole vector. Eq 14 predicts (i) the 'conservation rule' $\Delta\varepsilon = \Delta\varepsilon_\alpha + \Delta\varepsilon_\beta$ applies when (T,P) are varied, (ii) α and β processes coexist above T_g (iii) the merged (αβ) process is a continuation of the α-process to higher temperatures. This approach to multiple relaxations and the 'crossover' region has been debated recently (*40-42*). Schröter et al (*40*) and Gomez et al (*41*) applied both the Williams ansatz and an additive ansatz to the BDS behaviour of amorphous polymers. For PEMA and PBMA Schröter et al found:- (i) $\Delta\varepsilon_\alpha$ decreased with increasing T, approaching zero at the apparent crossover temperature T_x and (ii) for $T > T_x$ the merged 'a' process gave a separate locus for $\log f_{max}$ vs. T^{-1}, between those projected for α and β processes. They concluded that the 'a' process 'is a distinct and separate process' in accord with the conclusion of Williams (*27e*) for PEMA. In contrast, Gomez et al (*41*) found their BDS data for poly(epichlorohydrin), polyvinylmethyl ether, isopolymethyl methacrylate, i-polymethyl methacrylate, PVAc and tetramethyl bisphenol polycarbonate:- (i) were analysed satisfactorily using eq 14, (ii) while A_α decreased with increasing T, it did not become zero at T_x and (iii) the merged process was mainly determined by mechanisms responsible for the α-process. In this writer's view merging/coalescence of the α and β processes as T is raised gives an αβ process, or 'a' process, that is essentially the α-processs extrapolated to higher temperatures but modified slightly by local motions (see (*27e*) for an early discussion).

References

1. McCrum, N.G.; Read, B.E.; Williams G. *Anelastic and Dielectric Effects in Polymeric Solids*; J. Wiley: New York, 1967; Dover Publ.: New York, 1991.
2. (a) Williams, G.; *Adv. Polym. Sci.* **1979**, *33*, 59. (b) Williams, G. in *Comprehensive Polymer Science*; Allen, G.; Bevington J.C. Eds.; Pergamon Press: Oxford, 1989, Ch.18, pp 601-632. (c) Williams, G. in *Materials Science and Technology Vol.12 Structure and Properties of Polymers*; Thomas E.L. Ed.; VCH Publ.: Weinheim, 1992, pp 471-528. (d) Williams, G. in *Static and Dynamic Properties of the Polymeric Solid State, NATO ASI*; Pethrick, R.A.; Richards, R.W. Eds.; D. Reidel Publ. Co.: Dordrecht, 1982, pp 213-240.
3. *Dielectric Spectroscopy of Polymeric Materials. Fundamentals and Applications*; Runt, J.P.; Fitzgerald J.J. Eds.; American Chemical Society: Washington, DC, 1997.
4. *Broadband Dielectric Spectroscopy*; Kremer, F.; Schönhals, A. Eds.; Springer Verlag: Berlin, 2003.
5. Williams, G. in *Spec. Period. Rep. Chem. Soc.(London). Dielectric and Related Molecular Properties*; Davies, M.M. Ed.; 1975, pp 151-182.
6. *(a)* Ediger, M.D.; Angell, C.A.; Nagel, S.R. *J. Phys. Chem.* **1996**, *100*, 13200. (b) Ediger, M.D. *Ann. Rev. Phys. Chem.* **2000**, *51*, 99.
7. Angell C.A.; Ngai, K.L.; McKenna, G.B.; McMillan, P.F.; Martin, S.W. *Appl. Phys. Rev.* **2000**, *88*, 3113.
8. *J. Non Cryst. Solids* (a) **1991**, *131-133*, pp 1285, (b) **1993**, *172-174*. pp 1457 (c) **1998**, *235-237*, pp 814.
9. Matsuoka, S. *Relaxation Phenomena in Polymers*; Hanser Publ.: Munich, 1992.
10. Angell, C.A.; Wong, J. *Glass Structure by Spectroscopy*; Marcel Dekker: New York, 1976.
11. Götze, W. in *Liquids Freezing and the Glass Transition*; Levesque, D; Hansen, J.P.; Zinn-Justin, J. Eds.; Elsevier: New York, 1991 pp 292-502.
12. (a) Ngai, K.L. *J. Non Cryst. Solids* **2000**, *275*, 7. (b) Ngai, K.L. *J. Phys. Condens. Matter.* **2003**, *15*, 1.

13. (a) Williams, G.; Fournier, J. *J. Chem. Phys.* **1996**, *184*, 5690. (b) Williams, G. in *Keynote Lectures in Selected Topics of Polymer Science*; Riande, E., Ed.; CSIC: Madrid, Spain, 1996, pp 1-39.
14. (a) Schmidt-Rohr, K.; Spiess, H.W. *Phys. Rev. Lett.* **1991**, *66*, 3020. (b) Sillescu, H. *J. Non Cryst. Solids* **1999**, *243*, 81.
15. Böhmer, R. and 12 coathors. *J. Non Cryst. Solids* **1998**, *235-237*, 1.
16. (a) Diezemann, G. *J. Chem. Phys.* **1997**, *107*, 10112. (b) Diezemann, G. *J. Chem. Phys.* **1999**, *111*, 1126. (c) Diezemann, G.; Sillescu, H.; Hinze, G.; Böhmer, R. *Phys. Rev.E* **1998**, *57*, 4398.
17. Kivelson, D.; Kivelson, S.A.; Zhao, X.; Nussimov, Z.; Tarjus, G. *Physica A* **1995**, *219*, 27. (b) Kivelson, D.; Tarjus, G. *Phil. Mag B* **1998**, *77*, 245. (c) Kivelson, D.; Tarjus, G. *J. Non Cryst. Solids* **1998**, *235-237*, 86.
18. Lubchenko, V.; Wolynes, P.G. *J. Chem. Phys.* **2003**, *119*, 9088. (b) Xia, X.; Wolynes, P.G. *Phys. Rev. Lett.* **2001**, *86*, 5526.
19. (a) Vidal Russell, E.; Israeloff, N.E. *Nature* **2000**, *408*, 695. (b) Vidal Russell, E.; Israeloff, N.E.; Walther, L.E.; Alvarez Gomariz, H. *Phys. Rev. Lett.* **1998**, *81*, 1461. (c) Walther, L.E.; Israeloff, N.E.; Vidal Russell, E.; Alvarez Gomariz, H. *Phys. Rev. B* **1998**, *57*, R15112.
20. (a) Berne, B.J. in *Phys. Chem. An Advanced Treatise, Vol 8B, The Liquid State*; Eyring, H.; Henderson, D.; Jost, W. Academic Press: New York, 1971, pp 50-713. (b) Williams, G. *Chem. Soc. Rev.* **1978**, *7*, 89.
21. (a) Deschenes, L.A.; Vanden Bout, D.A. *J. Chem. Phys.* **2002**, *116*, 5850. (b) Deschenes, L.A.; Vanden Bout, D.A. *J. Phys. Chem.B* **2001**, *105*, 11978.
22. Miura, N.; MacKnight, W.J.; Matsuoka, S.; Karasz, F.E. *Polymer*, **2001**, *42*, 6129.
23. Leroy, E.; Alegria, A.; Colmonero, J. *Macromolecules* **2003**, *36*, 7280.
24. Lodge, T.P.; McLeish, T.C.B. *Macromolecules* **2000**, *33*, 5278.
25. (a) Ezquerra, T.A.; Balta-Calleja, F.J.; Zachmann, H.G. *Polymer* **1994**, *35*, 2600. (b) Nogales, A.; Ezquerra, T.A.; Garcia, J.M. *J. Polym. Sci. Polym. Phys.* **1999**, *37*, 7. (c) Ezquerra, T.A.; Liu, F.; Boyd, R.H.; Hsiao, B.S. *Polymer* **1997**, *38*, 5793. (d) Sies, I. et al *Macromol. Sci. Phys.* **2000**, *28*, 4516. (e) Sies, I. et al *Polymer* **2003**, *44*, 1045.
26. O'Reilly, J.M. *J. Polymer Sci.* **1962**, *47*, 429.
27. (a) Williams, G. *Trans. Faraday Soc.* **1964**, *60*, 1548. (b) Williams, G. *ibid* **1964**, *60*, 1556. (c) Williams, G. *ibid* **1965**, *61*, 1564. (d) Williams, G.; Watts, D.C. in *Dielectric Properties of Polymers*; Karasz, F.E. Ed.; Plenum Press: New York, **1972** pp.17-44. (e) Williams, G. *Trans. Faraday Soc.* **1966**, *62*, 2091.
28. (a) Sasabe, H.; Saito, S. *J. Polym. Sci. Part A2* **1968**, *6*, 1401. (b) Saito, S.; Sasabe, H.; Nakajima, T.; Yada, K. *ibid* **1968**, *6*, 1297.

29. Mierzawa, M.; Floudas, G. *IEEE Trans. Dielectrics EI*, **2001**, *8*, 359.
30. Floudas, G; Gravalides, C.; Reisinger, T.; Wegner, G. *J. Chem. Phys.* **1999**, *111*, 9847.
31. Hoffman, J.D.; Williams. G.; Passaglia, E. *J. Polymer Sci. Part C*, **1966**, 173.
32. (a) Roland, C.M.; Casalini, R.; Santangelo, P.; Secula, M.; Ziolo, J.; Palauch, M. *Macromolecules* **2003**, 36, 4954. (b) Casalini, R.; Roland, C.M. *J. Chem. Phys.* **2003**, *119*, 4052. (c) Roland, C.M.; Casalini, R. *ibid* **2003**, *119*, 1838.
33. (a) Palauch. M.; Roland, C.M.; Casalini, R.; Meier, G.; Patkowski, A. *J. Chem. Phys.* **2003**, *118*, 4578. (b) Palauch, M.; Roland, C.M.; Gaponski, J.; Patkowski, A. *ibid* **2003**, *118*, 3177. (c) Patkowski, A.; Palauch.; Kriegs, H. *ibid* **2002,** *117*, 2192. (d) Palauch, M.; Casalini, R.; Best, A.; Patkwoski A. *ibid* **2002,** *117*, 7624.
34. (a) Kivelson, D.; et al *J. Chem. Phys.* **1998**, *109*, 8010. (b) Alba-Simionesco. C.; Kivelson, D.; Tarjus, G. *ibid* **2002,** *116*, 5033.
35. Williams, G. *unpublished.*
36. (a) Albuquerque, L.M.P.C.; Reis, J.C.R. *Trans. Faraday Soc.* **1989,** *85*, 207. (b) Albuquerque, L.M.P.C.; Reis, J.C.R. *J. Chem. Soc. Faraday Trans.* **1991,** *87*, 1553.
37. (a) Garrington, D.C. *Ph.D Thesis, University of Wales* 1976. (b) Garrington, D.C.; Williams G. *unpublished.*
38. Struik, L.C.E. *Polymer* **1997**, *38*, 733.
39. Williams, G.; Watts. D.C. in *NMR, Basic Principles and Progress*; Diehl, P.; Fluck.; Kosfeld, R., Eds.; Springer-Verlag, Berlin, 1971; Vol. 4, pp 271-285.
40. Schröter, K.; et al *Macromolecules* **1998**, *31*, 8966.
41. Gomez, D.; et al *Macromolecules* **2001**, *34*, 503.
42. Beiner, M. *Macromol. Rapid Commun.* **2001**, *22*, 869.

Chapter 21

Viscometry and Light Scattering of Polymer Blend Solutions

M. J. K. Chee[1], C. Kummerlöwe[2], and H. W. Kammer[1,3,*]

[1]School of Chemical Sciences, University Sains Malaysia, 11800 Minden Penang, Malaysia
[2]University of Applied Sciences Osnabrück, Albrechtstrasse 30, D–49076 Osnabrück, Germany
[3]Current address: Mansfelder Strasse 28, D–01309 Dresden, Germany

Polymer blends comprising PHB in combination with PEO and PCL, respectively, were dissolved in common good solvents. Viscometry and light scattering served studying interactions between unlike polymers in dilute solutions of the blends. Specific viscosities obey the Huggins equation to an excellent approximation in the concentration range studied. Huggins coefficients display nonlinear dependencies on blend composition. – Apparent quantities of the ternary systems, molecular mass, radius of gyration and second virial coefficient, determined by light scattering, show also non-linear variation with blend composition. The second virial coefficients versus blend composition display positive and negative deviations from additivity those are indicative of repulsion and attractions, respectively, between the polymer constituents.

Introduction

We report on viscosity and light scattering studies of ternary polymer-blend solutions to reveal dominant interactions in polymer blends. Knowledge about interactions between constituents in polymer blends is important for understanding of their phase behavior. Evaluation of polymer-polymer miscibility is especially complicated in polymer blends comprising crystallizable constituents. Solution viscometry and light scattering might then be powerful tools in assessing miscibility of the components in the amorphous state.

The viscometric method utilizes the fact that polymer-polymer interactions are reflected in the viscosity of polymer-blend solutions consisting of chemically different polymers in a common solvent. In recent years, several attempts have been made using dilute-solution viscometry to predict polymer-polymer miscibility (*1-12*). Starting point is the equation for the viscosity of a polymer solution proposed by Huggins (*13*). According to Huggins' equation, the second-order term in the viscosity as a function of polymer concentration represents interactions between constituents of the system. This equation is extended to a polymer-blend solution consisting of two polymers in a common solvent. There are polymer-solvent and polymer-polymer interactions in the system. The interactions, including polymer-polymer interactions, are reflected in Huggins' equation via hydrodynamic interactions. In a common solvent, coils have a certain hydrodynamic volume. This volume will change owing to attractions or repulsions between chemically different chains, which in turn causes alteration of the Huggins coefficient. It means, we can deduce information about polymer-polymer interactions from the Huggins coefficient. Given the two polymers are dissolved in a common good solvent, at sufficiently low concentration of polymer they form separate swollen coils that behave like hard spheres and do not interpenetrate. Hence, one expects additive behavior of intrinsic viscosities in ternary solutions. When the concentration increases, the coils will interpenetrate and the Huggins coefficient reflects not only polymer-solvent but also polymer-polymer interactions. In that way, one can determine easily parameters from measurements of viscosity that are suitable to evaluate phase behavior of a polymer blend.

There are a number of light scattering studies on ternary solutions chiefly of immiscible polymers in a common solvent (*14 – 21*). Light scattering data yield the second virial coefficient. In Flory-Huggins approximation, the second virial coefficient for polymer blend solutions can be represented by an additivity term and an excess term with respect to blend composition (*22*). The excess term is proportional to the interaction parameter of the two polymer constituents. Positive deviation from additivity is caused by a positive interaction parameter and indicates immiscibility of the two polymers. Miscibility can be inferred when the second virial coefficient as a function of composition displays negative deviation from additivity. Determination of the excess term of the second virial coefficient in ternary solutions becomes especially simple under so-called optical Θ condition (*16*, *17*, *23*). This condition is characterized by refractive index increments of opposite sign for the two polymer components. Unfortunately, for the blends under discussion no common good solvent could be found that obeys this condition.

In this paper we present quantities provided by viscosity and light scattering measurements on polymer blend solutions comprising poly(3-hydroxy butyrate) (PHB) in combination with poly(ethylene oxide) (PEO) and poly(ε-caprolactone) (PCL), respectively, in chloroform and trifluoroethanol (TFE). We focus especially on the composition dependence of Huggins coefficient and second virial coefficient for the blend solutions.

Theoretical Background

Viscosity. We start with Huggins' equation for the viscosity of polymer solutions. Formulated for a polymer blend solution, it reads

$$\eta_{spec,b} = [\eta]_b c_b + K_{Hb}[\eta]_b^2 c_b^2 \tag{1}$$

where c is the mass concentration of the macromolecules in the solvent and K_H is the Huggins coefficient, subscript b refers to polymer blend. Specific viscosity, η_{spec}, and intrinsic viscosity, $[\eta]$, are defined as usual. Quantity $[\eta]$ of eq 1 comprises dependence of the terms on molecular mass. Hence, the Huggins coefficient is independent of molecular mass to a good approximation.

Equation 1 provides a linear relationship between reduced viscosity, η_{spec}/c, and concentration c with intercept $[\eta]$, and the slope yields the dimensionless Huggins coefficient K_H. Quantities $[\eta]_b$ and $K_{H\,b}$ that follow from eq 1 refer to the blend solution. These quantities might be represented by superposition of quantities of the respective binaries. With total polymer concentration $c_b = c_1 + c_2$, it follows for intrinsic viscosity and $K_{H\,b}$ of the blend solution

$$[\eta]_b = [\eta]_1 w_1 + [\eta]_2 w_2 \tag{2}$$

$$K_{H\,b}[\eta]_b^2 = \left(\sqrt{K_{H1}}[\eta]_1 w_1 + \sqrt{K_{H2}}[\eta]_2 w_2\right)^2 + 2\kappa[\eta]_1[\eta]_2 w_1 w_2 \tag{3}$$

with mass fraction $w_i = c_i / c_b$ (i = 1,2). Eq 2 is straightforward since one expects additivity of the intrinsic viscosities for dilute solutions. The Huggins coefficient can be expressed as a second order equation of the individual quantities of the constituents plus an excess term, which is ruled by parameter κ defined by

$$\kappa \equiv [K_{H12} - (K_{H1}K_{H2})^{1/2}] \quad (4)$$

We define a solution as ideal if the excess term of eq 3 disappears, i.e., $\kappa = 0$ or $K_{H12} = (K_{H1}K_{H2})^{1/2}$. Parameter κ reveals deviations from ideal behavior that are, to a good approximation, chiefly due to thermodynamic interactions between the different macromolecular species. At sufficiently high polymer concentration, coils will interpenetrate. Then, attractions between chemically different molecules will cause swelling of the coils leading to an excess increase in viscosity as compared to perfect behavior. Hence, positive deviations from ideal behavior are indicative of attractions between the different polymer species whereas negative deviations result from repulsions. Expressing these findings in terms of coefficient κ, it follows, $\kappa > 0$ reflects miscibility whereas $\kappa < 0$ stands for immiscibility.

Light scattering. Light scattering data are organized in Zimm plots (*24*). Extrapolations of the reduced scattering intensity to scattering angle $\Theta = 0$ and concentration $c = 0$ yield the quantities of interest, the molecular mass, M_w, the second virial coefficient, A_2, and the form factor, $P(\Theta)$. The latter quantity depends on the mean square radius of gyration, $\langle S^2 \rangle$. Moreover, the optical constant K of the scattering equations depends on the refractive index increment, which is symbolized by n_c. The scattering equations, applied to ternary blend solutions, yield apparent quantities of the ternary system (*14*, *22*). These quantities can be expressed analogously to eqs 2 and 3 in terms of the corresponding binaries. It follows

$$n_{ct}^2 M_{wt} \left\langle S^2 \right\rangle_t = \sum_{i=1,2} n_{ci}^2 M_{wi} \left\langle S^2 \right\rangle_i w_i \quad (5)$$

$$A_{2t}(n_{ct}M_{wt})^2 = \left(n_{c1}M_{w1}\sqrt{A_2^{(1)}}w_1 + n_{c2}M_{w2}\sqrt{A_2^{(2)}}w_2 \right)^2 + n_{c1}M_{w1}n_{c2}M_{w2}Xw_1w_2 \quad (6)$$

where subscript t refers to the ternary system. For the apparent molecular mass and the apparent radius of gyration, eq 5, we get simply additivity relations since these quantities are coefficients of first order terms in concentration. The second virial coefficient of the ternary system can be represented by a perfect and an excess part. The excess term characterizes deviations of the real blend solution from the perfect blend solution. Parameter X of eq 5 might be seen as an excess second virial coefficient. It is defined in a similar way as parameter κ of eq 4

$$X \equiv \frac{G_{12}}{V_S} - 2\sqrt{A_2^{(1)} A_2^{(2)}} \tag{7}$$

Quantities G_{12} and V_S are the derivative of Gibbs free energy per NkT with respect to concentrations of polymer components 1 and 2 in the solution and the molar volume of the solvent, respectively. Experimental results of light scattering are discussed chiefly in terms of eqs 5 and 6. In Flory-Huggins approximation and for small total volume fraction of polymer in the solution, quantities in eq 7 read $\rho_1\rho_2 G_{12} = 1 + \chi_{12} - \chi_{S1} - \chi_{S2}$ and $2V_S\rho_i^2 A_2^{(i)} = 1 - 2\chi_{Si}$ where ρ_i is the density of the amorphous component i, and χ_{ij} is the interaction parameter between components i and j. It follows for the interaction parameter χ_{12} from eq 7

$$\chi_{12} = V_S\left[\rho_1\rho_2 X - \left(\rho_1\sqrt{A_2^{(1)}} - \rho_2\sqrt{A_2^{(2)}}\right)^2\right] \tag{8}$$

Equation 8 demonstrates that negative excess X of the second virial coefficient leads to $\chi_{12} < 0$, which is indicative of miscibility of the two polymers 1 and 2.

Experimental

Materials. Molecular characteristics of the polymers used in this study are given in Table I. Polymers were purified by dissolution and precipitation in chloroform and methanol, respectively, and finally dried in vacuum at 50 °C for 48 h. Polymer blend solutions for viscometry and light scattering were prepared in chloroform and trifluoroethanol (TFE), respectively, as solvents. Solvents were used without further purification.

Viscosity measurement. Ternary solutions of PHB with PEO and PCL, respectively, were prepared by dissolving polymer mixtures having different weight ratios in chloroform at 50 °C. The solutions were diluted to the designated volume at room temperature to adjust desired concentrations. Concentrations ranged between 1 and 4g/l. Purification of solutions was done by filtration using nylon membrane filter prior to viscosity measurement. Ubbelohde viscometers of appropriate sizes were employed to determine the relative viscosities, η_r, of the blend solutions. In any case, the flow time of pure solvent exceeded 200 seconds so as to minimize experimental errors. Values of η_r fall within the range of 1.2 – 2.0. Under these experimental conditions the kinetic energy and shear corrections were negligible. Viscosity measurements were carried out at 298 K. Temperature control was recorded to ± 0.1 K.

Light scattering experiments. Ternary solutions of PHB and PEO or PCL, respectively, were prepared in TFE in the same way as described under viscosity measurement. Solutions of five different total polymer concentrations, ranging from 1 to 5 g/l, were studied. Solutions were filtered prior to the light scattering experiments.

Table I. Characteristics of the polymer samples and the solvent

Sample	M_w[a]/ kgmol^{-1}	M_w/M_n[b]	ρ/gcm^{-3} [c]	M_o/gmol^{-1}	*Source*
PHB	102	2.32	1.15 (*25*)	86	Aldrich
PEO	102	1.91	1.13 (*26*)	44	Acros
PCL	130	1.95	1.08 (*27*)	114	Union Carbide
TFE			1.373 (*28*)	100	Aldrich

[a] obtained from light scattering in this study
[b] obtained from GPC analysis
[c] amorphous density at 25 °C

Refractive index increments were measured directly in a differential refractometer (Brice-Phoenix). Unfortunately, polymer blend solutions in chloroform could not be used in light scattering experiments since the refractive index increment of PHB solutions in chloroform turned out to be very small. A photometer, equipped with a Helium-Neon Laser from Spectra-Physics (λ_o = 632.8 nm), a Thorn-Emi photo multiplier, Model 9863/100KB, which is attached to a goniometer SP-81 from ALV-Laser GmbH, Langen (Germany), allowed to monitor the scattering intensity.

Results and Discussion

Viscometry. Data of specific viscosity over blend concentration, η_{spec}/c, plotted versus concentration c, for the blend solutions in chloroform fit Huggins equation 1 with high correlation; typically, correlation coefficients amount to r = 0.9998. Regression analysis shows that intrinsic viscosities have errors of around 0.2 %. Selected results for intrinsic viscosities and Huggins coefficients are compiled in Table II. We note, both PCL and PEO display higher intrinsic viscosities than that of PHB. Moreover, intrinsic viscosities, $[\eta]$, vary with blend composition according to eq 2 in excellent approximation. The respective regression functions are also listed in Table II.

Table II. Selected intrinsic viscosities and Huggins coefficients for blends with PHB in chloroform

Sample	$[\eta]$/cm^{3}g^{-1}	K_H	$K_{H\,id}$
PEO/PHB			
100/0	173.2 ± 0.2	0.3347 ± 0.0035	
70/30	151.1 ± 0.2	0.3451 ± 0.0045	0.3373
50/50	140.4 ± 0.4	0.3529 ± 0.0092	0.3395
30/70	112.5 ± 0.4	0.364 ± 0.013	0.3424
0/100	91.2 ± 0.2	0.3487 ± 0.0082	
η_b(PHB/PEO)/cm^{3}g^{-1} = 173.5 (1 – w_{PHB}) + 92.9 w_{PHB}; correlation: 0.992			
PCL/PHB			
100/0	207.6 ± 0.2	0.3180 ± 0.0033	
70/30	162.3 ± 0.3	0.3146 ± 0.0065	0.3228
50/50	142.2 ± 0.2	0.3131 ± 0.0043	0.3272
30/70	119.4 ± 0.2		0.3334
η_b(PHB/PCL)/cm^{3}g^{-1} = 196.8 (1 – w_{PHB}) + 89.5 w_{PHB}; correlation: 0.992			

Experimental values of the Huggins coefficients, K_H were determined according to eq 1 from the slopes of functions η_{spec}/c versus c. Values of $K_{H\,id}$ were calculated after the first term of eq 3. Optical inspection of data in Table II shows that we observe outside experimental errors (K_H - $K_{H\,id}$) > 0 for PEO/PHB blends and (K_H - $K_{H\,id}$) < 0 for PCL/PHB blends. These results indicate miscibility of the two constituents in blends of PHB and PEO whereas immiscibility in blends with PCL. This can be seen more clearly in Figure 1 where experimental values of K_{Hb} are depicted versus blend composition. Positive and negative deviations from ideal behavior become evident. Relevant results are summarized in Table III. Parameters κ are the result of regression calculations.

For the blend PEO/PHB, it becomes obvious that the value of K_{H12} exceeds significantly the values of K_{Hi} for the parent constituents. The opposite is true for blends of PCL and PHB. This fact points towards miscibility of the components in the former case and immiscibility in the latter case. These conclusions are consistent with results published in references (*29* – *31*). Martuscelli et al. inferred miscibility of PHB and PEO from melting point depression (*29*). Measurements of glass transition temperatures supported this conclusion (*30*). On the other hand, detection of glass transition temperatures and morphological studies provided evidence for immiscibility of PHB and PCL (*31*).

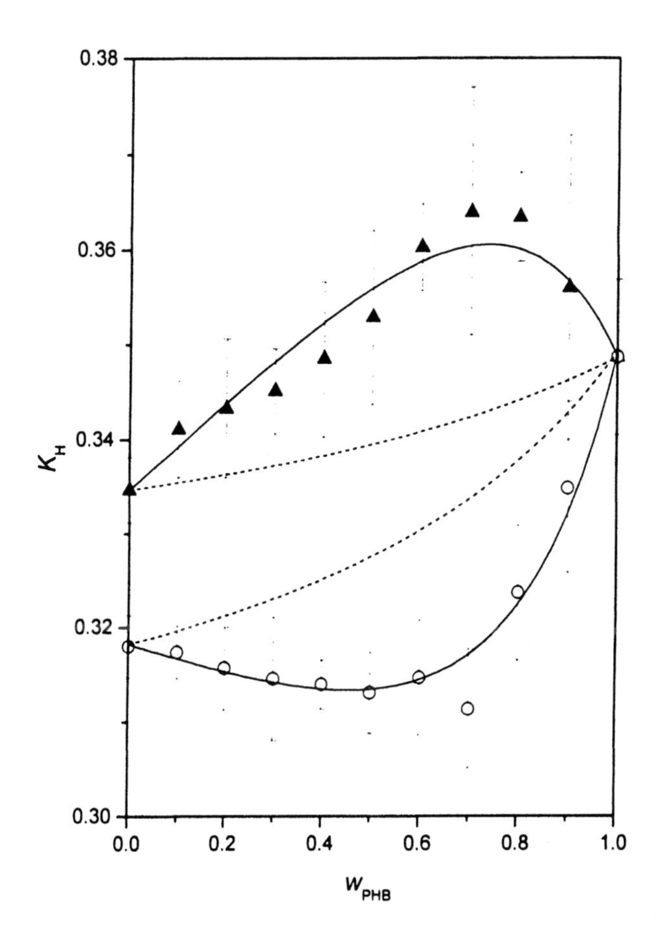

Figure 1. Huggins coefficient versus blend composition. The dashed curves represent perfect behavior according to the first term of eq 3; the solid curves are second order regression curves. Data refer to solutions in chloroform.
▲ *PEO/PHB;* ○ *PCL/PHB*

Table III. Huggins coefficients and parameter κ for the blends in chloroform

System	K_{H12}	$(K_{H1}K_{H2})^{0.5}$	κ
PEO/PHB	0.378	0.341	0.037
PCL/PHB	0.300	0.333	- 0.033

Light scattering. Scattering data of the blend solutions in TFE were organized in Zimm plots that provide quantities of interest. Refractive index increments, apparent molecular mass, radius of gyration and second virial coefficient were determined as functions of blend composition. It turned out that all quantities display non-linear dependencies on blend composition. Selected results are listed in Table IV. The refractive index increments show slightly positive deviations from additivity. Experimental data for the refractive index increments can be fitted by the following regression functions

$$\left(\frac{dn}{dc}\right)_{PEO/PHB} = \left[0.1680w_{PEO} + 0.1475(1-w_{PEO}) + 0.0271w_{PEO}(1-w_{PEO})\right]cm^3g^{-1} \quad (9)$$

$$\left(\frac{dn}{dc}\right)_{PCL/PHB} = \left[0.1661w_{PCL} + 0.1475(1-w_{PCL}) + 0.0214w_{PCL}(1-w_{PCL})\right]cm^3g^{-1} \quad (10)$$

Table IV. Results of light scattering for blend solutions in TFE at 25°C

PHB/PEO	(dn/dc)/ cm^3g^{-1}	M_{wt}/ $kgmol^{-1}$	A_{2t} 10^4/ cm^3molg^{-2}	$\langle S^2\rangle_t^{1/2}$ / nm	χ_{1S}
100/0	0.1480	102	32.2	29.5	0.20
70/30	0.1586	91	23.7	17.3	
50/50	0.1640	83	20.6	15.9	
30/70	0.1680	87	18.0	20.4	
0/100	0.1666	102	21.85	17.4	0.30
PHB/PCL					
70/30	0.1567	109	46.4	23.3	
50/50	0.1626	114	51.1	34.1	
30/70	0.1648	129	52.0	39.5	
0/100	0.1671	130	50.6	46.1	0.07

The second virial coefficients for the blend constituents yield interaction parameters $\chi < ½$, which confirms that TFE is a good solvent for PHB, PEO and PCL. Moreover, Table IV reveals that the experimentally determined apparent molecular masses, M_{wt}, display in blends both negative and positive deviations from eq 5 (with $\langle S^2\rangle = 1$).

Figure 2 presents the molar coil volume $N_A\langle S^2\rangle^{3/2}$ for the blend solutions. We observe that this volume does not change markedly in PHB/PCL solutions as long as PHB is in excess. For lower content of PHB ascending coil volume occurs. In PHB/PEO blend solutions the volume descends and is at blend composition 50/50 around six times smaller than in PHB solutions (2.4/15.5). In both cases, we observe negative deviation from behavior after eq 5.

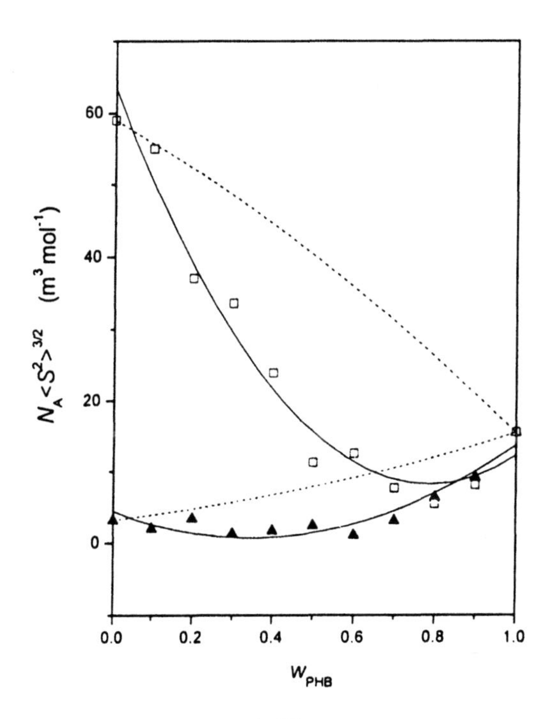

Figure 2. Molar coil volume, $N_A<S^2>^{3/2}$, versus blend composition for blend solutions in TFE at 25°C. The solid curves indicate regression functions; the dashed curves give the molar volume after eq 5.
▲ – PEO/PHB, □ – PCL/PHB

The dependence of the second virial coefficient times square of apparent molecular mass, A_2M^2, on blend composition is shown in Figure 3. When one defines after eq 6 perfect behavior of the blend solutions by $X = 0$, we see positive and negative deviations from perfect behavior that can be characterized by parameter X. The regression analysis yields

$$X_{PEO/PHB} = -0.0033 \text{ molcm}^3\text{g}^{-2} \qquad X_{PCL/PHB} = 0.0023 \text{ molcm}^3\text{g}^{-2} \qquad (11)$$

The first equation immediately allows after eq 8 the conclusion $\chi_{PEO/PHB} < 0$, which is indicative of miscibility of the two polymers. The interaction parameters between the polymers can be estimated from eq 8 and use of eq 11. It follows for the PHB blend with PEO $\chi = -0.32$ and with PCL $\chi = 0.20$.

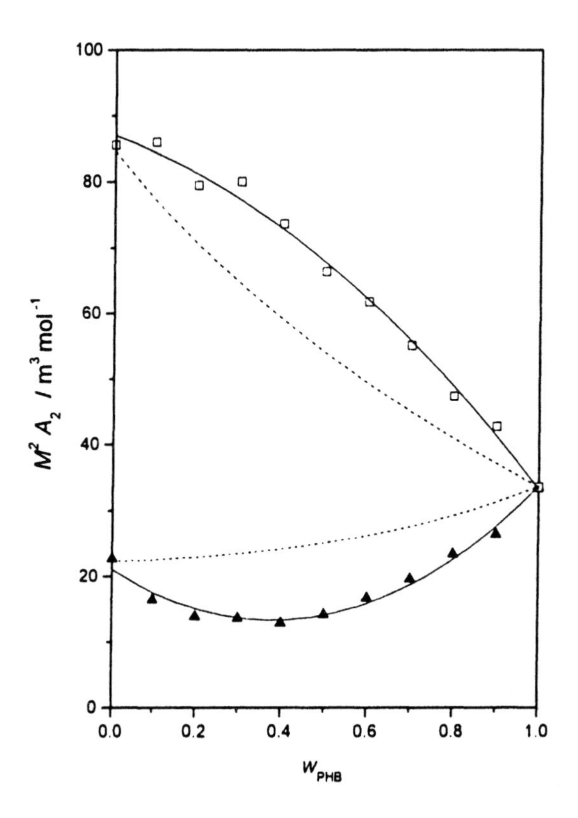

Figure 3. Second virial coefficient times square of molecular mass versus mass fraction for PEO/PHB (▲) and PCL/PHB (□) in TFE at 25 °C. The solid curves represent the regression functions of second order whereas the dashed curves were calculated after eq 6 with X = 0 and use of the regression functions for (dn/dc).

We note that the second virial coefficient itself displays qualitatively the same behavior as shown in Figure 3. This result we may discuss qualitatively as follows. The 2nd virial coefficient of PHB decreases with addition of PEO. One may say the solution approaches apparently closer to Θ - conditions or to ascending ability of the coils for interpenetration. Negative deviations from perfect behavior occur which means ability of the coils towards interpenetration is favored more than in the perfect blend solution. This effect is seen as attraction between PHB and PEO or miscibility of the two polymer constituents. The opposite effect occurs with addition of PCL to a PHB solution; the 2nd virial coefficient increases. It means the solvent becomes increasingly better or the coils repel more vigorously each other. This might be seen as reflection of immiscibility of the polymer constituents. These results are in perfect qualitative agreement with the results from viscosity studies of the same polymers in chloroform, given above.

We add here that dependencies shown in Figures 2 and 3 are consistent. Adopting the relationship between $N_A<S^2>^{3/2}$ and M^2A_2 proposed by Yamakawa (*32*) also for the ternary system, we have

$$M^2A_2 = 4\pi^{3/2}N_A<S^2>^{3/2}\Psi \tag{12}$$

where Ψ is the so-called interpenetration function which disappears under Θ - conditions. For discussion of eq 12, we formulate quantity M^2A_2 as in eq 6. For simplicity, we ignore the composition dependence of quantity n_{ct}. Then, the excess coefficient $(M^2A_2)^E$ of eq 6 reads

$$(M^2A_2)^E = 2\left[\left(A_2M^2\right)_{12} - M_1M_2\sqrt{A_2^{(1)}A_2^{(2)}}\right] \tag{13}$$

for the binary mixture of polymers 1 and 2. Analogously, we can define the coefficient of the excess volume $V^E = 2\left[V_{12} - \sqrt{V_1V_2}\right]$ with $V \equiv N_A<S^2>^{3/2}$. Figures 2 and 3 show $V^E < 0$ for the blends under discussion and $(M^2A_2)^E > 0$ for PCL/PHB as well as the opposite for PEO/PHB blends, respectively. If we define after eq 12 (ignoring numerical factors)

$$\Psi_{12} \equiv \frac{\left(A_2M^2\right)_{12}}{\sqrt{V_1V_2}}$$

it follows for PCL/PEO

$$\Psi_{12} - \sqrt{\Psi_1\Psi_2} > 0 \tag{14}$$

The inequality of eq 14 reveals that the interpenetration function for PCL/PHB increases as compared to the perfect system, i.e. ability for interpenetration of the chains decreases. The opposite conclusion can be drawn for the PEO/PHB blends. With eqs 13 and 14 and the result of Figure 3, it follows for PCL/PHB

$$1 < \frac{\left(A_2 M^2\right)_{12}}{\sqrt{V_1 \Psi_1 V_2 \Psi_2}} = \frac{(V\Psi)_{12}}{\sqrt{V_1 \Psi_1 V_2 \Psi_2}} > \frac{(V)_{12}}{\sqrt{V_1 V_2}}$$

where the last inequality follows according to eq 14. Hence, positive excess in quantity M^2A_2 does not necessarily generate positive excess in coil volume $<S^2>^{3/2}$.

Conclusions

Viscometric studies on polymer blend solutions of PHB with PEO and PCL, respectively, in chloroform show that intrinsic viscosities of the blends are linear functions of blend composition to an excellent approximation. Huggins coefficients display non-linear dependencies on blend composition. Both positive and negative deviations from perfect behavior can be observed. Blend solutions of PEO and PHB display positive deviations that originate from attractions between the polymer species. Opposite deviations could be detected for blends of PHB and PCL indicating repulsions between the coils that lead to immiscibility.

Light scattering studies on blend solutions with PHB yield qualitatively the same results on miscibility of the constituents as the viscosity studies. Blend solutions of PHB and PEO display negative deviations of the second virial coefficient from perfect behavior, which is indicative of miscibility of the two constituents. The opposite is observed for PHB/PCL blend solutions. Generally, we note that all quantities characterizing the blend solutions display non-linear dependencies on blend composition.

References

1. Dondos, A.; Benoit, H. *Makromol. Chem.* **1975**, *176*, 3441.
2. Pierre, E.; Dondos A. *Eur. Polym. J.* **1987**, *23*, 347.
3. Kulshreshtha, A. K.; Singh, B. P.; Sharma,Y. N. *Eur. Polym. J.* **1988**, *24*, 29.
4. Chee, K. K. *Eur. Polym. J.* **1990**, *26*, 423.
5. Kent, M. S.; Tirrell, M. *Macromolecules* **1992**, *25* , 5383.
6. Moszkowicz, M.J.;Rosen, S.L. *J. Polym. Sci.,Polym.Phys.Edn.* **1979**, *17*, 715.

7. Cragg, L. H.; Bigelow, C. C. *J. Polym. Sci.* **1955**, *16*, 177.
8. Zhu, P. P.; Wang , S. *Eur. Polym. J.* **1997**, *33*, 411.
9. Haiyang, Y.; Pingping, Z.; Shiqiang, W.; Yiming, Z.; Qipeng, G. *Eur. Polym. J.* **1998**, *34*, 1303.
10. Pingping, Z.; Haiyang, Y.; Yiming, Z. *Eur. Polym. J.* **1999**, *35*, 915.
11. Neiro, S. M.; Dragunski, D. C.; Rubira, A. F.; Muniz, E. C. *Eur. Polym. J.* **2000**, *36*, 583.
12. Shanfeng, W.; Gaobin, B.; Pingping, W.; Zhewen, H. *Eur. Polym. J*. **2000**, *36*, 1843.
13. Huggins, M. L. *J. Am. Chem. Soc.* **1942**, *54*, 2746.
14. Kratochvil, P.; Vorlicek, J.; Strakova, D.; Tuzar, Z. *J. Polym. Sci., Polym. Phys. Ed.* **1975**, *13*, 2321.
15. Kratochvil, P.; Strakova, D.; Stejskal, J.; Prochazka, O. *Eur. Polym. J.* **1983**, *19*, 189.
16. Fukuda, T.; Nagata, M.; Inagaki, H. *Macromolecules* **1987**, *20*, 654.
17. Fukuda, T.; Nagata, M.; Inagaki, H. *Macromolecules* **1987**, *20*, 2173
18. Kuleznev, V. N. *Poly. Sci. Ser.B* **1993**, *35*, 1156
19. Sun, Z.; Wang, C. H. *Macromolecules* **1996**, *29*, 2011
20. Sun, Z.; Wang, C. H. *J. Chem. Phys.* **2000**, *112*, 6844
21. Posharnowa, N.; Schneider, A.; Wünsch, M.; Kuleznew, V.; Wolf, B. A. *J. Chem. Phys.* **2001**, *115*, 9536
22. Stockmayer, W. H. *J. Chem. Phys.* **1950**, *18*, 58
23. Fukuda, T.; Nagata, M.; Inagaki, H. *Macromolecules* **1984**, *17*, 548
24. Zimm, B. *J. Chem. Phys.* **1948**, *16*, 1093
25. Barham, P.J.;Keller,A.;Otun,E.L.;Holmes, P.A. *J. Mater. Sci.* **1984**, *19*, 2781
26. van Krevelen, D. W. *Properties of Polymers*; Elsevier: Amsterdam, 1976
27. Crescenzi, V. G.; Manzini, G.; Calzolari, G.; Borri, C. *Europ. Polym. J.* **1972**, *8*, 449
28. *Aldrich Information*; 2002
29. Avella, M.; Martuscelli, E. *Polymer* **1988**, *29*, 1731
30. Goh, S.H.; Ni, X. *Polymer* **1999**, *40*, 5733
31. Reeve, M.S.; McCarthy, S.P.; Gross, R. A. *Macromolecules* **1993**, *26*, 888.
32. Yamakawa, H. *Modern Theory of Polymer Solutions*; Harper and Row: New York, 1971, p 172.

Chapter 22

Spectroscopic Studies of the Diffusion of Water and Ammonia in Polyimide and Polyimide–Silica Hybrids

P. Musto[1], G. Ragosta[1], G. Scarinzi[1], and G. Mensitieri[2]

[1]Institute of Chemistry and Technology of Polymers, National Research Council of Italy, Via Campi Flegrei 34, 80078 Pozzuoli (NA), Italy
[2]Department of Materials and Production Engineering, University of Napoli Federico II, P.le Tecchio 80, 80125 Napoli, Italy

Mass transport of low molecular weight penetrants in polyimide and silica/polyimide hybrids has been investigated using *time-resolved* FTIR spectroscopy and gravimetric analysis. In particular, transport of reacting (ammonia) and non-reacting (water) penetrants has been studied as a function of penetrant concentration, evidencing peculiar features related to the presence of the inorganic phase in the hybrid systems. For the case of water, diffusivity and sorption equilibrium have been evaluated in an activity range between 0.1 and 0.75. Free water as well as molecular aggregates have been detected in both systems. In the case of ammonia, its reactivity with polyimide has been directly observed, and the rection mechanism elucidated. Furthermore, it has been possible to discriminate diffusion and reaction phenomena due to the different time scales of the two processes.

Polyimides (PI's) represent an important class of high-performance polymeric materials, widely used because of their outstanding thermal-oxidative stability, excellent mechanical properties and very high glass transition temperature. These materials find applications as electrical and electronic insulators, and in separation-membrane technology *(1-3)*. More recently, they have been proposed as matrices for organic/inorganic (O-I) hybrid systems prepared by the sol-gel route *(4-7, 9)*. Hybrid systems are an interesting class of new-generation materials which combine the relevant properties of a ceramic phase (heat resistance, high-temperature mechanical performances, low thermal expansion) and those of organic polymers (toughness, ductility and processability). Polyimides are among the few matrices well suited for use in connection with sol-gel technology, which is a complex process accomplished in aqueous media and with formation of water and alcohols as by-products.

In the present study the diffusion behaviour in a polyimide has been compared with that of a polyimide-silica hybrid having a nanoscale morphology. As penetrants, two low molecular weight compounds were selected, namely water and ammonia. Water sorption has obvious implications with respect to the performances of the materials as insulators, as well as in their aging behaviour. Of particular relevance, in this respect, is the state of aggregation of the absorbed water molecules and the molecular interactions they form with the polymeric matrix and/or with the inorganic phase in the case of the hybrid system. These issues have been addressed by a careful analysis of the infrared spectra of the penetrant in the ν_{OH} frequency range.

Ammonia was selected as the second molecular probe to be investigated since it has been reported in earlier investigations *(8)* that the polyimide, which is otherwise highly stable to solvents, displays a certain reactivity when contacted with basic reagents. These findings were based on gravimetric measurements which showed a marked non-Fickian behaviour characterized by a prolonged sorption process without approaching sorption equilibrium. The subsequent desorption evidenced that a substantial amount of ammonia was irreversibly retained in the polymer as a consequence of reaction of ammonia with imide linkages. Information on the reaction and its possible mechanism and kinetics were tentatively inferred from a Danckwerts analysis of sorption data *(8)*. *Time-resolved* FTIR spectroscopy applied to diffusion studies offers the unique opportunity to monitor in real time the reagent diffusion and the chemical reaction of the penetrant with the polymeric substrate. The data presented here confirm the propensity of the polyimide to undergo a chemical attack by ammonia and the reaction mechanism is elucidated. The same measurements carried out on the hybrid system revealed that the inorganic phase accelerates the reaction.

Results and Discussion

Sorption of Water in Polyimide and the Hybrid

In figures 1 and 2 are reported the water spectra in the ν_{OH} stretching region collected at sorption equilibrium (a = 0.4) for the polyimide and the hybrid, respectively.

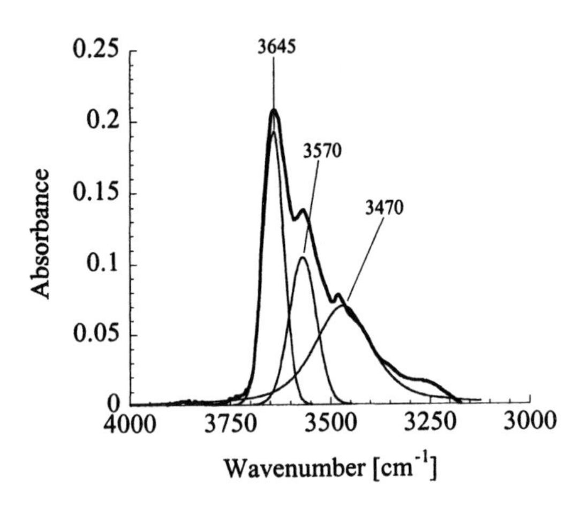

Figure 1: Curve fitting analysis of the spectrum representative of water absorbed at equilibrium (a = 0.4) in neat polyimide. The figure displays the resolved components (thin lines) and the experimental profile (thick line).

The complex spectral profiles were resolved by curve fitting analysis and the resulting components are reported in the same figures. These spectra have been interpreted on the basis of a simplified association model (*10, 14, 15*), whereby three different water species (S_0, S_1 and S_2) could be spectroscopically distinguished. In particular, with S_0 we indicate water molecules which do not establish any H-bond, while with S_1 and S_2 we designate, respectively, water molecules with one H atom or two H atoms involved in H-bonding with a proton acceptor. In the spectra reported in figs. 1 and 2 the peaks at the highest frequency (3645, 3638 cm^{-1}) can be associated with S_0 molecules, those located at the next lower frequency (3570, 3593 cm^{-1}) to S_1 molecules, while the various components appearing below 3570 cm^{-1} can be associated to S_2 species. The observation that in the spectrum of Fig. 1 there is only one S_2 component, while in Fig. 2 three distinct peaks are identified, implies the presence of a single S_2 adduct in pure polyimide and of a series of H-bonding complexes in the polyimide/silica system, characterized by different interaction strengths (the lower the frequency of the component, the higher the strength of the H-bond).

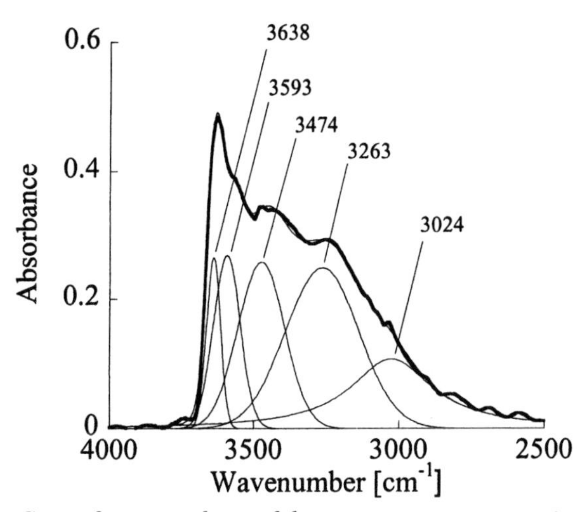

Figure 2: Curve fitting analysis of the spectrum representative of water absorbed at equilibrium (a = 0.4) in the hybrid sample. The figure displays the resolved components (thin lines) and the experimental profile (thick line).

It is likely that S_0 and S_1 species are characterised by high molecular mobility. S_0 molecules should be confined into excess free volume (microvoids) or molecularly dispersed with no H-bonding interactions (bulk dissolution), while S_1 may either interact moderately by H-bonding or may represent self-associated dimers. On the other hand, S_2 molecules are expected to be characterized by a much lower mobility; these species should be firmly bound to specific sites present in the matrix or be involved in clusters of more than two water molecules. When comparing the two investigated samples, significant differences emerge in the species distribution at sorption equilibrium. A higher amount of sorbed water is evident in the case of the hybrid sample, as well as an increase of the contribution of strongly interacting species. These findings can be interpreted by assuming that, in the case of the hybrid, the inorganic phase, which contains a significant amount of interacting OH groups, promotes strong H-bonding interactions between penetrant molecules and the silica phase, which results in an enhanced solubility. Conversely, in the case of pure polyimide, due to the low tendency of the matrix to form H-bonding, the number of water molecules involved in molecular interactions is much lower and most likely related to water clustering, which increases as the concentration of sorbed water increases.

Sorption isotherms evaluated gravimetrically and reported in figure 3 are consistent with the proposed interpretation. In fact, solubility in the hybrid is higher and a significant difference in the shape of the isotherms is observed. In the case of polyimide the upward concavity at higher activities can be related to the occurrence of water clustering which contributes to an increase of solubility with activity. Conversely, in the case of the hybrid, the slightly downward concavity of the isotherm suggests water adsorption on specific sites present in the inorganic phase. No upturn is evident in this case, owing to the increased hydrophilicity of the substrate which prevents water from clustering.

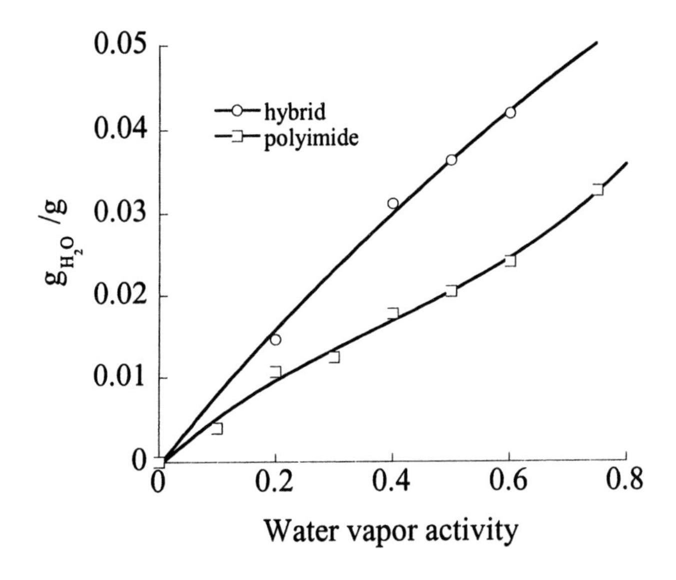

Figure 3: Sorption isotherms at 30 °C expressed as grams of sorbed water per gram of matrix vs water vapour activity for neat polyimide (open squares) and hybrid sample (open circles).

Time-resolved FTIR spectroscopy allowed a precise monitoring of sorption/desorption kinetics: examples of such curves, evaluated from the absorbance area of the stretching band of water are reported in Figure 4a for the polyimide and in Figure 4b for the hybrid. For both materials at all the investigated activities, desorption kinetics is slower than sorption, which is a typical feature of Fickian systems where mutual diffusivity is an increasing function of penetrant concentration. When mutual diffusivity increases with concentration, the shape of the sorption curve does not depend significantly on the functional form of the diffusion coefficient.

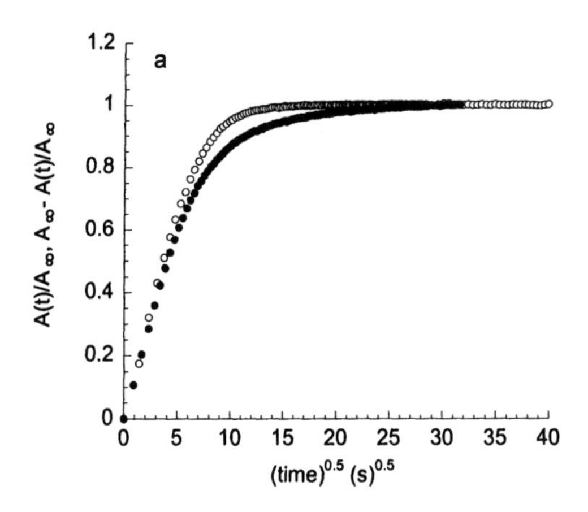

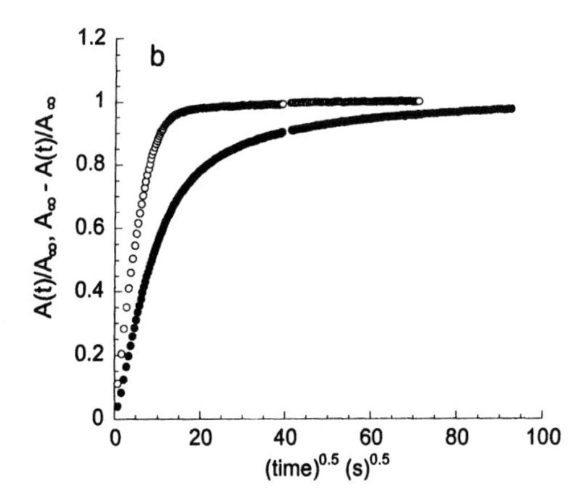

Figure 4: Sorption and desorption kinetics (a = 0.4) in terms of $A(t)/A_\infty$ for sorption, and of $[A_\infty - A(t)]/A_\infty$ for desorption.(a): pure polyimide; (b):hybrid.

The diffusion coefficient, D, evaluated from the Fick's relationship *(16)* is reported, as a function of water vapour activity, in Figure 5, which shows that diffusivities are very close in both materials. In this analysis D represents a diffusivity averaged over the whole range of water concentrations established during the sorption test inside the sample. Thus, although comparison of sorption and desorption kinetics points to a slight dependence of diffusivity on water concentration, these averaged values result to be independent.

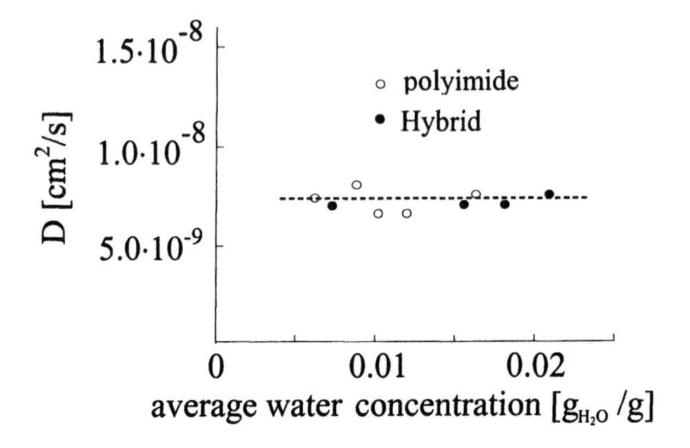

Figure 5. Diffusion coefficient, D, as a function of average water concentration for pure polyimide and the hybrid.

Sorption of Ammonia in Polyimide and the Hybrid

Time-resolved FTIR spectroscopy is particularly useful whenever the diffusion phenomenon is accompanied by changes in the molecular structure of the polymeric substrate since, in principle, both phenomena can be concurrently monitored. As an example of this particular application, we report preliminary results obtained for the case of sorption of ammonia in polyimide and polyimide hybrids. This analysis is also relevant in techonlogical applications in view of possible use of the investigated materials in aggressive environments.

Time evolution of the sample spectrum in the 3480–3000 cm^{-1} range is reported in figure 6 with reference to the sorption test conducted on the polyimide at 100 Torr and 30°C. The diffusing ammonia is clearly discernible, giving rise to two well resolved peaks at 3400 and 3306 cm^{-1} due, respectively, to the asymmetric and symmetric stretching vibrations of the diffusing molecule. A gradual absorbance increase in the whole 3480–3000 cm^{-1} range, which is observed in the same time frame, is due to concurrent reaction of ammonia with imide moieties. After a short time (about 5 minutes) the intensity of both ammonia peaks reaches a plateau value, indicating the attainment of an apparent sorption equilibrium.

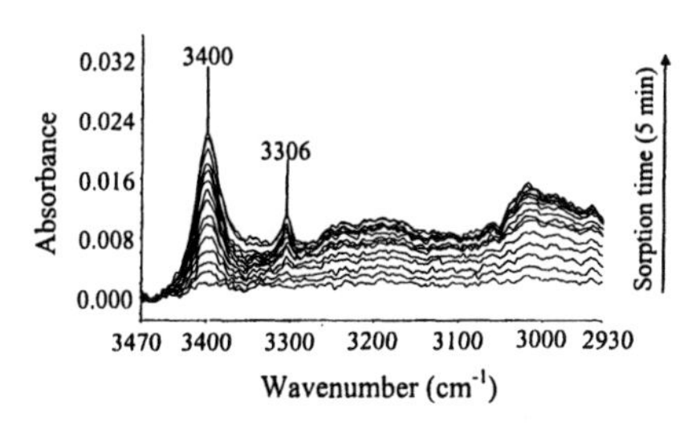

Figure 6: FTIR transmission spectra in the 3480-3000 cm^{-1} range, collected at increasing times in the first 5 minutes of the sorption of ammonia in neat polyimide [p(NH_3)=100 Torr, T=30°C].

At longer times, the spectrum of the matrix displays conspicuous changes, clearly indicating the reactivity of the system. As an example, in figure 7, are compared the spectra collected at 0, 14 and 28 hours, in the range 2000 – 500 cm^{-1} for the test carried out on pure polyimide at 30°C and 760 Torr. This comparison evidences an extensive reduction of the peaks arising from the imide moiety (1780, 1725, 1377 and 1170 cm^{-1}) and the concurrent increase of absorptions at 1670, 1606 and 1408 cm^{-1}. The latter can be attributed to the normal modes of the amide functionality which represents the reaction product .

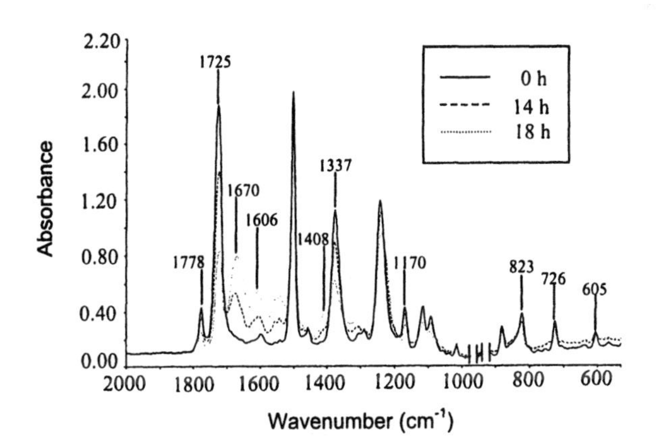

Figure 7: FTIR transmission spectra in the 2000-1000 cm^{-1} range, collected at three times (0, 14 and 28 hours) during the sorption of ammonia in neat polyimide [p(NH_3)=760 Torr, T=30°C].

According to the above evidences, the likely mechanism is the following:

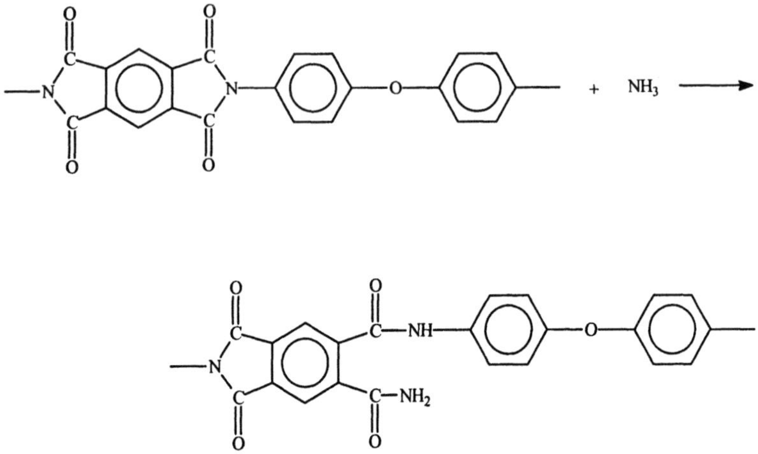

The evolution with time of the peaks at 3400 and 3306 cm^{-1} in terms of normalized absorbance (i.e. $A(t)/A_\infty$, with A(t) absorbance at time t and A_∞ absorbance at sorption equilibrium) is reported in figure 8. The interpretation of sorption kinetics using ideal Fickian behavior (*16*) gave a calculated ammonia diffusivity equal to $2.2 \cdot 10^{-9}$ cm^2/s.

A gravimetric analysis of ammonia diffusion in Kapton® polyimide at 30°C has been previously reported by Iler et al. (*8*) for the same experimental conditions adopted in the present study. These investigators interpreted sorption kinetics curves using an approach proposed by Danckwerts (*17*) for the case of diffusion-reaction processes with a pseudo-first order reaction kinetics. They calculated diffusivity values for ammonia which were at least two orders of magnitude lower than those expected on the basis of the size of the diffusing molecule. This is shown in figure 9, where the diffusion corefficients at infinite dilution for various penetrants in Kapton® are reported as a function of the van der Waals volume, b, of the penetrant (*8*). Iler et al. attributed the deviation to strong molecular interactions occurring between ammonia and the matrix. Actually, the value we have estimated in this study using *time-resolved* FTIR spectroscopy is very close to that expected on the basis of the molecular size of the penetrant, the slight positive deviation being reasonably related to the increase of diffusivity with penetrant concentration.

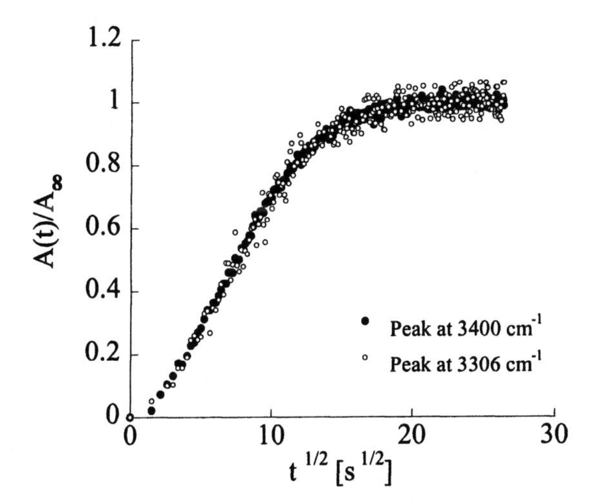

Figure 8: $A(t)/A_\infty$ for the sorption test of ammonia in neat polyimide [$p(NH_3)$=100 Torr, T=30°C].

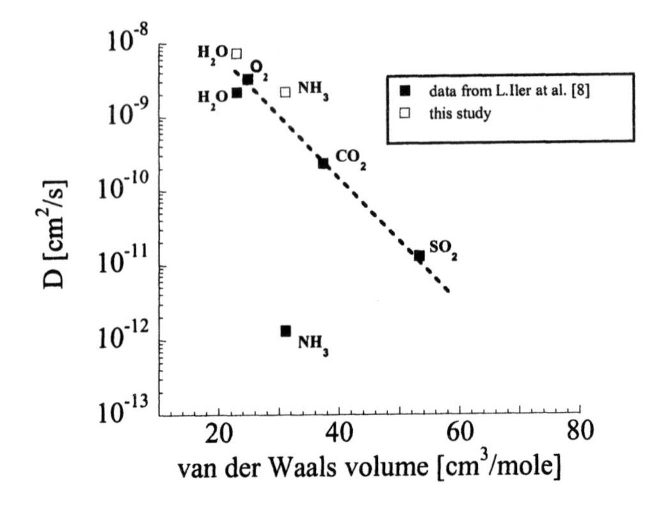

Figure 9: Correlation of diffusion coefficient at infinite dilution for various penetrants in Kapton® at 30°C with van der Waals volume, b, of the penetrant. Full squares refer to data from Iler et al. (8), open squares refer to ammonia and water diffusivities evaluated in the present study.

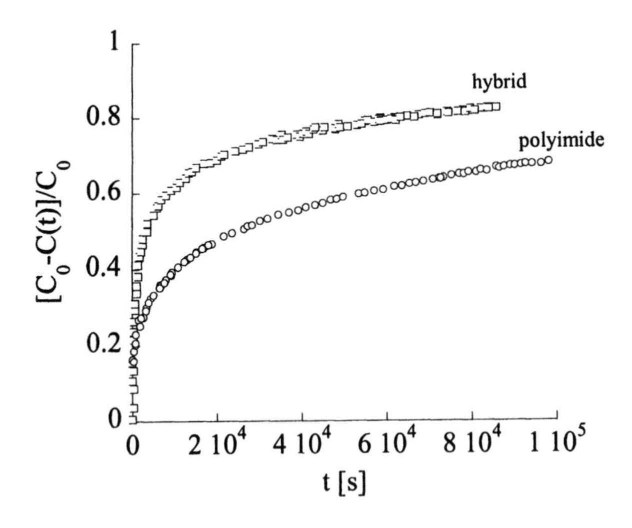

Figure 10: Relative conversion of the imide groups at 30 °C as a function of time for the neat polyimide and the hybrid sample. [$p(NH_3)$=760 Torr, T=30°C]

The kinetics of the chemical reaction with time can be followed by monitoring the absorbance of several peaks related to the imide ring. In particular we have measured the peak centered at 726 cm^{-1}, since it is well resolved both for the neat polyimide and the hybrid.

The behavior of the two materials is compared in figure 10 at 770 Torr. It is found that the polyimide reactivity in the examined conditions is conspicuous, leading to final conversion of imide rings exceeding 60 %. The reaction rate and the final conversion are considerably higher in the case of the hybrid, and this effect may be related to a catalytic activity of the inorganic phase towards the ammonolysis of imide rings. A similar effect has been documented in a recent publication *(9)* in terms of the increased rate of imidization of poly(amic acid) in the presence of silica nanophase.

Concluding Remarks

Water sorption and transport in polyimide and polyimide/silica hybrids has been investigated by combining time-resolved FTIR spectroscopy measurements and gravimetric analysis. For the neat polyimide, the spectropic analysis evidenced that the prevailing penetrant species is monomeric water along with minor contribution from self-associated species (dimers and multimers). Conversely, in the hybrid system the amount of sorbed water increases with respect to the control material due to a substantial increment of H-bonded water, likely because of molecular interaction occurring with the inorganic phase.

The transport behavior is substantially Fickian in all cases investigated with diffusivity essentially independent of concentration.

The same experimental approach has been employed to study transport of ammonia in both types of materials. A significant reactivity of the penetrant with the organic phase has been evidenced by time-resolved FTIR spectroscopy. Analysis of the time evolution of the IR spectra allowed us to propose a likely reaction mechanism. an appropriate selection of experimental conditions allowed a wide difference between the characteristic times for diffusion and reaction of ammonia, thus making it possible to monitor concurrently both processes. A considerable catalytic activity of the inorganic phase towards the ammonolysis was evidenced and its possible origin was discussed.

References

1. *Polyimides; Fundamentals and Applications,* Gosh M.K.; Mittal K.L., Eds., Marcel Dekker, New York, 1996.
2. *Polyamic Acids and Polyimides: Synthesis, Transformation and Structure,* Bessonov M.I.; Zubkov V.A., CRC Press, Boca Raton, FL, 1993.
3. *Polymers for Microelectronics: Resists and Dielectrics,* Thompson, L.F.; Wilson, C.G.; Tagawa S. Eds., ACS Symposium Series 537, ACS, Washington DC, 1994.
4. Mascia, L. *Trends Polym. Sci.,* **1995**, *3*, 61.
5. Morikawa, A.; Iyoku, Y.; Kakimoto M.; Imal, Y. *Polym. J.,* **1992**, *24*, 107.
6. Nandi, M.; Conklin, J.A.; Salvati L. Jr.; Sen, A. *Chem. Mater.,* **1991**, *3,* 201.
7. Kioul A.; Mascia, L. *J. Non-Cryst. Solids,* **1994** , *175,* 169.
8 *Polyimides. Synthesis, Characterization and Applications,* Mittal K.L. Ed., Volume 1, p.443. Plenum Press, New York, 1984.
9. Musto, P.; Ragosta, G.; Scarinzi G.; Mascia, L.; *Polymer,* **2004**,*45,* 1697.
10. Cotugno, S.; Larobina, D.; Mensitieri, G.; Musto, P.; Ragosta, G. *Polymer,* **2001,** *42,* 6431.
11. *Perry's Chemical Engineers Handbook,* 7th edition, Perry, R.H.; Green D.W.; Maloney J.O. Eds, McGraw Hill, New York, 1999.
12. Marquardt, D.W. *J. Soc. Ind. Appl. Math.,* **1963**, *11,* 441.
13. Maddams, W.F. *Appl. Spectroscopy,* **1980**, *34,* 245.
14. Musto, P.; Ragosta, G.; Mascia, L. *Chem. Mater.,* **2000**, *12*, 1331.
15. Musto, P.; Ragosta, G.; Scarinzi, G.; Mascia, L. *J. Polym. Sci. Part B: Polym. Phys. Ed.,* **2002**, *40,* 922.
16. *The Mathematics of Diffusion* Crank,. J. 2nd edition, Clarendon Press, Oxford, 1975.
17. Danckwerts, P.V. *Trans. Farad. Soc.,* **1951**, *47*, 1014.

Chapter 23

Connectivity of Domains in Microphase Separated Polymer Materials: Morphological Characterization and Influence on Properties

Samuel P. Gido

Department of Polymer Science and Engineering, University of Massachusetts, Amherst, MA 01003

The spatial arrangement and connectivity of materials has been demonstrated to be a controlling factor in a range of material properties including, mechanical properties, transport properties, and electro-optical properties. Materials composed of two or more microphase separated domain materials may either display long range lattice order, as is typical for block copolymers, or they may be microphase separated but randomly ordered. The morphological characterization of phase connectivity in microphase separated materials is described. This includes materials with long range order such as block copolymers, as well as more randomly ordered microphase separated materials which can still display varying degrees of spatial connectivity among their domains. The connectivity information thereby obtained is then used to model composite material properties, transport properties in the present chapter.

This chapter discusses micro or nanodomain connectivity in microphase separated polymer materials. It is dedicated to Prof. Frank Karasz on the occasion of his 70th birthday.

Newnham[1] showed that for a biphasic, ordered material that there are, in general, ten possible connectivity patterns. Those relevant to block copolymer morphology are illustrated in Figure 1. The connectivity pattern is designated by two numbers, the connectivities of each of the domain materials. In this formalism, the connectivity is an integer equal to the number of dimensions or component directions in which there are sample spanning paths in the material of interest. The 3-0 connectivity pattern corresponds to the block copolymer case of spheres of one dispersed component in a continuous matrix of another; the 3-1 pattern corresponds to the cylinder morphology; the 3-3 pattern corresponds to the bicontinuous gyroid morphology; and the 2-2 pattern corresponds to the lamellar block copolymer morphology. This scheme provides a qualitative classification of composite structures. It would, however, be desirable to develop a quantitative measure of domain continuity useful for modeling structure-property relationships.

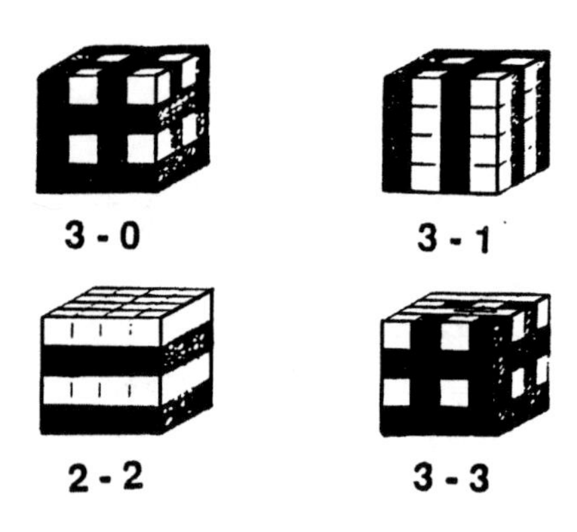

Figure 1: Simple Continuity Classification for Two Component Materials

Morphological characterization of microphase separated materials, is a crucial step in understanding domain connectivity and continuity. Such analysis generally employs transmission electron microscopy (TEM) and small angle X-ray scattering (SAXS).

Transmission Electron Microscopy of Ordered Morphologies

As compared to scattering techniques, TEM has the advantage of producing real space images which may be directly interpretable provided one keeps in mind the mechanism by which contrast arises as well as the how thin sectioning and projection of electrons through the sample generate the image.

Due to the extremely strong interaction between electrons and matter, samples for TEM observation must be very thin, typically 100 nm or less for block copolymers. Samples preparation for TEM frequently involves ultramicrotome sectioning of block copolymers with a diamond knife, frequently at cryogenic temperature. The total areas of thin sample typically observed while working in the magnification range commonly employed (about 5,000x to 50,000x) are miniscule. Thus the data obtained is representative of only a very small volume of the sample, and should only be trusted when its conclusions are consistent with scattering data.

A central problem of electron microcopy of polymers is obtaining contrast between different regions of the sample, i.e. between the microphase separated domain materials in block copolymers. In typical block copolymers in which both blocks are amorphous, the only possible mechanism of contrast in the TEM is mass-thickness contrast.[2] Most electron scattering by the atoms composing the sample is in the forward direction or at very small angles. As the masses of the scattering atoms increase, the probability of scattering at higher angles increases. Additionally, increasing the path of the electrons through the sample increases the probability of higher angle scattering. Use of an objective aperture to block out higher angle electron scattering from regions of the sample which are either thicker or contain heavier elements than other regions results in a lack of electron intensity reaching the parts of the magnified image corresponding to the more highly scattering material. Thus these regions appear darker in TEM images. In mass thickness contrast the intensity, I, at a given point in the image is given by $I = I_o\ exp(-S\rho t)$, where I_o is the intensity of the electron beam incident on the sample, and ρt is the product of average density and sample thickness along a ray tracing path through the region of the sample corresponding to the point on the image.[3] S is an instrument- and sample-specific constant of secondary importance to the relative contrast between regions of the sample. The relationship between a block copolymer morphology and its image in the TEM can be understood by a ray tracing exercise in which the thickness and average density along many paths through a sample are evaluated and used to plot a simulated image. In doing this calculation in a block copolymer with two domain materials, it is helpful to approximate the equation for intensity by its series expansion: $I \approx I_o\ (1 - S\rho_1 t_1 - S\rho_2 t_2 + ...)$.

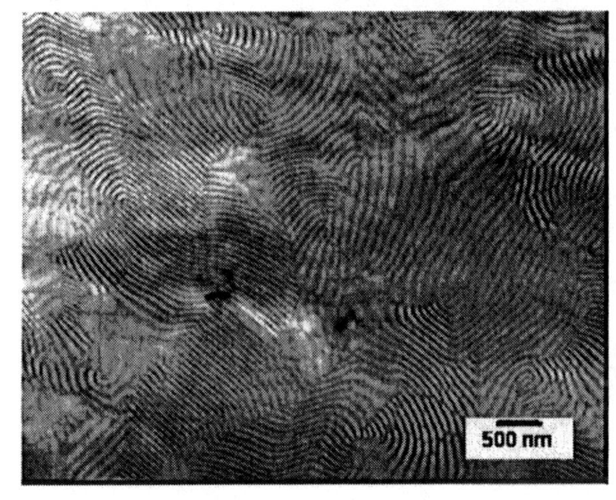

Figure 2: Lamellar Block Copolymer Morphology

Figure 2, shows a TEM image of a lamellar forming poly (styrene - b -butadiene) block copolymer which has been stained with OsO_4 rendering the polybutadiene domains dark. On a 100 nm length scale distinctive lamellar stripes are visible. These result from the higher average density of the osmium stained polyisoprene domains. The lamellar bands visible in the image appear to have a range of thicknesses even though we know that the lamellar repeat distance of a near-monodisperse block copolymer material is constant. This is a projection artifact, as illustrated through a ray tracing exercise in Figure 3, due to the orientation of the lamellar layers within the thin microtomed section.[4] When the lamellar layers are exactly perpendicular to the thin section, the image shows their true thickness and contrast is at a maximum. As the lamella become tilted in the film, the repeat spacing of banding in the projected image increase and image contrast decreases. The projected repeat distance observed in the image, L^*, is given by the following relationship: $L^* = L/\cos\theta$. Here L is the true lamellar repeat distance and θ is the angle between the lamellae and the surface of the microtomed section. Contrast disappears entirely when the lamella are lying parallel to the surfaces of the thin section and thus parallel to the electron beam.

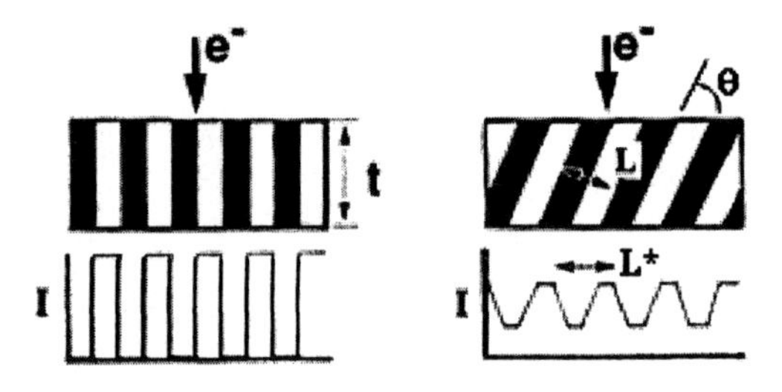

Figure 3: Ray Tracing of Lamellar Morphology

TEM images of other morphologies such as bcc spheres, hexagonally pack cylinders and cubic bicontinuous gyroid structures, are much more complex but can be understood by keeping the principles of ray tracing in mind. Cylinders for instance produce images that look like a hexagonal arrangement of circles if the projection direction is down the cylinder axis, and they produce images which look like alternating stripes if the projection is perpendicualr to the cylinder axis. This can be confusing since the stripe projection of cylinders could be mistaken for lamella and the circle projection could be mistaken for spheres. Cubic packings of spheres look like a hexagonal arrangement of circles if the projection direction is down the body diagonal of the cubic unit cell. Projections parallel the edge of the cubic sphere unit cell produces an image with a square arrangement of circles. More irregular projections of these structures are often very difficult to interpret. Thus care must be taken in assigning a morphology to a sample based on TEM images. If possible one should also consult small angle scattering data (SAXS or SANS).

Additionally, the sample may be tilted relative to the electron beam, thus changing the projection geometry. With a high tilt TEM instrument one can actually obtain, for instance, both square and hexagonal projections of the same volume of bcc spherical morphology by tilting the approximately 60° between the *[100]* projection and the *[111]* projection, as shown in Figure 4.[5-7]

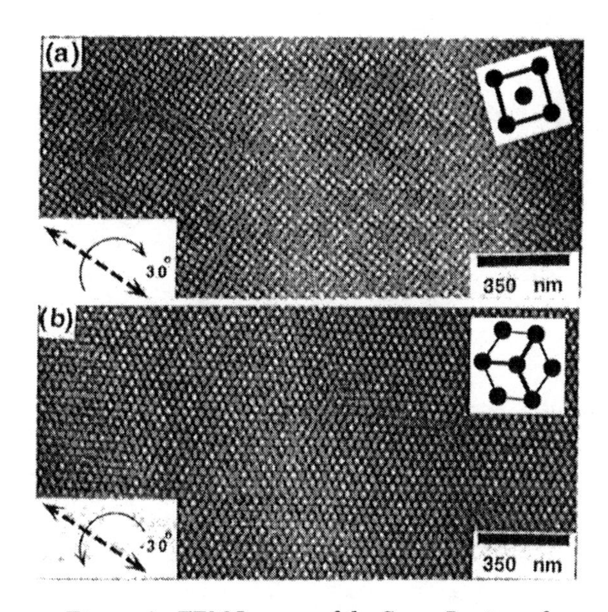

Figure 4: TEM Images of the Same Region of Spherical Morphology at 60° Relative Tilt

This section has discussed some basic features of determining morphology and phase continuity in ordered block copolymers using TEM. It is important that these conclusions be supported by scattering data, which should agree with TEM with respect to approximate length scale as well as crystallographic symmetry. Principles of ray tracing to understand electron projection in complex geometries can and have been extended to analyze the more complex bicontinuous morphologies,[8] as well as the morphologies found in ABC triblock terpolymers.[9] Additionally, electron tomography (on beam stable materials) can be used to reconstruct three dimensional computer models of morphologies as a way to analyze their structural connectivities.[10]

Randomly Ordered, Microphase Separated Materials

In addition to the highly ordered microphase separated morphologies discussed in the previous section, some materials have microphase separated structures that are random and chaotic.[11-13] As illustrated in Figure 5a, a bicontinuous structure without long range order consists of separated domains which differ in composition. These domains are mutually interwoven, both forming network structures that branch and extend to span three dimensional space. The two domain networks completely fill all space in the material so that the domains of one type of material are bounded by their interface with the other type of material, and *vice versa*.

Imaging such a randomly ordered bicontinuous structure can be very difficult if the length scale of the morphology is very small, on the order of 10 nm or less. At the high magnifications required to resolve such small structures in TEM, a dappled or speckled texture is present in images due to the limitations of the instrument optics as expressed through the *contrast transfer function.*[2] This texture is pure artifact, but it also looks very much like the type of image one would expect from a projection of a thin section of a randomly ordered bicontinuous structure, as shown in Figure 5. So how does one tell the real structure from the artifact?

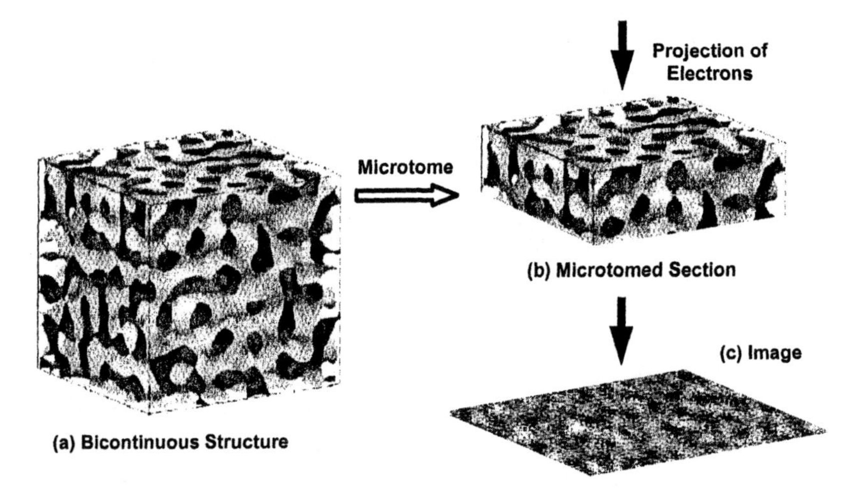

Figure 5: Disordered Bicontinuous Structure, Microtomed Thin Section and Projected Image

If the TEM is in true focus then the contrast transfer function artifact will be minimized, and the real sample morphology will be imaged. However, determining true focus is potentially very difficult when the morphology visible in focus and the artifact visible out of focus, appear similar. If one looks at the sample and turns the focus control of the TEM until the "best looking image" is obtained, there is a very good chance that it will be an artifact. The best method to ensure correct focus is to find an edge of the thin section or a hole in the section. Focusing to minimize the Fresnel fringes at this edge will give proper focus. Then translate the electron beam to a nearby area of sample and record TEM micrographs without further adjustments to the focus.

An interesting, and commercially important class of materials that have a bicontinuous morphology of the type shown in Figure 5, are silicone hydrogels used to make extended wear contact lenses. These materials are composed of microphase separated domains of a siloxane rich material, which form a co-continuous morphology with domains of a hydrophilic carbon based polymer material. The silicone domains provide oxygen permeability, while the hydrophilic domains allow the lens to remain moist. This dual oxygen and moisture permeability resulting from two different domain types, requires the continuity of both domains throughout the structure of the lens, i.e. it requires a bicontinuous morphology. Additionally, the microphase separated, bicontinuous morphology must be on a nanometer length scale in order to preserve optical clarity.

Figure 6a shows a TEM image at 100,000 times magnification, of a silicone hydrogel, contact lens material: CIBA Vision Focus Night and Day. This image is properly focused, and a chaotic microphase separated morphology is visible. The morphological length scale, which is properly measured from the center of a dark domain to the center of the next dark domain (or the center of a light domain to the center of the next light domain), is roughly 20 nm. This image was taken without sample staining, the presence of heavier Si atoms in the siloxane rich domains, which appear dark, provides adequate mass-thickness contrast. Figure 6b shows a TEM image of the same CIBA silicone hydrogel out of focus (underfocus). The dappled texture of the contrast transfer function induced artifact is a bit more pronounced and somewhat smaller than the real morphology. Otherwise, however, it appears quite similar. This artifact arises from a mechanism known as phase contrast which operates concurrently with the mass-thickness contrast that we wish to use in imaging the contact lens materials.[2] The contrast transfer function is dependent upon fixed factors such as the spherical aberration coefficient of the objective lens and the energy spread of the electron beam, and it also depends on operator controlled factors, most notably the focus condition of the microscope. The contrast transfer function results in some spatial frequencies being passed more effectively to the final image than others. In Figure 6b, the different efficiencies with which information about different spatial frequencies is passed to the image results in a speckle structure with an average length scale corresponding to the spatial frequency favored by the contrast transfer function.

Another technique, scanning transmission electron microscopy (STEM), does not use the objective lens for imaging and thus is not influenced by the contrast transfer function. In a STEM instrument, an extremely small probe of electrons is rastered across the sample in a rectangular pattern. Each probe position in the raster pattern corresponds to a pixel in the final STEM image.

Detectors collect the electrons that pass from the fine probe through a given point on the sample. Because this technique does not image with the objective lens, it is not influenced or limited by the contrast transfer function. The contrast transfer function speckle, as shown in Figure 6b, is absent in STEM images.

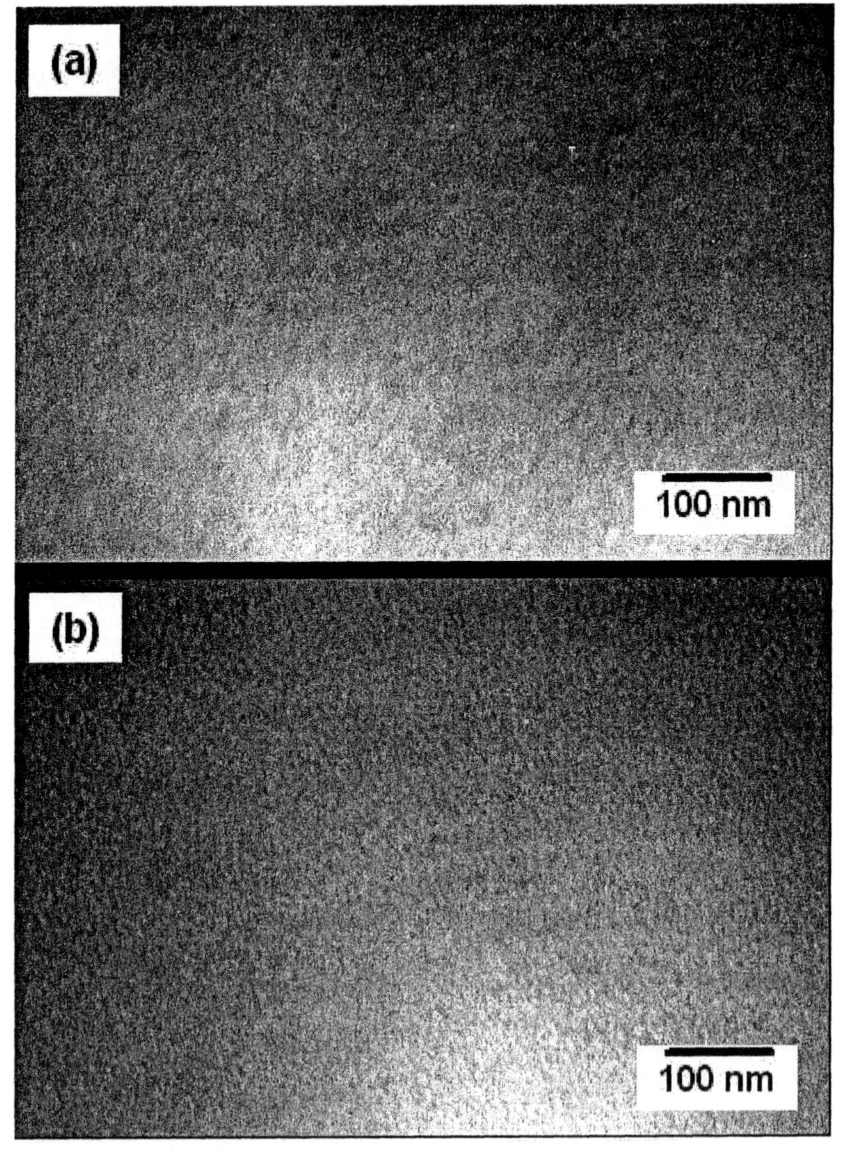

Figure 6: TEM of Silicone Hydrogel. (a) In Focus, (b) Out of Focus, Artifact.

Figure 7 shows a dark field (DF), field emission gun (FEG) STEM image of the silicon hydrogel lens material which was imaged by conventional TEM in Figure 6a. The image was obtained with a JEOL 2010 F in FEGSTEM mode at, 200 kV accelerating voltage, and a probe size of 0.5 nm. The DF image is produced by scattered rather than transmitted electrons and thus the more strongly scattering Si-rich domains appears brighter in this image than the more weakly scattering Si-poor domains. This the opposite contrast to that observed in the conventional TEM image of Figure 6a. In Figure 7, a microphase separated, bicontinuous structure of light (Si-rich) and dark (Si-poor) domains on a length scale of about 20 nm is clearly visible.

Figure 7: Dark Field FEGSTEM image of Silicone Hydrogel Contact Lens Material

Both TEM and FEGSTEM work by projecting electrons through a very thin, typically about 50 to 100 nm thick, section of material. Figure 5b illustrates such a thin section removed, by microtoming, from the original bicontinuous structure. Some of the three dimensional interconnectivity of the bicontinuous structure is lost in this process since it occurs in regions of the material which are excluded from the nearly two dimensional thin section. However, an isotropic bicontinuous structure will show a significant amount of quasi-two dimensional interconnectivity in the thin microtomed section. Since the bicontinuous structure is the same in all directions throughout the material, any randomly chosen thin section should display a significant amount of domain interconnectivity. Figure 5c illustrates the FEGSTEM imaging process on the thin section of such a bicontinuous material. As the electron probe passes through each point of the section, it interacts with the relative amounts of the two types of material on a trajectory directly through the thin section. The electron intensity detected, depends on the relative amounts of the two domain types encountered.

Phase Continuity

The morphological analysis of the three dimensional arrangement and connectivity of domains in microphase separated materials, as discussed above, provides the information for a quantitative description of phase continuity which can also be related to materials properties.

Consider a lamellar "bicrystal" with a 90° twist grain boundary, as shown in Figure 8. This grain boundary preserves continuity of both domain materials from one grain into the next.[4,14] Each of the lamellar grains separately has 2-2 connectivity in the Newnham classification. However, the presence of a continuous grain boundary between them gives the two-grain system a higher connectivity. A probe molecule that starts in a lamella of the light material in the left-hand grain and is constrained to move only in the light domain material can move freely in a plane as required by 2-2 connectivity. It can also move out of this plane into the third dimension by moving across the grain boundary into the perpendicularly oriented neighboring grain. This provides some measure of connectivity in the third dimension.

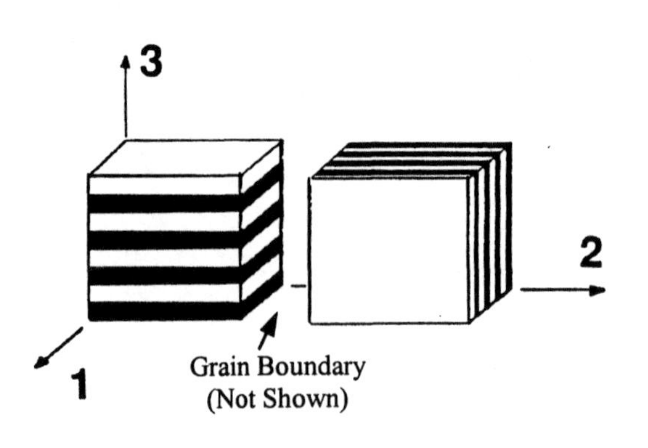

Figure 8: Lamellar Bicrystal

Newnham's approach would have to classify the lamellar bicrystal as a 3-3 structure. However, it is clear that the path into the third direction is indirect. The continuity parameters thus should be formulated in such a way as to take into account the tortuosities of connected paths in different dimensions. The component materials in the lamellar bicrystal should each have continuity numbers somewhere between 2 and 3.

A procedure for calculating average microscopic tortuosities for probe motion in a single component of a two component material (i.e. assuming one component is permeable and the other component is total impermeable) has been developed.[15] Tortuosity is the ratio of the contour length of a particle's path to the net, straight line motion in a particular direction. This method is similar to that recently reported by Slater and co-workers[16,17] for the calculation of diffusion coefficients in a periodically heterogeneous material. For a three dimensional structure (d = 3), these microscopic tortuosities (τ_i) are calculated for motion in three component (i = 1-3) directions. Once the component microtortuosities are determined, the phase continuity, C, of a microphase separated material may be calculated as follows:

$$C = \sum_{i=1}^{d} \frac{1}{\tau_i}$$

The more tortuous the path a probe must take to move in a given direction the less connectivity the material has in that direction. Thus the reciprocal of a component direction tortuosity provides a measure of connectivity in that direction. The phase continuity parameter is thus the sum over all dimensions, d, of the reciprocals of the microtortuosities. Figure 9, shows how the standard block copolymer morphologies can be placed on a scale of phase continuity from 0 to 3. Matrix phases in spheres and cylinders as well as cylinders and lamella forming phases in multigrain systems generally have non-integral continuity numbers.

Phase Continuity and Transport Properties

The basic problem in modeling structure property relationships in composite materials is to describe composite properties in terms of the properties of the component materials. The continuity of these components can have an important effect on these properties independent of their relative volume fractions. Gas permeability, P, relates the flux, J, of the transported gas to the driving pressure gradient: $J = P \nabla p$. The permeability is also equal to the product of the diffusion coefficient, D, and the gas solubility, S: $P = DS$. Experimental data for CO_2 permeability in polystyrene-polybutadiene (SB) diblock copolymers is available in the literature.[18] The PB domains are about 100 times more permeable to CO_2 than the PS domains. As a first approximation, the permeablity of the SB diblock may be modeled as a volume fraction weighted average of the PS and PB permeabilities: $P_{SB} = \phi_S P_S + \phi_B P_B$. Figure 10a shows a plot of P_{SB} vs. ϕ_B, the volume fraction of the more permeable component, corresponding to different block copolymer morphologies. The straight line represents the volume fraction weighted average approach of the above equation. The experimental data obviously deviates considerably from this line. The volume fraction weighted average is an inadequate model because it only specifies the relative amounts of the two domain materials but does not account for their spatial arrangements and connectivities.

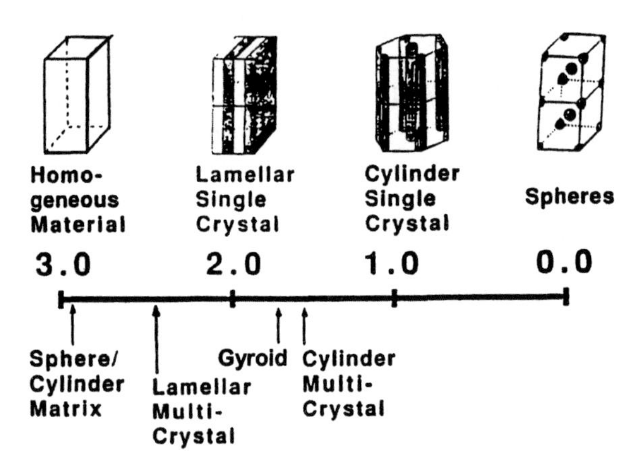

Figure 9: Scale of Phase Continuity

In order to incorporate continuity information into our modeling of a block copolymer nanocomposite, a result is used from percolation theory[19] for the effective permeablity, P_{eff}, of a composite of a conducting material and nonconducting obstacles: $P_{eff} = (\phi_c/\tau)P_c$. Here ϕ_c and P_c are the volume fraction and permeablity of the the conductive microphase. The effective permeablity is inversely proportional to the tortuosity, τ, of transport paths in the conductive domains. The permeablity in a particular domain material is therefore proportional to its continuity *(C)*, and thus $P_{SB} = (1/3)C_S\phi_S P_S + (1/3)C_B\phi_B P_B$. The factors of *1/3* result from the three dimensional nature of the systems of interest. Figure 10b shows a plot of the same permeability data plotted vs. the product $\phi_B P_B$, and the straight line approximates the continuity and composition weighting embodied in the above equation. The line is not an exact representation of the equation because, unlike the case were only volume factions were used as weighting coefficients, $C_S\phi_S + C_B\phi_B$ is not a constant. The actual predictions of the phase continuity number weighting are represented by square symbols while the experimental data is represented by triangles. The agreement between the experimental and phase continuity weighted symbols is excellent.

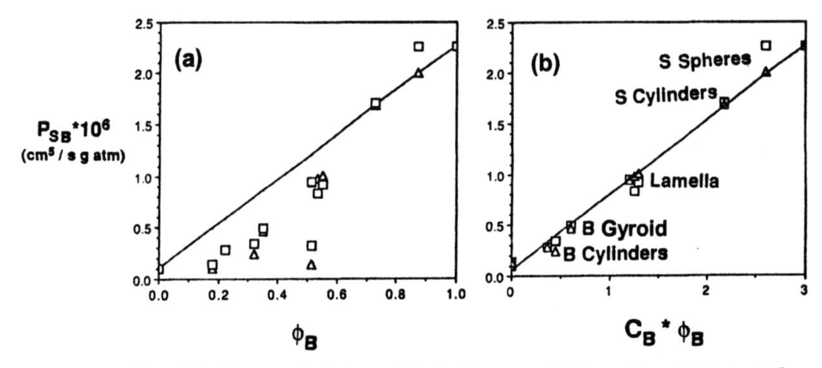

Figure 10: CO_2 Permeabilities of Poly(Styrene-b-Butadiene) Materials vs. (a) Polybutadiene Volume Fraction, and (b) Product of Polybutadiene Domain Continuity and Volume Fraction

The phase continuity approach, as applied to calculating gas permeabilities in block copolymers, may be compared to previous work using effective medium theory,[18,20,21] and to a computational model for transport in block copolymers developed by Cohen and co-workers.[22-24] For a multigrain lamellar structure, the material encountered by a permeant is modeled as a column of cubic grains with random lamellar orientations. This approach does incorporate the tortuous transport paths which are not captured by the effective medium theory. While good agreement between modeling and experimental data could be obtained with the random column model, it was achieved by using an adjustable parameter.

Investigations of transport properties in block copolymer materials with various morphologies and degrees of long range order have been conducted using a gas and vapor sorption balance device.[18,21] The permeability of non-interacting gases such as N_2, Ar, O_2, and CO_2 in PS-polydiene materials was investigated. In these materials, at room temperature, the rubbery polydiene microdomains are as much as 100 times more permeable to these gases as the glassy PS domains. Therefore, the sample spanning connectivity or percolation paths in the polydiene domains control the transport behavior.

Figure 11 shows a schematic comparison of possible transport pathways in lamellar materials with small isotropic grains (Fig. 11a) and large anisotropic grains (Fig. 11b). Studies by Gido and Thomas,[4,25] and others,[26-28] have shown that the boundaries between adjacent grains in Figure 11 allow for connectivity across the boundaries, and thus allow for transport from one grain into the next. The tortuosity of transport paths is potentially larger in samples with large grains oriented in a direction perpendicular to the direction of transport. Transport data from studies on the influence of morphology type as well as on grain size and degree of long range order can all be unified through the connectivity results: $P_{eff} = (\phi_c/\tau)P_c = + (1/3)C_c\,\phi_c P_c$. Here ϕ_c, C_c and P_c are the volume fraction, phase continuity, and permeablity of the conductive microphase. Figure 12 shows the permeability data for three gases obtained in studies of both morphology type and grain size and orientation. Measured effective material permeability is plotted vs (ϕ_c/τ) where ϕ_c is known from the polydiene volume fraction of the block copolymers and τ is estimated from the TEM and SAXS analysis of morphology, grain size, and degree of orientation. For the three gases studied one obtains a linear relationship between permeability and (ϕ_c/τ), with the slope giving the permeability of the polydiene domains.

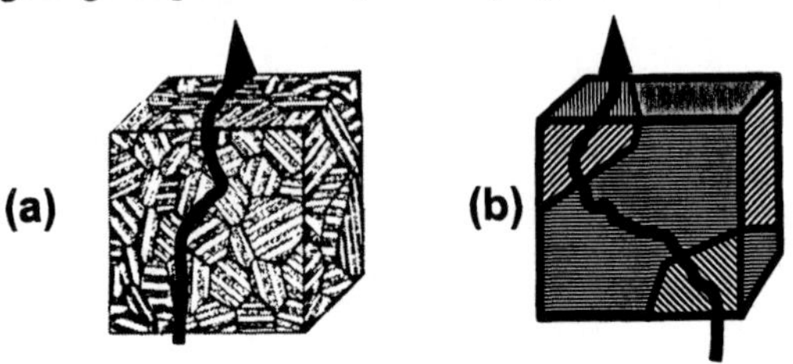

Figure 11: Transport Paths in Small vs. Large Grain Morphologie. (a) Small, Isotropic Grains, (b) Large, Anisotropic Grains.

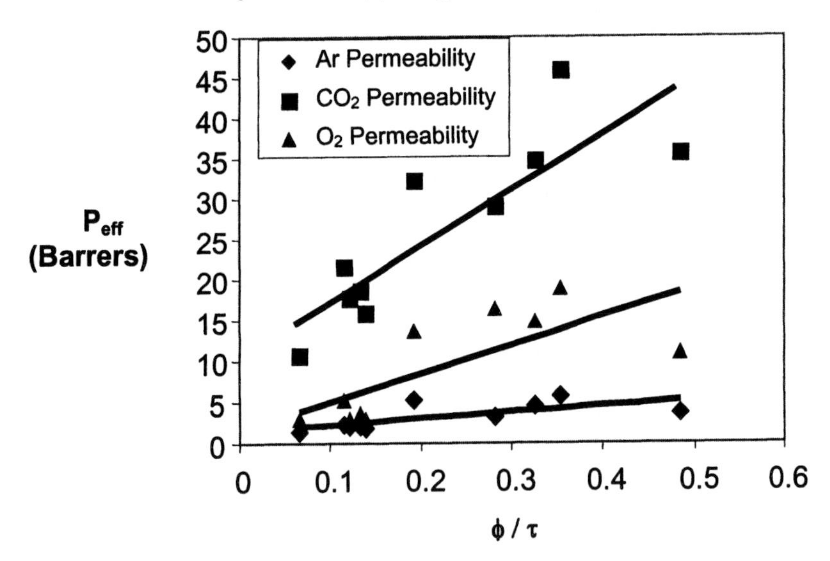

Figure 12: Gas Permeability in PS-polydiene block Copolymers vs. ϕ/τ. The Unit of Permeability, Barrers = $10^{10}cm^3(STP)cm/(cm^2s\ cm(Hg))$.

Conclusions

The morphological analysis of bicomponent materials using electron microscopy can be used to obtain information about microdomain or nanodomain connectivity. In doing so, one must be mindful of how the structural projections observed in TEM are related to the three dimensional structure. In the case of microphase separated materials that lack long range order, one must also be careful not to confuse the structure with the contrast transfer function artifact, which appears very similar. Use of STEM imaging techniques avoids this artifact. A method is present in which domains of a given type in a material are assigned a continuity number between *0* and *3* which can then be used to effectively model transport properties.

References

1. Newnham, R. E.; Skinner, D. P.; Cross, L. E. *Materials Research Society Bulletin* **1978, 13**, 525.
2. Williams, D. B.; Carter, C. B., *Transmission Electron Microscopy*. ed.; Plenum Press: New York, 1996; 'Vol.' III Imaging, p.
3. Anderson, D. M.; Bellare, J.; Hoffman, J. T.; Hoffman, D.; Gunther, J.; Thomas, E. L. *Journal of Colloid and Interface Science* **1992,** 148, 398.
4. Gido, S. P.; Gunther, J.; Thomas, E. L.; Hoffman, D., *Macromolecules* **1993,** 26, 4506.
5. Gido, S. P.; Lee, C.; Pochan, D. J.; Pispas, S.; Mays, J. W.; Hadjichristidis, N. *Macromolecules* **1996,** 29, 7022.
6. Lee, C.; Gido, S. P.; Poulos, Y.; Hadjichristidis, N.; Beck Tan, N.; Trevino, S. F.; Mays, J. W. *Polymer* **1998,** 19, 4631.
7. Thomas, E. L.; Kinning, D. J.; Alward, D. B.; Henkee, C. S. *Macromolecules* **1987,** 20, 2934.
8. Hajduk, D. A.; Harper, P. E.; Gruner, S. M.; Honeker, C. C.; Kim, G.; Thomas, E. L.; Fetters, L. J. *Macromolecules* **1994,** 27, 4063.
9. Auschra, C.; Stadler, R. *Macromolecules* **1993,** 26, 2171.
10. Spontak, R. J.; Fung, J. C.; Braunfeld, M. B.; Sedat, J. W.; Agard, D. A.; Kane, L.; Smith, S. D.; Satkowski, M. M.; Ashraf, A.; Hajduk, D. A.; Gruner, S. M. *Macromolecules* **1996,** 29, 4494.
11. Fredrickson, G. H.; Bates, F. S. *J. Polm. Sci: B. Polym. Phys.* **1997,** 35, 2775.
12. Bates, F. S.; Maurer, W. W.; Lipic, P. M.; Hillmeyer, M. A.; Almdal, K.; Mortensen, K.; Fredrickson, G. H.; Lodge, T. P. *Physical Review Letters* **1997,** 79, 849.

13. Beyer, F. L.; Gido, S. P.; Bueshl, C.; Iatrou, H.; Uhrig, D.; Mays, J. W.; Chang, M. Y.; Garetz, B. A.; Balsara, N. P.; Beck Tan, N.; Hadjichristidis, N. *Macromolecules* **2000,** 33, 2039.
14. Gido, S. P.; Thomas, E. L. *Macromolecules* **1994,** 27, 849.
15. Gido, S. P. Ph.D. Thesis. Chemical Engineering, Massachusetts Institute of Technology, 1993.
16. Mercier, J. F.; Slater, G. W.; Guo, H. L. *Journal of Chemical Physics* **1999,** 110, 6050.
17. Mercier, J. F.; Slater, G. W. *Journal of Chemical Physics* **1999,** 110, 6057.
18. Kinning, D. J.; Thomas, E. L.; Ottino, J. M., *Macromolecules* **1987, 20,** 1129.
19. Sahimi, M., *Applications of Percolation Theory.* ed.; Taylor & Francis: London, 1994.
20. Mohanty, K. K.; Ottino, J. M.; Davis, H. T. *Chemical Engineering Science* **1982, 37**, 905.
21. Saxs, J.; Ottino, J. M., *Polymer Engineering and Science* **1983, 23,** 165.
22. Csernica, J.; Baddour, R. F.; Cohen, R. E. *Macromolecules* **1987, 20,** 2468.
23. Csernica, J.; Baddour, R. F.; Cohen, R. E., *Macromolecules* **1989, 22,** 1493.
24. Rein, D. H.; Csernica, J.; Baddour, R. F.; Cohen, R. E., *Macromolecules* **1990, 23,** 4456.
25. Gido, S. P.; Thomas, E. L. *Macromolecules* **1994,** 27, 6137.
26. Matsen, M. W. *Journal of Chemical Physics* **1997,** 107, 8110.
27. Netz, R. R.; Andelman, D.; Schick, M. *Physical Review Letters* **1997,** 79, 1058.
28. Villain-Guillot, S.; Netz, R. R.; Andelman, D.; Schick, M. *Physica A* **1998,** 249, 285.

Chapter 24

Dual Detection High-Performance Size-Exclusion Chromatography System as an Aid in Copolymer Characterization

Tatjana Tomić[1], Marko Rogošić[2], Zvonimir Matusinović[2], and Nikola Šegudović[1]

1Strategic Development, Research and Development Sector, INA Industrija nafte d.d., Lovinčićeva bb, HR–10000 Zagreb, Croatia
[2]Faculty of Chemical Engineering and Technology, University of Zagreb, Marulicev Trg 19, HR–10000 Zagreb, Croatia

Compositional heterogeneity is one of the essential factors controlling the physical and physico-chemical properties of copolymers; its determination is in most cases a rather difficult and challenging scientific task. The overall composition of a copolymer comprising two kinds of repeating units can be determined as a function of molecular mass by the dual detector high performance size exclusion chromatography (HPSEC) technique. The common combination of the refractive index (RI) detector with the UV spectrophotometer can be applied if at least one of the monomer units absorbs at a suitable wavelength and if the UV-spectra of both components are sufficiently different. In this paper, synthesized random copolymers of styrene and methyl methacrylate were dissolved in tetrahydrofuran and characterized by the RI-UV HPSEC technique, at the UV wavelength of 254 nm. Home made software and a corresponding hardware interface were used for the equipment control, data acquisition, data processing and calculation. The obtained results show that the technique is capable for fast, routine determination of average copolymer composition. There is no significant variation of chemical composition with molecular mass for investigated copolymers with lower polystyrene content. For copolymers rich in polystyrene, chemical composition varies with molecular mass. In this case, calculated apparent mass fractions of styrene may exceed 100%, and this is explained by the presence of small quantities of low-molecular-mass styrene homopolymer.

Introduction

The properties of a copolymer are affected not only by its average composition and molecular mass, but they also depend on the sequential arrangement of monomeric units along the chain, i.e. on the chain architecture. From the viewpoint of the chain architecture, copolymers composed of chemically different monomers can be classified into four main types, i.e. alternating, statistical, block and graft copolymers [1]. The composition of graft copolymers generally correlates with the molecular mass, but this may not be true for statistical and block copolymers [2]. In particular, compositional heterogeneity may be the essential factor controlling the physical and physico-chemical properties of statistical copolymers.

In the last twenty years, size exclusion chromatography (SEC, former GPC, gel permeation chromatography) became the most common method for the determination of molecular mass and molecular mass distribution (MMD) of polymers [3]. SEC separates the solutes in solutions based on the size and hydrodynamic volume of molecules. This kind of sorting takes place by passing the solution through a column containing a separation packing. For simple homopolymers, the apparatus can be calibrated to relate size (as measured by the volume of mobile phase eluted between injection and elution of the molecule) to molecular mass using simple concentration detector, such as refractive index detector (RI).

For homopolymers, the size of a molecule (as well as other molecular properties) depends not only on its molecular mass but on other variables as well, such as the degree of branching. For copolymers, even more structural properties (chemical composition distribution, functionality distribution, details of molecular architecture, etc.) may affect the size of a molecule in solution, although elution volumes are often given only as functions of molecular mass [4]. By taking into account above mentioned effects, other analyses, like those of composition and configuration of copolymer molecules are also possible by using modified SEC.

In other high performance liquid chromatography (HPLC) techniques, common in analysis of small molecules, like normal (NP) and reverse (RP) phase chromatography it is common to extend separation and identification by coupling different detectors [5-7] to chromatographic columns. Moreover, multidimensional HPLC coupled with a range of detectors may be a method of choice for performing difficult separations and analyses of complex petrochemical samples according to the size, volatility and polarity of constituent molecules [6, 8].

Using various detection systems with SEC, one may characterize copolymers according to different properties, such as chemical composition distribution, grafting efficiency, graft ratio, graft frequency, etc. In addition, the conversion of monomers during copolymerization may be monitored on-line using SEC [9-11].

When spectroscopic detectors (UV-Vis, DAD, IR) are used, it is expected that at least one component of a copolymer is an UV or IR absorbing medium, and that its spectrum differs as much as possible from that of another component. Wavelengths to be monitored should be chosen to maximize the difference of molar absorbance of the parts. At the same time, corresponding mobile phase (solvent) must possess low UV cut-off properties.

The aim of this paper was to describe the simple case of copolymer compositional heterogeneity characterization, using SEC dual detection system equiped with UV spectrophotometer at 254 nm as a selective detector and RI detector as a concentration detector. The model copolymers to be studied were statistically prepared styrene – methyl methacrylate copolymers.

Home made software and a corresponding hardware interface were used for data acquisition, processing and manipulation.

Experimental

Sample preparation

Statistical copolymers of styrene and methyl methacrylate were prepared by the batch radical solution copolymerization at 60°C in toluene, under nitrogen atmosphere. Total monomer concentrations were kept approximately constant (3 mol dm^{-3}), but monomer mole fractions as well as initiator (azobisisobutyronitrile, AIBN) concentrations and reaction times were varied in order to produce copolymers having different average compositions and different molecular mass distributions. Yet, overall monomer conversions were kept under 10% by mass to prevent the copolymer composition shift. Synthesized copolymers were purified by the three times repeated dissolution/precipitation process, using toluene as a solvent and cold methanol as a non-solvent. Copolymer compositions were determined by the elemental analysis using evaporative combustion/GC coupled CHNS analysis system (Perkin Elmer 2400, ser. 2). The resulting C/H/O mass fractions were converted to styrene/methyl methacrylate molar fractions using simple stoichiometric relations. The data related to the sample preparation and elemental analysis are given in Table I.

Table I. Homo and copolymerization conditions

Sample ID	*[M]*	*[I]*	*t*	*X*	*f*	*F*
PS	3.00	$1.0\ 10^{-2}$	190	9.4	100	100
SMMA 0.2x2	3.00	$2.5\ 10^{-3}$	240	4.1	84.2	78.6
SMMA 0.2x4	3.00	$6.3\ 10^{-4}$	480	4.7	84.2	80.1
SMMA 0.5x2	3.00	$2.5\ 10^{-3}$	160	2.9	45.3	46.5
SMMA 0.5x4	3.00	$6.3\ 10^{-4}$	390	3.3	45.3	50.0
SMMA 0.8x2	3.00	$2.5\ 10^{-3}$	240	6.9	11.3	17.5
SMMA 0.8x4	3.00	$6.3\ 10^{-4}$	360	4.5	11.3	17.7
PMMA	2.58	$1.0\ 10^{-2}$	210	34.8	0	0

[M] total monomer concentration, mol dm^{-3}; *[I]* initiatior concentration, mol dm^{-3}; *t* reaction time, min; *X* total monomer conversion, wt%; *f* styrene mole fraction in feed, mol%; *F* styrene mole fraction in copolymer as determined by elemental analysis, mol%

SEC measurements

All measurements were performed on a modular SEC system comprising the Waters 6000A high pressure pump, Waters U6K manual injector capable for injection of up to 2.0 ml of sample solution, three PL-gel mixed B columns (30.0 × 0.76 cm, 10 μm), Waters 440 UV fixed wavelength detector (at 254 nm) and Waters 401 RI detector. All sample solutions were prepared in concentrations of 0.52±0.02 g / 100 mL, by dissolving the samples in freshly distilled tetrahydrofuran (THF). Sample solutions were left to rest overnight to allow the dissolution process to reach its equilibrium. Before injection, all solutions were filtered through 0.5 μm membrane filters (Millipore MX) and 50μL of solutions were injected. Flow rates were kept constant, at the value of 1.0 mL min^{-1}. The separations were performed at ambient temperature. For the calibration of both detectors, narrow molecular mass polystyrene standards (molecular mass ranging from 1.3×10^3 to 2.9×10^6 were used. This procedure is rather common, but the calculated molecular masses of SMMA copolymers may be considered only as relative values. The time lag between detectors was 9 seconds (150 μL).

Equipment control, data acquisition, reduction, processing and reporting were performed using home made software and a corresponding hardware interface. The interface is able to monitor both the injection event and the evolution of chromatogram. Analog signals from both detectors are amplified and forwarded to AD connector (16 channels and 12 bit resolution) to obtain digital values. The software is divided into three main parts: 1) on-line data acquisition and plotting of the chromatogram, 2) calibration part and 3) data reduction procedures and post-run manipulation. In the acquisition and plotting part, a sequence of digitalized raw data is plotted against time as a chromatogram, starting from the injection event and ending at any appropriate time chosen by the operator. In the calibration part, chromatogram peaks are recognized and converted to corresponding retention times. By the input of corresponding molecular mass values, individual calibration data are then converted to calibration curves, described by 1st or 3rd order polynomials. The data reduction part enables post-run calibration and manipulation of the acquired data. The following features are included:

- data filtering using Savitzky-Golay filters
- automatic recognition of onsets and offsets of chromatogram peaks
- baseline deduction
- calculation of differential and integral molecular mass distributions using previously determined or externally supplied calibration curves
- calculation of average molecular mass values (M_n, M_w, M_z)
- fitting of experimental distribution curves data with some theoretical functions (Flory, Schulz, Tung, Wesslau)
- generation of final reports

Results and discussion

Before running of SEC analyses of copolymers, UV-spectra of corresponding homopolymers (polystyrene, PS and polymethyl methacrylate, PMMA) were recorded (Fig 1). The results showed the wavelength of 254 nm to be appropriate for subsequent measurements, because at that wavelength the

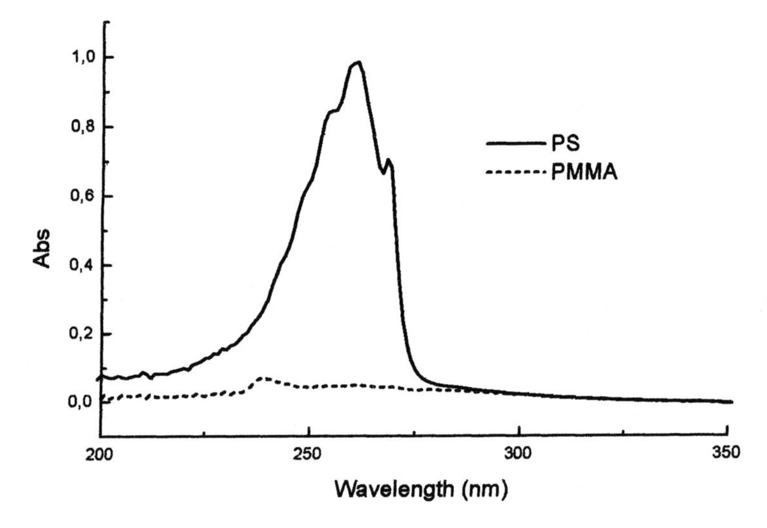

Figure 1. UV spectra of PS and PMMA

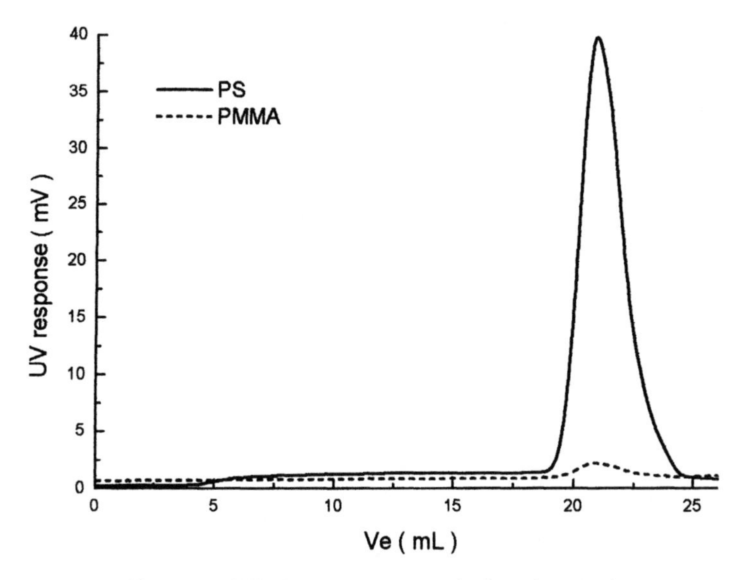

Figure 2. UV chromatograms of PS and PMMA

absorbance of PMMA could be neglected compared to the absorbance of PS, and UV detector could be taken as a selective detector. The UV cut-off of THF was below 254 nm.

SEC-UV chromatograms of the same samples at equal concentration are shown in Fig 2, confirming that the UV signal of PMMA can be neglected compared to the PS signal. The chromatograms clearly indicate that the almost complete UV signal in copolymers belongs to PS.

Figs 3 and 4 present chromatograms of PS with different injection volumes, obtained with IR and UV detector, respectively. 25, 50 and 75 μL of the stock solution in THF were injected in the system to obtain response factors for both detectors. The RI response factor for PMMA was determined in very much the same way.

The elemental analysis of prepared copolymers has shown that three different copolymer compositions were obtained (molar fraction of MMA component of approximately 0.2, 0.5 and 0.8). Taking every composition, there were two samples with the expected difference in molecular mass, based on the variations in the initiator concentration. Fig 5 shows UV chromatograms of the three samples (differing in composition) that are prepared at a higher initiator concentration. Chromatograms of the samples prepared at a lower initiator concentration look very much the same (Fig 6). Corresponding molecular mass averages (M_n and M_w) determined by RI and UV specific calibration curves are presented in Table II.

Table II. Molecular mass averages and average sample composition determined by SEC

Sample ID	M_n *(RI)*	M_n *(UV)*	M_w *(RI)*	M_w *(UV)*	*F (UV) / wt%*
PS	24500		43400		
SMMA 0.2x2	43800	38700	85600	84500	78.3
SMMA 0.2x4	71400	43900	142800	139600	78.8
SMMA 0.5x2	53300	40400	94400	103700	51.3
SMMA 0.5x4	78400	50800	146500	142500	51.0
SMMA 0.8x2	78300	58800	144000	139300	20.0
SMMA 0.8x4	87200	67500	172000	169500	20.0
PMMA	55100		91300		

M_n and M_w number and mass average molecular mass, respectively, as determined by *UV* or *RI* detector response, *F* styrene mole fraction in copolymer as determined by *UV* response at a peak

Upon the comparison of the raw UV chromatograms of copolymer and homopolymer (PS) samples, it becomes obvious that the height of the chromatogram is related to the overall content of PS component. From this observation, taking the height of the PS homopolymer chromatogram as a fixed 100% (by mass) point, mass fractions of PS in copolymer samples were calculated. The results are included in Table II as the sixth column; calculated average compositions correspond well with the results obtained by the different method (Table I, column 7).

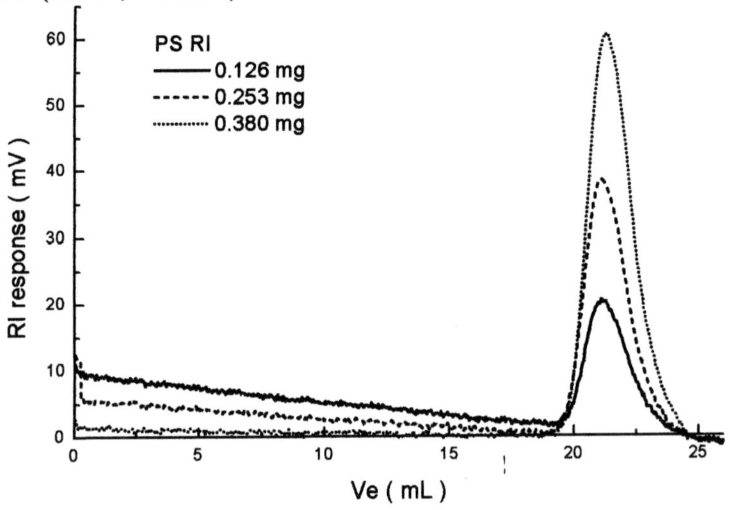

Figure 3. RI chromatograms of PS with different concentrations

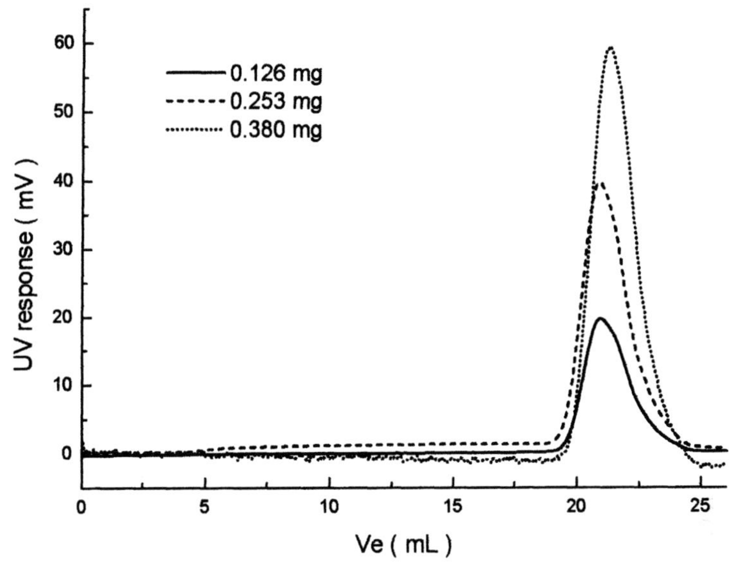

Figure 4. UV chromatograms of PS with different concentrations

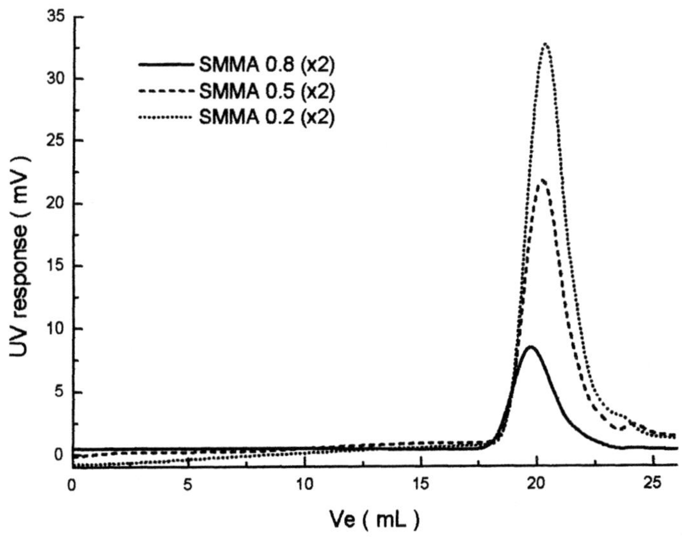

Figure 5. UV chromatograms of copolymer samples with different compositions (higher initiator concentration)

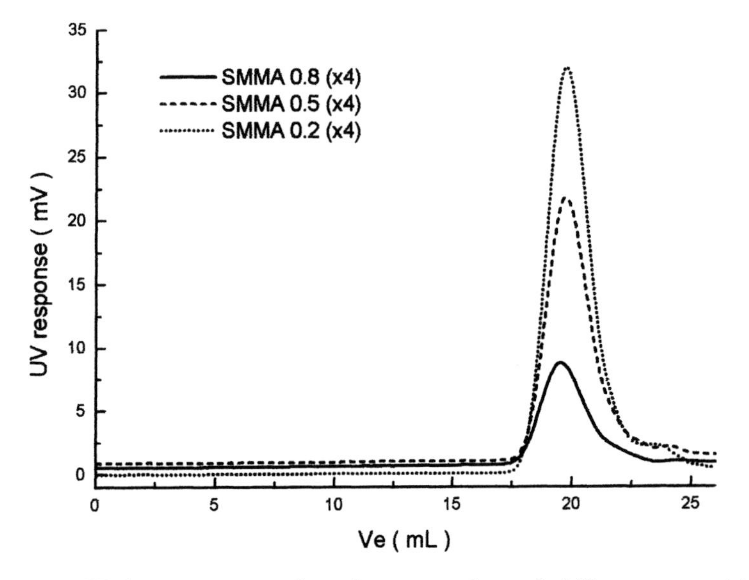

Figure 6. UV chromatograms of copolymer samples with different compositions (lower initiator concentration)

There is significant doubt in the application of the height of chromatogram in HPLC, when the chromatogram is not a narrow one [12]. In the first instance, the height may be used for symmetric (that is, relatively narrow) peaks at M_w and only if characteristic detector values (refractive index increment and molecular absorptivity, respectively) ar not M_w-dependent. However, it seems that the usage of the height as a representative of chromatogram has given qualitatively correct results in the determination of average copolymer composition. The observed quantitative deviations deserve further discussion.

Molecular masses for copolymer samples, calculated from the RI and UV detector responses using corresponding specific polystyrene calibration curves, agree well in the higher molecular mass range, i.e. in M_w-values. The discrepancy is found primarily in the lower molecular mass range, i.e. in M_n-values, because the tailing parts of the UV chromatogram are more pronounced. Even more, in the case of copolymer samples with high styrene contents (SMMA 0.5 and particularly SMMA 0.2) there is some indication of a bimodal distribution, the second peak appearing in an oligomer molecular mass range. The indices of homogeneity (M_w/M_n-values), calculated according to the RI response are approximately 2.0, but those calculated according to the UV response are much higher, approaching the value of 3.0, depending on the average composition of copolymer.

A possible explanation for the low molecular weight peak may be found in the secondary polymerization reaction, that takes place either simultaneously or prior to the main polymerization reaction. The relative height of the second peak seems to increase with the content of styrene comonomer units; hence, one immediately suspects the styrene homopolymerization to be the secondary reaction. It is well known that styrene is prone to thermally initiated polymerization (see e.g. [13]). This may occur during the warm-up period of the reactor, i.e., before the injection of the initiator solution, but also during the storage of styrene monomer after the removal of inhibitor. The second peak is missing in RI detector responses and it is probably unimportant in determining the overall copolymer properties, except Mn (see Fig 6 for the three samples prepared at the higher initiator concentration). UV detector is much more sensitive than RI detector in the oligomer molecular mass range, due to the high molecular absorptivity.

In dual detection SEC systems, chemical composition along the peak is calculated from the ratio of detector signal intensities (X_1 / X_2, in this case UV response / RI response [9]). The results of the calculations of chemical composition distribution (CCD) data are shown in Fig. 7, together with the corresponding RI detector responses. The data are given for the three samples prepared at the higher initiator concentration, but the results for the set of samples prepared at the lower initiator concentration look very similar. The data indicate that there is no variation in composition across the copolymer elution volume range in case of the copolymer with lower styrene content.

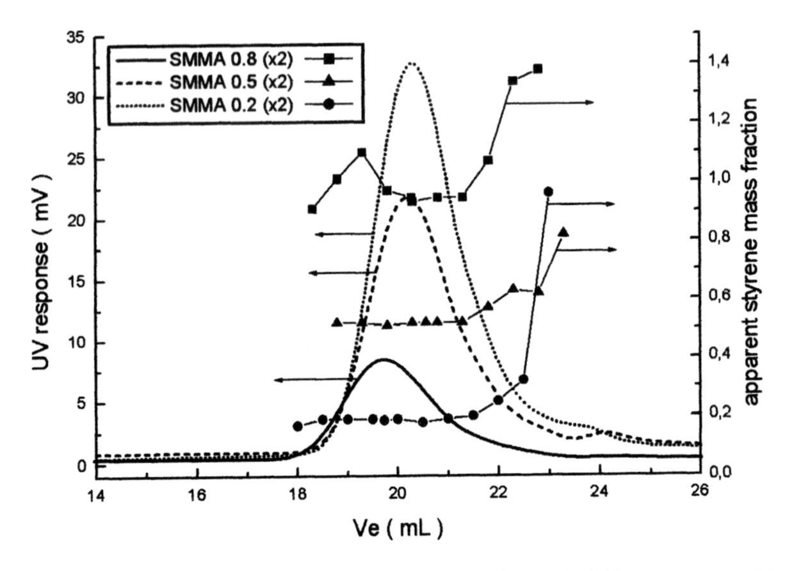

Figure 7. UV chromatograms of copolymer samples with different compositions (curved lines) and corresponding apparent styrene mass fractions (line connected symbols)

However, there is a strong variation in cases of the copolymers with higher styrene content; in some cases apparent mass fractions exceed the value of 1.0.

The possible explanation for the observed, obviously unrealistic styrene mass fraction values may be given as follows. In this case, investigated samples (especially at higher styrene content) may be regarded as polymer blends, consisting of a major component, SMMA copolymer and a minor component, styrene homopolymer. The chemical composition distributions of the main sample polymer, i.e. SMMA copolymer have to be considered as uniform, since overall monomer conversions were kept below 10% by mass to prevent the copolymer composition shift. Thus, it may be suspected that a stronger UV signal, obtained at higher elution volumes (lower molecular mass range), may be a consequence of the minor polystyrene homopolymer quantities, present in the samples due to the thermal homopolymerization of styrene. Upon decreasing of the molecular mass (i.e. increasing of the elution volume), the samples are gradually enriched with the PS homopolymer component. Therefore, the discrepancies of the average and the fractional styrene content increases in the same direction, thus producing styrene mass fraction values above 1.0, as a consequence of the applied calculation procedure. As mentioned before, the quantities of styrene homopolymer in investigated copolymer samples remain rather low, thus not affecting the average sample properties.

The effect of variation of the apparent composition distribution in both the high and low molecular mass ranges was observed earlier [9], and two reasons were suggested for the explanation [14]: in dual detection SEC systems the chemical composition data will be influenced by the peak dispersion between the detectors as well as by the incorrect interdetector volume (time lag).

The obtained results have also pointed out the possible role of the more sensitive UV detector, not only in the copolymer but also in the homopolymer (PS) characterization. Due to the high molecular absorptivity, UV detector is more sensitive in the low molecular mass range and is therefore capable to detect low molecular mass tails or secondary peaks in SEC chromatrograms, not observable by the RI detector. This feature may partially explain the big discrepancies in MMD (especially in M_n-values) amongst different laboratories, obtained previously on selected test samples [15].

Conclusion

Dual detection SEC technique comprising RI and UV detection in the characterization of statistical styrene – methyl methacrylate copolymers was described.

The obtained results show that the technique is capable for the fast, routine determination of average copolymer composition. The results are comparable to those obtained by other technique (elemental analysis). Calculated molecular mass distributions and averages vary according to the copolymerization conditions. However, the results are dependent on the type of the detector used; lower number average molecular mass values and broader distributions are obtained using the UV calibration curve. Preliminary results show that there is no significant variation of chemical composition with molecular mass for copolymers with lower PS content. For copolymers rich in styrene monomer units, chemical composition varies with molecular mass. In this case, calculated apparent mass fractions of styrene may exceed 100%.

References

1. Hamielec, A.E.: *Pure Appl.Chem.* **54**(29) (1982) 293.

2. Inagaki, H; Tanaka, T: "Developments in Polymer Characterization" J. V Dawkins, Ed.; Applied Science: London 1982; Vol.3.
3. Yau, W.W.; Kirkland, J.J.; Bly, D.D.: Modern Size Exclusion Liquid Chromatography Wiley- Interscience: New York, 1979.
4. Meister, J.J.; Nicholson, J.C.; Patil, D.R.; Field, L.R.: *Macromolecules* **19** (1986) 803.
5. Pasadakis, N.; Gaganis, V.; Varotsis, N.: *Fuel* **80** (2001) 147.
6. Schoenmakers, P.; Blombery, J.; Keskuliet, S.: *LC/GC Intern.* **11** (1996) 718.
7. Sarowha, S.L.S.; Sharma, B.K.; Sharma, C.D.; Bhagat, S.D.: *Energy and Fuels* **11** (1997) 566.
8. Gratzfeld-Huesgen, A.: HPLC Appl. (87-5) Hewlett Packard note
9. Mori, S.; Suzuki, T.: *J. Liq. Chromatog.* **4**(10) (1981) 1685.
10. Fodor, Zs.; Fodor, A.; Kennedy, J.P.: *Polym. Bull.* **29** (1992) 689.
11. Huang, N.J.; Sundberg, D.C.: *Polymer* **35**(26) (1994) 5693.
12. Kipiniak, W.: *J. Chromatog. Sci.* 19 (1981) 332.
13. Hui, A.W.; Hamielec, A.E.: *J. Appl. Polym. Sci.* **16** (1972) 749.
14. Trathnigg, B.; Ferchtenhofer, S.; Kollroser, M.: *J. Chromatogr. A.* **786** (1997) 75.
15. Berek, D.: Results of the Round Robin test of IUPAC WP 1V.2.2 project No. 4., personal communication

Chapter 25

Polymer Permeability Measurements via TGA

Brandi D. Holcomb[1,2] and Harvey E. Bair[1,3,*]

[1]Lucent Technologies, Bell Laboratories, 600 Mountain Avenue, Murray Hill, NJ 07974
[2]Current address: North Carolina State University, Raleigh, NC 27607
[3]Current address: HEB Enterprises, 17 Seminole Court, Newton, NJ 07860

A patented metal capsule small enough to fit inside a commercial TGA was fashioned with an opening that was more than 200 times smaller in area than used in the standard ASTM test method for water vapor transmission of materials. Unfortunately, in the latter method the weight determinations are not made in the controlled environment and the size of the opening is recommended to be at least 3000 mm^2 (4.65 in^2) or greater. In an attempt to gain water vapor transmission rates in-situ and rapidly as a function of temperature and humidity a new thermogravimetric approach was created. 'Wet' and dry gas streams were mixed to control the humidity inside the TGA chamber to any value between 0 and 90%. In this manner a number of measurements could be made reproducibly and quickly at a variety of temperatures and humidities. Water vapor transmission measurements, WVT, were made on a variety of polymer films ranging from polyethylene terephthalate to a permeable, Goretex® membrane.

Introduction

If the amount of a new film for testing is quite small, it may be impractical to use one of the standard ASTM methods to track the egress of water vapor from the water rich side of a film to its drier or lower humidity side (1). Furthermore, in our work we wanted, if possible, to monitor the movement of moisture out of a tiny optoelectronic device directly. Hence, we started to think of using our TGA instrument to carry out these types of measurements. Immediately, we discovered that if one attempts to design a new container to conduct water vapor transmission (WVT) studies in a TGA environment a leak-proof sealing technique must be developed to hold the film in-place since none of the ways ASTM suggested for sealing a film to a test dish appeared to work on any liquid container that was small enough to fit inside a TGA instrument. Once this hurdle was cleared the TGA method proved to be superior in terms of time to any of the prior ASTM ways to determine the water vapor permeability of polymer films as a function of temperature and humidity. In this paper our patented WVT capsule is described and used in a TGA to determine the water permeability of polymer films (2).

Experimental

Our invention relates to methods and apparatus for securing a solid film or a liquid sample for TGA or photo-dsc thermal analysis (2). In accordance with the invention a film of the polymer to be tested is positioned inside a circular lid and held in place via an o-ring that is typically a deformable material such as a rubber. The lid with film in-place is positioned atop an impermeable, metal receptacle that has a bottom and side walls as shown in Figure 1. The top edges of the side walls are bent towards the center of the receptacle. The capsule's lid is placed across the top edges of the container's side walls and the two parts are squeezed together. The lid compresses the o-ring onto the lower half of the capsule, sealing it against the bent top edges of the receptacle and the polymer film above it. In the case, of water transmission tests about 50 mg of water are sealed inside our WVT capsule. Another use for this capsule is in the confinement of volatile liquids that undergo photo-induced polymerization inside a photo-dsc. In the latter application a UV transparent polymer such as a copolymer of polyhexafluoropropylene/polytetrafluoroethylene or poly (vinylidene fluoride), PVF, can be employed. In the photo-dsc work the film

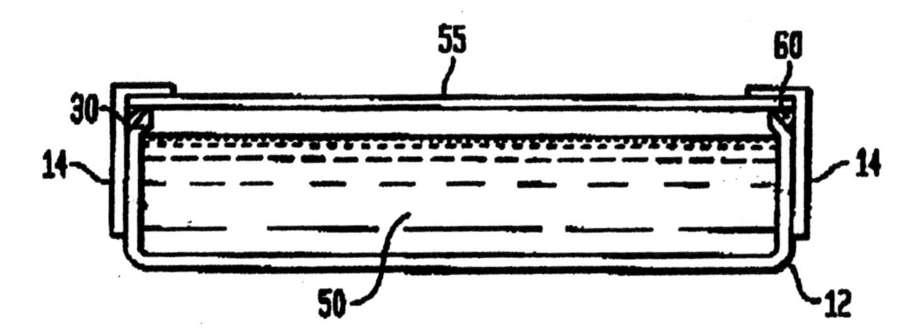

Figure 1 Schematic of WVT capsule showing: (12) receptacle, (14) lid, (30) top edges of side wall, (55) polymer film, (50) liquid permeant or water and (60) o-ring.

should have a leak rate of less than about 1.5 weight percent/hr. Also, in order to provide a uniformly thick sample on the bottom of the receptacle, the inside surface of the container is coated with a material not wetted by the sample such as a flurorcarbon. An example of the use of this capsule for photocalorimetric work can be found elsewhere (3,4).

The WVT capsule was assembled with a 0.05 mm thick PVF film and 67 mg of water and held isothermally at temperatures up to 80°C. The capsule's lid had a hole with a diameter of 4.21 mm. The sealed system lost 0.0001, 0.0087, 0.024, 0.083 and 0.315 wt% per 10 minutes at 23, 36, 46, 60 and 80°C, respectively. In the prior experiment the level of humidity inside the TGA was held close to 0% with nitrogen flowing at 100 cc/min. If the humidity inside the TGA is raised from 0 to 62 % the rate of water loss at 23°C is reduced by a factor of five over the dry state rate. Typically in a photo-dsc free radical polymerizations are completed in less than 10 minutes (5,6). Hence, this capsule could be used in a photo-dsc to study the reaction of aqueous solutions at temperatures approaching 80°C. Recently this capsule was used to contain a blend of an epoxy and alcohol at 100°C for an 8 minute irradiation period with only the loss of a few micrograms of the volatile blend (4).

In these studies the capsules were used inside a PerkinElmer TGA7 that was equipped with a high temperature furnace. The humidity was controlled inside the TGA unit by mixing two streams of wet (>95% humidity) and dry nitrogen gas. Dynamic scans of materials of interest were performed on a power compensated Perkinelmer DSC7 in a manner described elsewhere (7).

Results and Discussion

A polycarbonate film (Lexan®, General Electric Co.) with a thickness of 0.00305 cm was seated in the lid of a WVT capsule. The latter had a 4.25 mm diameter hole. The capsule was sealed with 53 mg of water present. The TGA

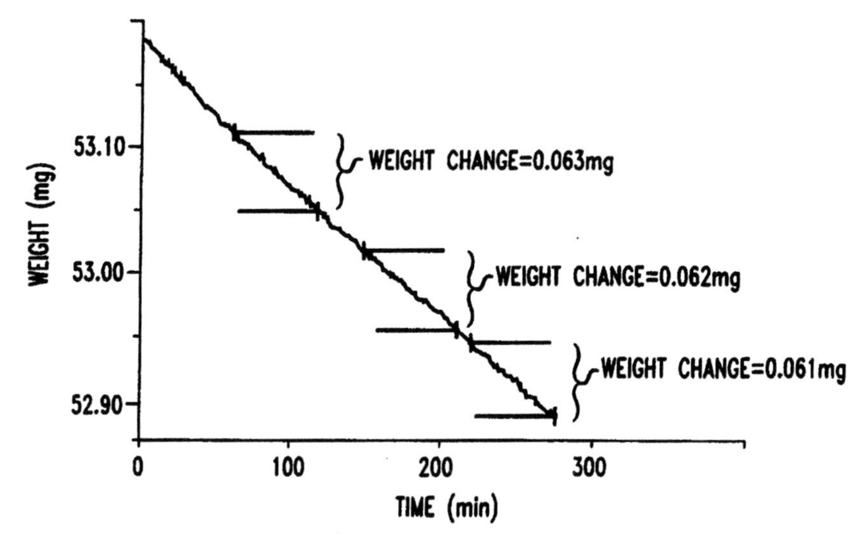

Figure 2 Weight loss of water vapor through a polycarbonate film at 23°C (weight changes are in one hour intervals).

was maintained at 23°C for the duration of the weight loss measurement while the vapor pressure difference across the PC film was held at 21.1 mm of Hg. In Figure 2 the graph of the weight change of the water in the sealed receptacle is plotted as a function of time. From the data in Figure 2 it follows that water vapor is transmitted through the amorphous Lexan polycarbonate film at a constant rate of 0.062±0.001 mg/hr or 0.45 mg/cm^{2}hr for a period of 275 minutes.

Based on this water vapor transmission rate, GE Lexan polycarbonate has a water permeability at 23°C of approximately 2.2 x 10^{-6} cm^3-mm/cm^2-sec-cmHg at STP. In 1963 Norton measured a value of 1.4 x 10^{-6} cm^3(STP)mm/cm^{2}sec-cmHg on polycarbonate using a mass spectrometer as an analyzer (8). Permeability (P) is calculated as

$$P = \Delta W d / t\, A(p_1 - p_2) \qquad (1)$$

where ΔW is the weight change for the sealed receptacle, d is the thickness of the polymer film, t is the time, A is the area of the hole in the receptacle lid, p_1 is the vapor pressure in the sealed capsule, and p_2 is the vapor pressure inside the TGA chamber.

A drop of water was placed in a stainless steel WVT capsule via a syringe. Then a 0.32 mm thick film of cured GE silicone RTV-615 was fitted over a 4.25 mm diameter opening in the container's lid. The sealed receptacle was positioned atop the Pt sample cup inside the TGA chamber. The rate of water vapor transmission through the silicone film was determined to be 1.2, 9.3, 4.5 and 16.7 mg/cm^2hr at 23, 48, 39 and 57°C, respectively (Figure 3). Note how quickly the sample adjusts to temperature changes and steady state water loss values are obtained. The partial pressure of the water in this experiment was 21.1, 52.4, 83.7 and 129.8 mmHg at 23, 39, 48 and 57°C, respectively.

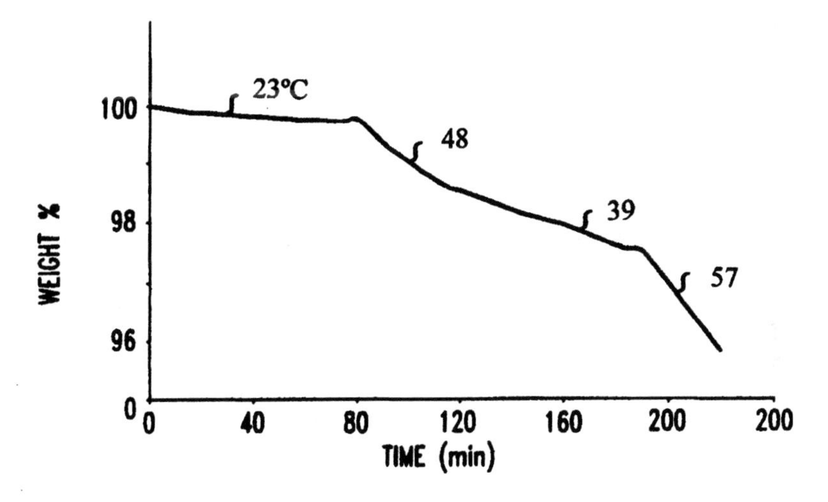

Figure 3 Water vapor loss through a silicone film at 23, 48, 39 and 57°C.

The permeability of the GE 615RTV silicone is plotted in Figure 4 as function of temperature and is shown to increase linearly with increasing temperature. Note that when temperature rose from 23 to 48°C permeability doubled.

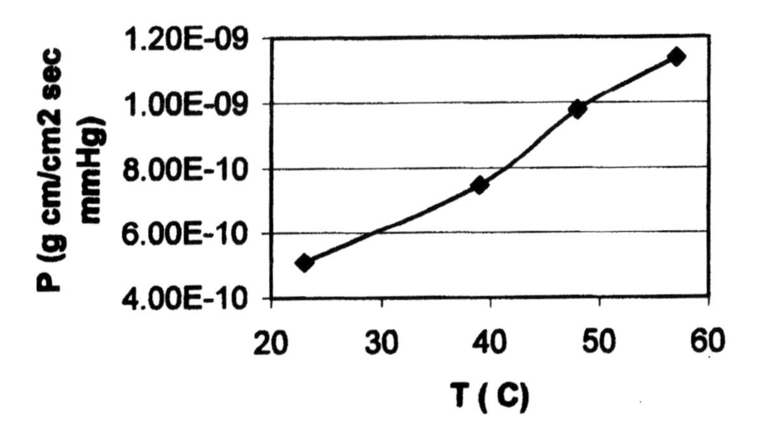

Figure 4 Permeability of cured RTV-615 silicone film as a function of temperature.

In Figure 4 at room temperature P equals 5.1 x 10^{-10} g-cm/cm^{2}sec-mmHg or 6.35 x 10^{-7} cm^3(STP)cm/cm^{2}sec-mm Hg. At 23°C the solubility of water in RTV-615 is 0.1wt%. Since the product of the diffusion coefficient D and the solubility coefficient, S equals P, the estimated value of D is 9.0 x 10^{-6}cm^2/sec

Permeability of a polymer depends not only on its chemical structure but also on its morphology, crystallinity and orientation. In an attempt to examine the effect of crystallinity on the rate of water vapor transmission though a film of polyethylene terephthalate (PET), two PET samples of were prepared. One was made amorphous by quenching and the other was annealed at 175°C for 24 minutes to induce partial crystallinity.

The DSC scan from 20 to 275°C at 15°C/min of the quenched PET film is shown in Figure 5 (dash-dot line). Note that a broad glass transition occurred beginning at about 45° and ending near 75°C with the magnitude of change in Cp associated with the glass transition, ΔCp, equal to 0.39 Jg^{-1}°C^{-1}. Tg occurred at ½ΔCp or 69 °C. Above Tg and near 100°C crystallization began and proceeded to about 212°C with the evolution of 36.5 J/g. Above this temperature crystals of PET melted with the absorption of 36.5 J/g. Since the heat evolved during crystallization equaled the subsequent heat lost during melting, one can conclude that the initial PET film was amorphous. In contrast to this behavior the film that was annealed for 24 minutes at 175°C showed no sign of crystallization

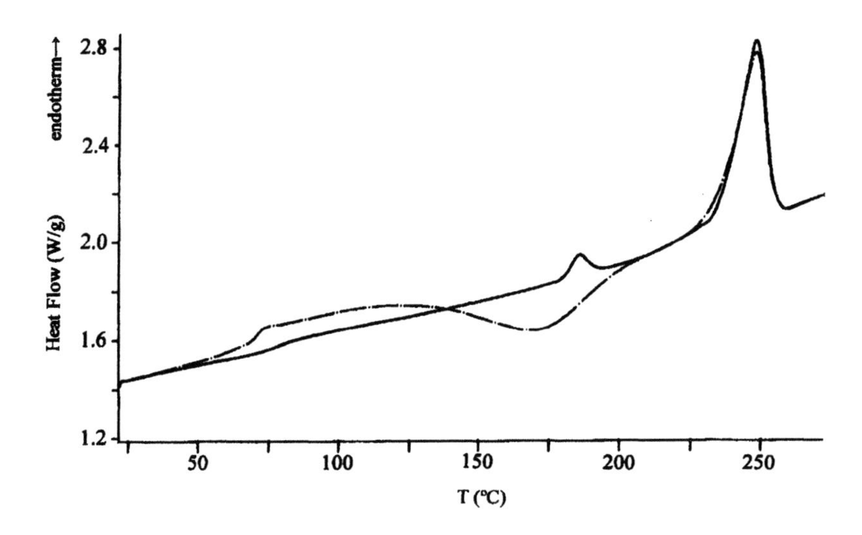

Figure 5 Comparative DSC scans of quenched and annealed PET films.

when heated above Tg (solid line, Figure 5). In the latter case, ΔCp is diminished to 0.13 J/g°C and Tg is increased to 81°C due to restrictions placed on the glass by the crystal phase (9). In this way the polymer chains are kept from attaining their liquid-like amorphous configurations and the accompanying liquid like Cp above Tg. A small group of crystals associated with the annealing melted near 185°C while the more perfect and thicker crystals melted between 212 and 256°C with an apparent heat of fusion of 36.5 J/g. Hence, crystallinity equals 36.5/122 or 30% for the annealed PET film.

A 0.241 mm thick film of the amorphous PET was sealed in a WVT capsule that contained a 4.25 mm diameter hole in its lid and placed in the TGA instrument. The interior of the TGA closure was maintained near 0% relative humidity and the WVT rate was monitored at a series of temperatures ranging from 20 to about 86°C. This data is displayed in Figure 6. Note that two regimes exit above and below 47°C. The latter temperature coincides with the onset of the glass transition that was observed by DSC for the amorphous PET as shown in Figure 5. Note that the rate of water vapor movement through PET increased sharply above the glass transition temperature as indicated in Figure 6.

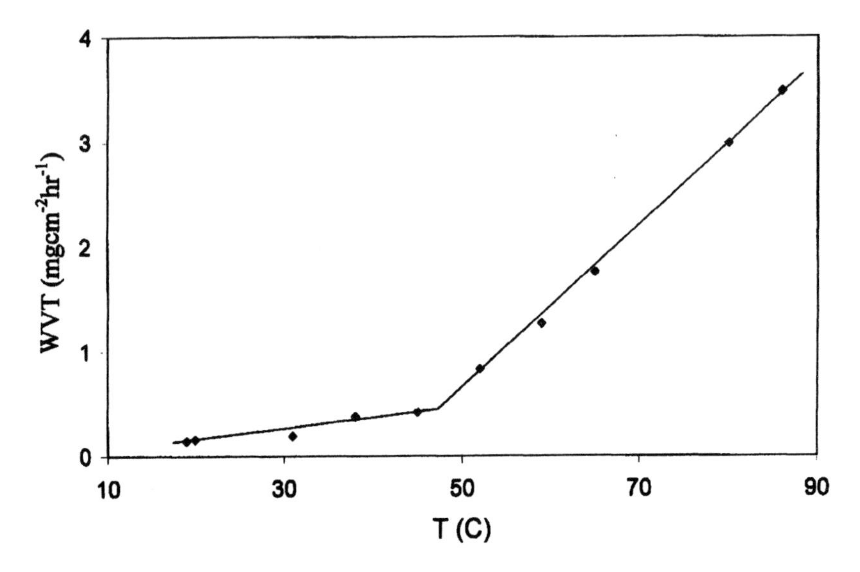

Figure 6 Rate of water vapor transmission through an amorphous PET film.

Below Tg water movement through the partially crystalline PET was reduced by a factor of about three from the rate that was found for the amorphous PET. For example near room temperature the WVT rate was about 0.15 and 0.05 mg/cm^{2}hr for the amorphous and semi-crystalline films, respectively.

A lot of clothing and footwear that is designed for outdoor activity allows water vapor to escape from the article but at the same time repels liquid water from the membrane's outer surface. In certain situations these same kinds of materials can be used to allow electronic devices to breathe while keeping liquid water out of the hardware. One manufacturer's product line using this special type of microporous membrane is Goretex®. The supplier reports the membrane is made from a fluorocarbon material.

A 5 mg sample of this membrane was scanned in a DSC as shown in Figure 7 (solid line). The material exhibited a melting endotherm with a peak at 328°C and fusion required 19.9 J/g. A second sample of polytetrafluoroethylene (PTFE), Teflon, is shown to have nearly an identical melting pattern (dash-dot line, Figure 7). The latter polymer melting peak is at 328.7°C with the absorption of 19.4 J/g of heat. Note that endothermic areas under each melting peak are defined by the solid lines with circles that stretch from about 275 to 340°C for each scan. In addition, both the PTFE and membrane samples were found to undergo two first order transitions near 19 and 31°C with a heat of transition of

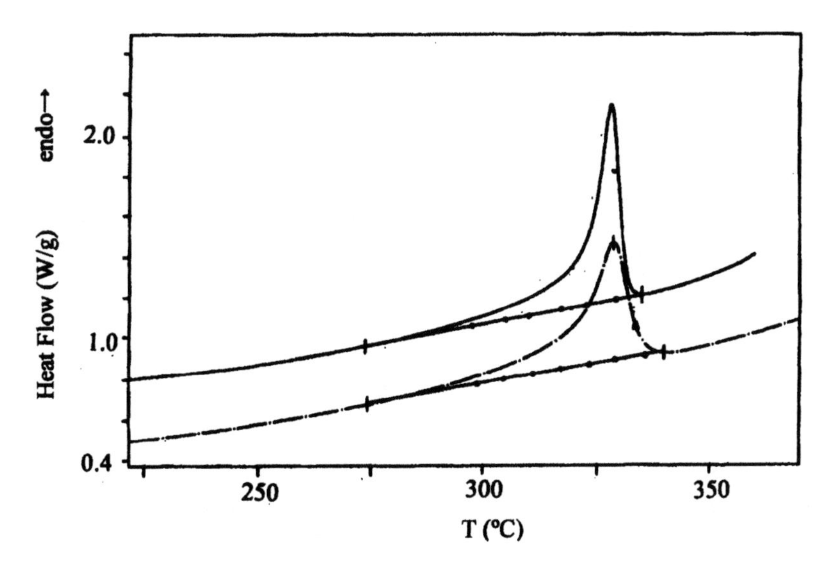

Figure 7 Two DSC scans of an unknown and polytetrafluoroethylene samples at 15°C/min.

about 5 J/g. These solid-solid phase transitions are associated with the onset of internal motion in the polymer's crystal phase (10,11).

From the similarities of the two material's melting temperatures and apparent heats of fusion it is concluded that the Goretex material is nearly pure PTFE. The scanning electron micrographs of the Goretex membrane in Figure 8 reveal the secret behind the product's ability to repel liquid water but allow water vapor to move through it quickly.

The lower half of Figure 8 shows a PTFE structure that appears to be composed of islands of solid PTFE measuring about 10 microns across and showing nearly infinite length in the opposite direction. Between the islands, many PTFE strands are stretched out into thin fibrils. These fibrils are about 1 μm in diameter and 20 μm in length. The spacing between strands of PTFE is clearly defined to be about 2μm as shown in the upper half of Figure 8. Hence, liquid water droplets will simply sit on the hydrophobic PTFE's surface but the vapor can freely pass through the grating created by the fibrillar PTFE structure.

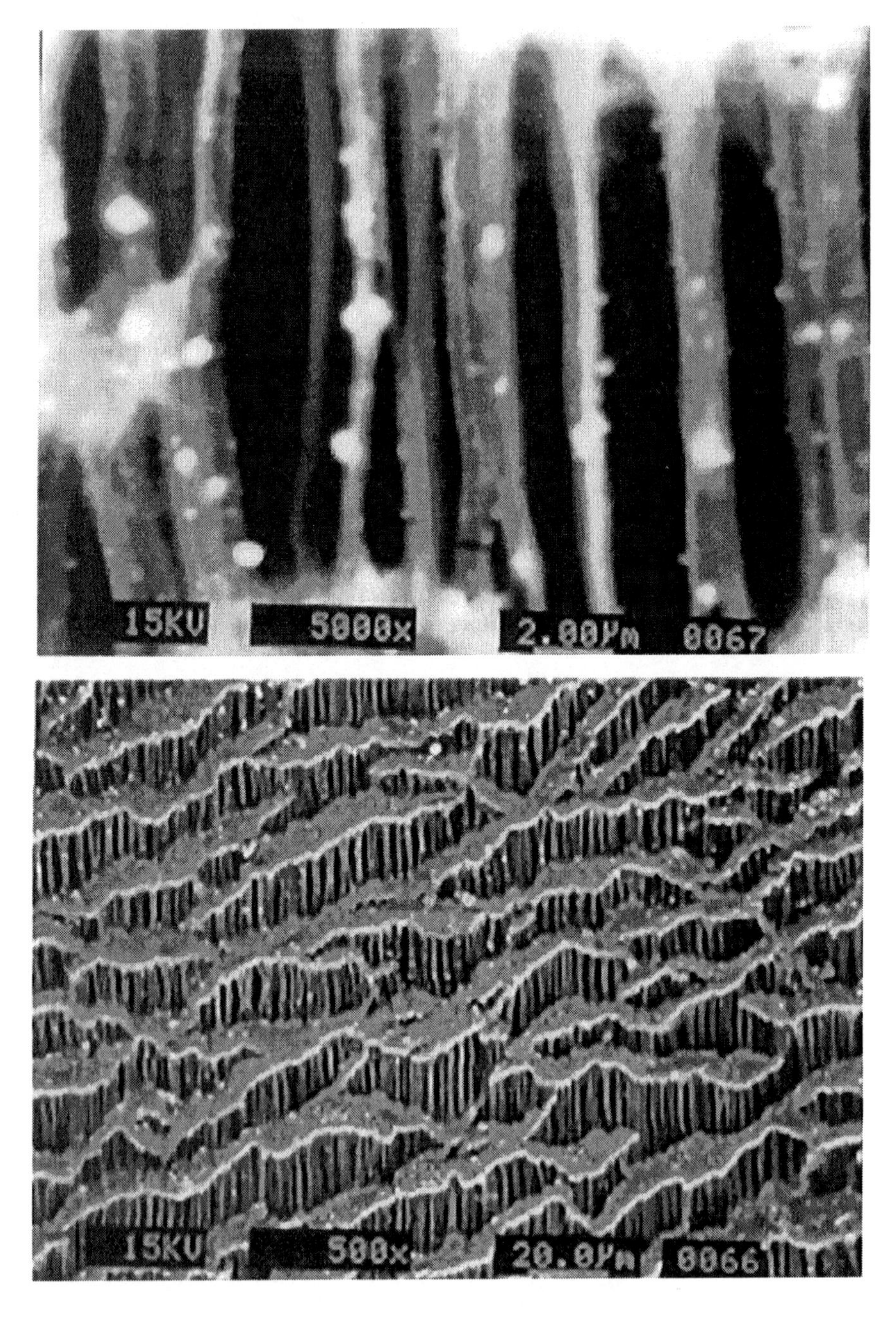

Figure 8 Scanning electron micrograph of a Goretex membrane

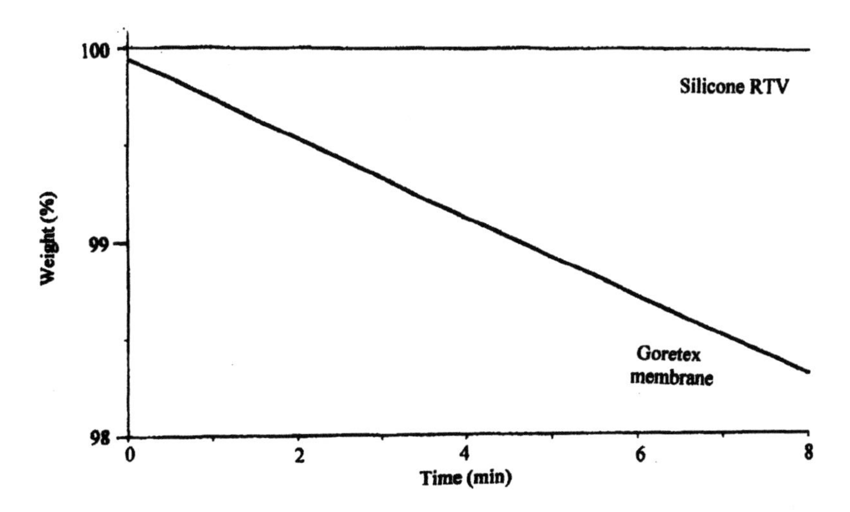

Figure 9 Water vapor loss through films of a RTV silicone (upper curve) and a Goretex membrane (lower curve) at room temperature.

Figure 9 demonstrates the effectiveness of the Goretex membrane in allowing water vapor to pass through its fibrillar structure as compared to the rubbery GE silicone, RTV-615. At 23°C the rate of water movement through the silicon RTV is 1.2 mg/cm^2hr whereas the rate through the permeable PTFE membrane is 65.7 mg/cm^2 hr. From this data P equals 1.94 x 10^{-4} cm^3(STP) cm/ cm^2sec-cmHg for the membrane. In a separate experiment the microporous membrane was aged for 3 weeks in an accelerated dust chamber. Reportedly this treatment is roughly equivalent to 10 years of aging in a city environment such as Newark, NJ. The aging only reduced the rate of water transmission through the membrane slightly or about 7% lower than found for the unaged, as received membrane.

Conclusions

A new hermetically sealeable capsule was created for use as a container for making water vapor transmission rate measurements via a TGA instrument. In this way WVT data can be collected accurately and quickly as a function of temperature and humidity. Permeability data was measured not only on glassy, rubbery and semi-crystalline polymer films but also on a microporous Goretex® membrane. With a slight modification of the WVT capsule it is also possible to confine volatile reactive liquids such as used in photo-dsc work at elevated temperatures.

Acknowledgement

This work is dedicated to Prof. Frank E. Karasz on the occasion of his 70th birthday for his many contributions to the field of polymer science, One of us, namely Harvey Bair, was fortunate to have Frank as his mentor when he was a research training fellow at the General Electric Research Laboratory in Schenectady, NY during the early 1960s. The authors would also like to acknowledge helpful discussions with Dr. Lee Blyler, of Chromis Fiberoptics, on permeability issues. In addition, we thank Dr. Arturo Hale for the SEM photos.

References

1. Annual Book of ASTM Standards, Standard E 96-94, American Society for Testing and Materials: Philadelphia, PA, **1994**; Vol 04.06, pp.696-1055.
2. Bair, H. E.; Hale, A.; Popielarski; S. R. US Patent 6,586,258 B1, **2003**, July 1.
3. Olsson, R. T.; Bair, H. E.; Kuck, V.; Hale A. Polymer Preprints **2000**, 42 (2), 797.2
4. Olsson, R. T.; Bair H. E.; Kuck, V.; Hale, A. "Thermomechanical Studies of Photoinitiated Cationic Polymerization of a Cycloalipathitc Epox Resin", In *Photinitiated Polymerization* K. D. Belfield and J. V. Crivello, Eds.; ACS Symposium Series **2003**, 847, p. 317.
5. Bair, H. E.; Blyler, L. L. Proc. 14th NATAS Conference, September **1985**, p. 392.
6. Bair, H. E.; Blyler, L. L.; Simoff, D. A. ACS Polymeric Materials Sci.&Eng. **1993**, 69, 268.
7. Bair, H. E. "Glass Transition Measurements by DSC", In *Assignment of the Glass Transition*, ASTM STP, Seyler, R. J., Ed.; American Soc. For Testing and Materials, Philadelphia, PA **1991**, p. 50.
8. Norton, F., J. Appl. Polymer Sci. **1963**, 7, 1649.
9. DiMarzio, E. A.; Dowell, F. J. App.Phys. **1979**, 50, 6061.
10. Bunn, C. W.; Howells, E. R. Nature **1954**, 174,549.
11. Kimmig, M.; Strobl, G. ; Stuhn, B. Macromolecules **1994**, 27,2481.

Chapter 26

Thermal Characterization and Morphological Studies of Binary and Ternary Polymeric Blends of Polycarbonate, Brominated Polystyrene, and Poly(2,6-dimethyl-1,4-phenylene oxide)

A. Z. Aroguz[1], Z. Misirli[2], and B. M. Baysal[3,4,*]

[1]Department of Chemistry, Faculty of Engineering, University of Istanbul, 34850, Avcilar, Istanbul, Turkey
[2]Advanced Technologies R&D Center, Bogazici University, 80815, Bebek, Istanbul, Turkey
[3]Department of Chemical Engineering, Bogazici University, 80815, Bebek, Istanbul, Turkey
[4]TUBITAK Marmara Research Center, P.O. Box 21, 41470 Gebze, Turkey

The thermal behaviors of binary blends of polycarbonate (PC)/brominated polystyrene (PBrS) and ternary blends of PC/PBrS/poly(2,6-dimethyl-1,4-phenylene oxide) (PPO) were investigated. The compatibilizing effect of PPO on the miscibility of the PC/PBrS blends was examined. The miscibility of binary and ternary blends was studied by using differential scanning calorimeter (DSC). The results of DSC indicate that the binary blends of PC/PBrS are immiscible but ternary blends of PC/PBrS/PPO in certain limits are miscible. The microstructural properties of the blends was characterized by environmental scanning electron microscope (ESEM) where the properties are determined in the natural state of the structural role of individual phases (PC/PBrS/PPO) and their effect on the overall microstructure of the products. DSC and ESEM results were supported by FT-IR measurements.

Poly(2,6-dimethyl-1,4-phenylene oxide)(PPO) an engineering plastic is miscible in all proportion with a commodity plastic polystyrene (1-6), due to the favorable specific interactions of the repeat units of these polymers (7,8). Mixtures of PPO and PS give amorphous blends of significant commercial importance (9). Limited compatibility was observed for the blends of PPO and halogene-substituted-polystyrenes (9). The phase behavior of PS, PPO, and brominated-polystyrenes was studied (7). The mean field theory of phase behavior was used to discuss, in detail, the interactions and compatibility in polymeric/copolymer systems of these blends (8,10,11).

In an earlier work, we studied the miscibility of the two binary polymeric systems: PS/brominated PS, and PPO/brominated PS. Only limited miscibility was observed for a binary PPO/brominated PS blends. The blends of PS/brominated PS are immiscible in all compositions (12). However, compatible compositions of ternary blends of PPO/PS/brominated PS were obtained due to the compatibilizing effect of PPO in this system.

In this work, first we studied the thermal behaviors of binary blends of an engineering plastic polycarbonate (PC) and brominated polystyrene (PBrS). Limited miscibility was observed for binary PC/PBrS blends. We have also studied the ternary blends of PC/PBrS/PPO and we examined the compatibilizing effect of Poly(2,6-dimethyl-1,4-phenylene oxide) on the miscibility of PC/PBrS blends.

Experimental

Materials

The polymers used in this study were obtained from commercial sources. PPO was purchased from Polysciences, Inc. (Warrington,PA; M_w=50x10^3 g mol^{-1}; M_n=20x10^3 g mol^{-1}; high softening point, 90 °C).

PBrS was also purchased from Polysciences, Inc. Gel permeation results obtained for PBrS were M_w=63x10^3 g mol^{-1}; M_n=19x10^3 g mol^{-1}. The structural characterization of this polymer was performed using IR, NMR and chemical analysis. We found that pendant phenyl groups of the PS backbone chain contain, on avarage, 2.66 Br atoms (i.e., two of each three pendant phenyl groups were *ortho-para*-tribromo, whereas the remaining one was dibromo-substituted).

PC was purchased from Bayer AG. The molecular weights of PC are M_w=45x10^3 g mol^{-1}; M_n=25x10^3 g mol^{-1}. All materials were dried at least 4h at 80°C under vacuum.

Formulas of polymers used in this work

Polycarbonate(PC):

Tg = 140 °C

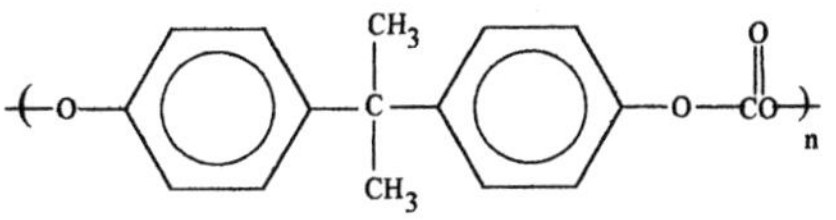

Brominated polystyrene (PBrS):

Tg=177 °C

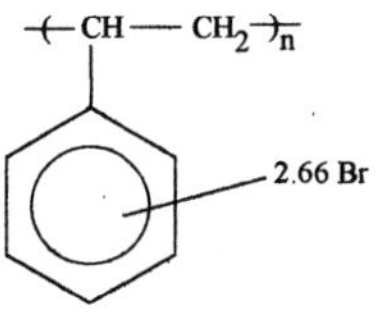

Poly(2,6-dimethyl-1,4-phenylene oxide)(PPO):

Tg=216 °C

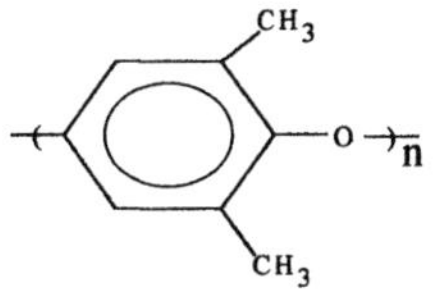

Methods of Characterizations

Preparation of the Blends

Nine binary blends of PC/PBrS were prepared by dissolving in chloroform with PC weight fractions of 90,80,70,60,50,40,30,20,10 at room temperature for DSC analysis. The solutions were clear for all blend compositions. In order to examine the compatibilizing effect of PPO on the miscibility of PC/PBrS blends, ternary blends were prepared in different compositions. It was known that the binary blends of PPO/PC (9,13) and PPO/PBrS (12) are partly miscible. The compositions of ternary blends PC/PBrS/PPO were 40/20/40, 40/40/20, 20/40/40.

The polymer concentration in the solution was kept below 2 wt % to obtain uniform mixing. The polymer solution was coprecipitated in acetone drop by drop with vigorous agitation and the white precipitates were repeatedly washed with acetone. The blends were dried under vacuum at 50°C for 2-3 days and used for DSC analysis.

FT-IR Measurements

In general, the miscibility in polymeric blends arises from specific interactions between the two polymers. Infrared spectroscopy can be used to establish the nature and the level of interactions in polymer blends. Fourier transform infrared (Mattson 1000 FT-IR spectrometer) spectra of samples were recorded in the range 400-4000 cm^{-1} with KBr pellet as IR transmitting material. Spectra were obtained at 8 cm^{-1} resolution.

Thermal Properties

The thermal properties of all samples were measured calorimetrically using Shimadzu Differential Scanning Calorimeter Computerized Thermal Analysis System 40-1 Model. Blend samples were heated from 298 to 532 K at a heating rate of 20 K min^{-1} under stream of nitrogen with a sample size between 15 and 25 mg using standard aluminum sample pans.

The DSC curves taken for analysis were obtained from the second run and glass transition temperature was taken as the initial onset of the change of slope in DSC curve. Temperature calibration was performed using indium (T_m =156.6°C H_e=28.5 J g^{-1})

Characterization of Microstructural Behavior by ESEM

Binary and ternary polymer films were prepared by solution casting for morphologic analysis at ESEM. The polymer mixtures were dissolved in a given compositions in chloroform at room temperature (3.0 w/v solution) for at least 2 days. Blends were cast on glass plates and all film samples were dried under vacuum for 15 days at room temperatures before use.

The surface morphology of the solution cast films was examined by ESEM-FEG/EDAX. The samples were fractured by immersing them in liquid nitrogen and then breaking them between two grips while at low temperature and then the sections should be free from dust or loose particles, were then stained with osmium tetroxide vapor for eight hours at room temperature. These surfaces were examined in their natural, uncoated state at low vacuum around 0.2-0.8 torr and 10 KeV. ESEM micrographs were obtained by secondary (SE) or backscattered electrons (BE) or mixed (SE+BE) images for effectively indicated at each phase detail of interest.

RESULTS AND DISCUSSION

FT-IR Characterization

Figure 1 shows the full FT-IR spectra of polymers, PC, PBrS, and PPO used in this work, respectively. Figure 2 shows the FT-IR spectra of the binary blend of PC/PBrS (50 /50).

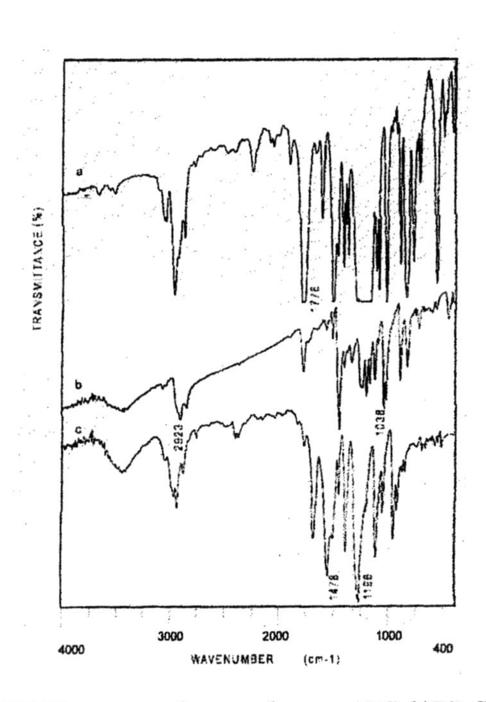

Figure 1. FT-IR spectra of pure polymers a)PC, b)PBrS, c)PPO

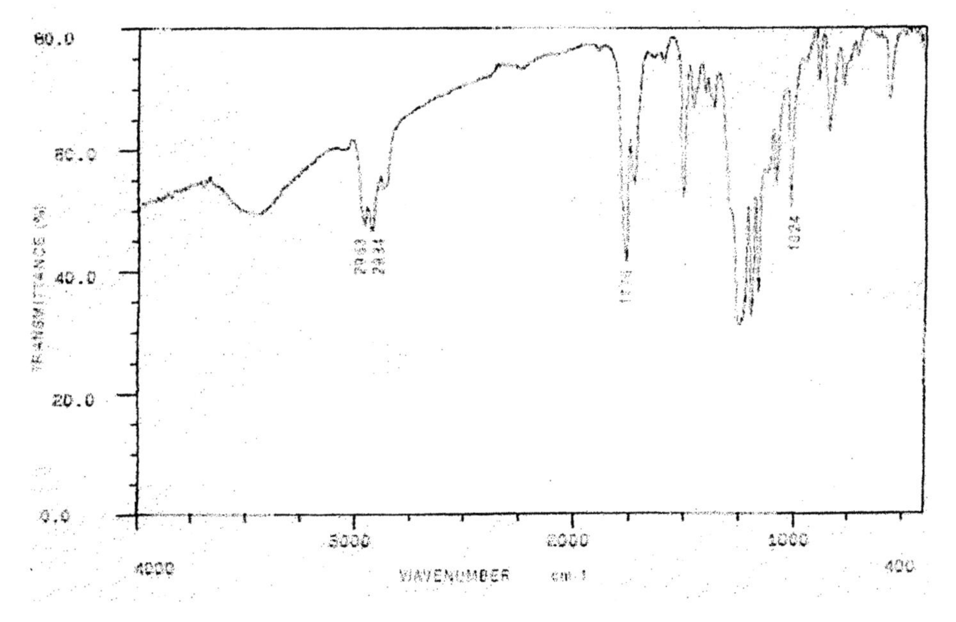

Figure 2. FT-IR spectra of binary blend PC/PBrS (50/50)

Carbonyl (C=O) adsorption peak is seen 1776 cm^{-1} in pure PC. Since the carbonyl frequency shift is very small this results emphasizes the weakness of the interactions, leading to thermodynamic miscibility in these blends. This peak was splitted at 10 wt% and 20 wt% of PC in the blend compositions as 1776 cm^{-1}, 1730 cm^{-1} and 1776 cm^{-1}, 1741cm^{-1}.

C-Br aromatic peaks were seen at 1038 cm^{-1}. This peak is shifted to lower wavenumbers as the composition of PC increased to 50wt%.

Aliphatic C-H streching is seen in pure PBrS at 2923 cm^{-1}. It changes from this value to 2934 cm^{-1} when adding PC to the blend. This peak is splitted at PC 50 wt% in the blend compositions as 2963 and 2934 cm^{-1}.

FT-IR spectra of ternary blends of PC/PBrS/PPO, 40/20/40, 40/40/20 are also seen in Figure 3. In the ternary blends of (PC/PBrS/PPO) (40/40/20) aliphatic C-H streching peaks are shifted at 2980 and 2940 cm $^{-1}$. If we compare PC/PBrS (50/50) and PC/PBrS/PPO (40/40/20) carbonyl peak were shifted from 1776 to 1788 cm^{-1}.

Thermal Characterization

Figure 4 shows the composition dependence of glass transition temperature of PC/PBrS binary blends. The glass transition temperature of PC was decreased from 140 °C to 129 °C as PBrS was added. This is due to the degradation of PC. When each of the polymer weight fraction in binary blend is very high (80 wt% or more) they are miscible otherwise immiscible.

From the DSC thermographs it can be seen that the ternary blends of PC/PBrS/PPO in the compositions of 40/40/20, 40/20/40, 20/40/40 are miscible (Figure5).

Microstructural Examinations

As shown in typical ESEM Figures (6a-6i) morphology studies of these binary and ternary multiphase polymer blends have been concerned with controlling the size and shape of dispersed phase. These morphological structures were basically described as being composed of a matrix (PC is a continuous phase) and second phase (PBrS is a dispersed phase). The dispersed and continuous phases can be seen in all compositions of the blends in different sizes as shown in the micrographs. The mean sizes of different phases were measured by atomatic image analysis using software program and results were given in Table I and Table II.

Figure 6a shows the internal structure of the pure PC. The surface morphology migrograph of the binary blend PC/PBrS (80/20) was given in Figure 6b. The spherical PBrS particles are dispersed and adherent to the matrix. Dispersed phase becomes elongated. This phenomenon was confirmed by the

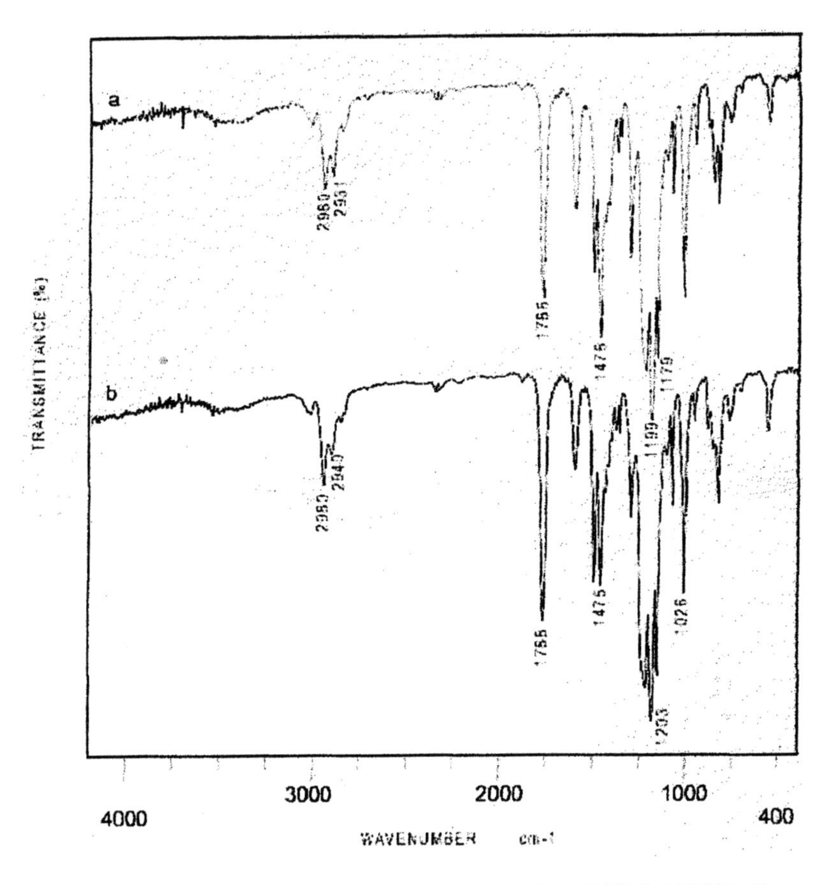

Figure 3. FT-IR spectra of ternary blends a)PC/PBrS/PPO (40/20/40), b)PC/PBrS/PPO (40/20/40)

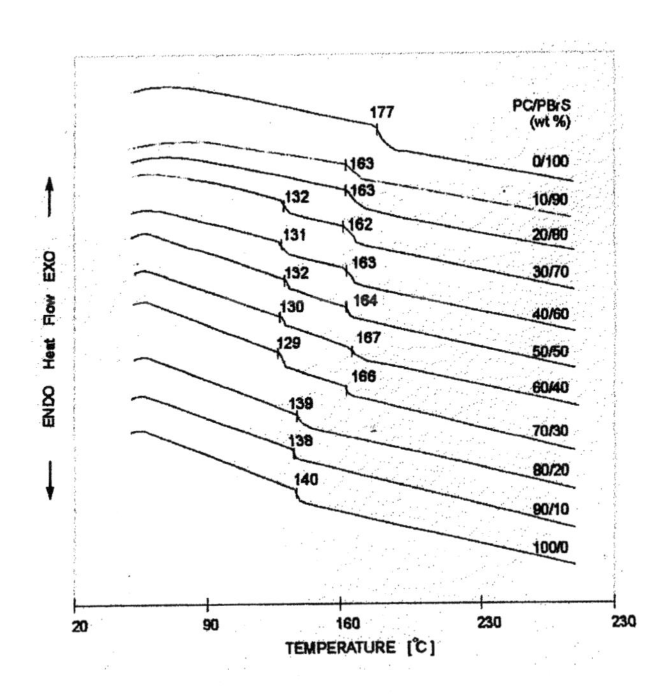

Figure 4. DSC Thermograms of binary blends of PC/PBrS.

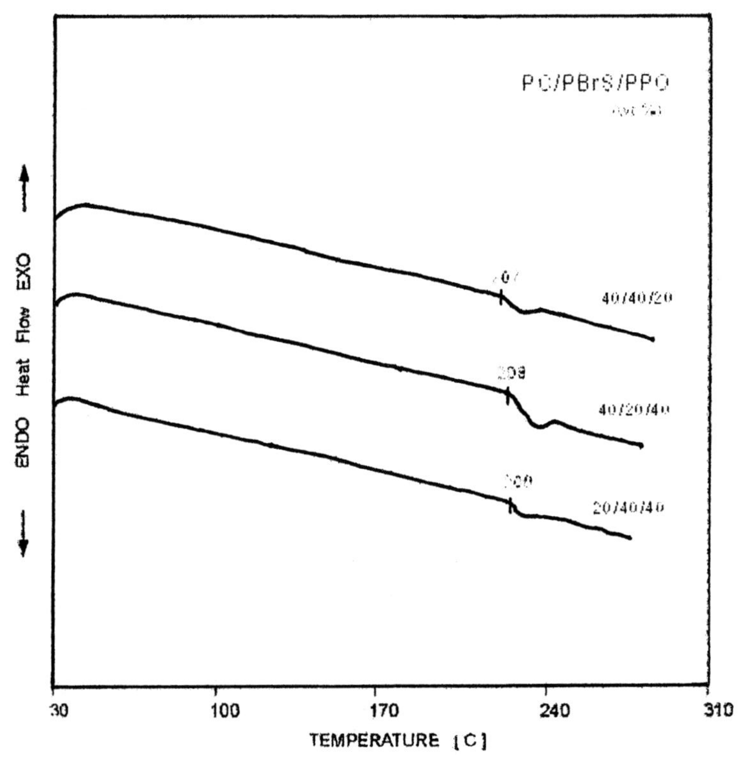

Figure 5. DSC Thermograms of ternary blends of PC/PBrS/PPO.

Table I . Statistical results of dispersed spherical particles. The sizes* of various binary blends given as μm.

PC(wt %)	10	20	30	40	50	60	70	80	90
Size	0.14	15	60	525	88	837	256	1175	1180

*mean size of dispersed phases was taken in cross-sectional view

Table II. Statistical results of dispersed spherical particles. The sizes* of various ternary blends given as μm.

Dispersed phases	Compositions		
	PC/PBrS/PPO 40/20/40	PC/PBrS/PPO 40/40/20	PC/PBrS/PPO 20/40/40
PPO	0.9	3	3
PBrS	20	24	34

*mean size of dispersed phases was taken in cross-sectional view

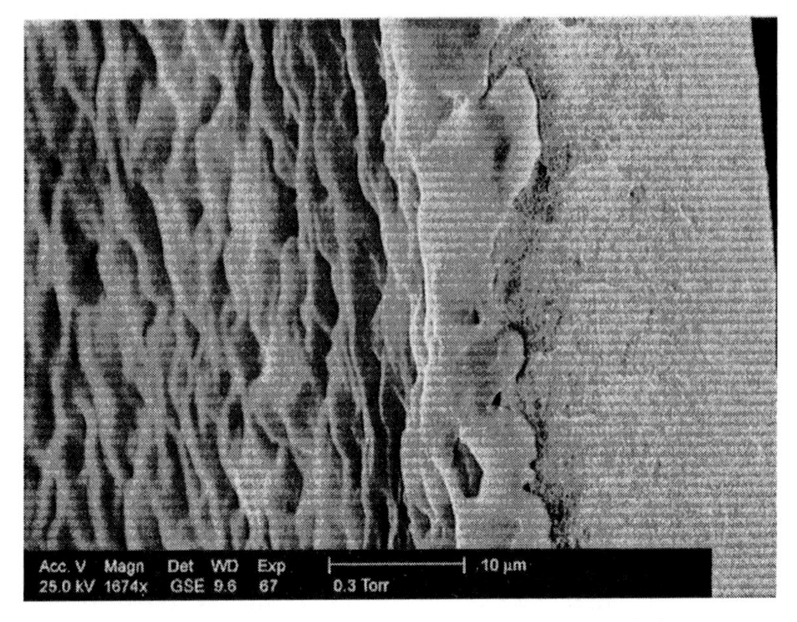

Figure 6a. SEM micrograph of pure PC surface morphology.

Figure 6b. SEM micrograph of surface morphology of PC/PBrS (80/20).

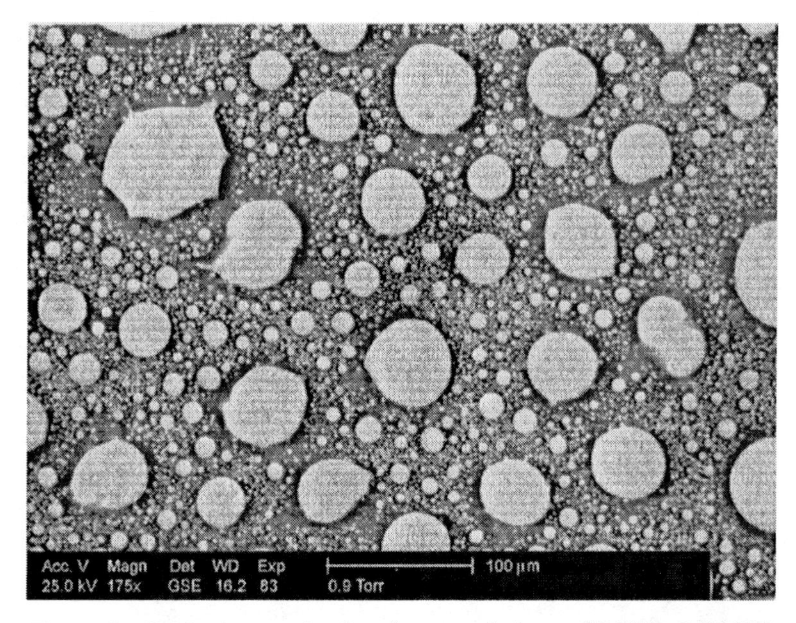

Figure 6c. SEM micrograph of surface morphology of PC/PBrS (50/50).

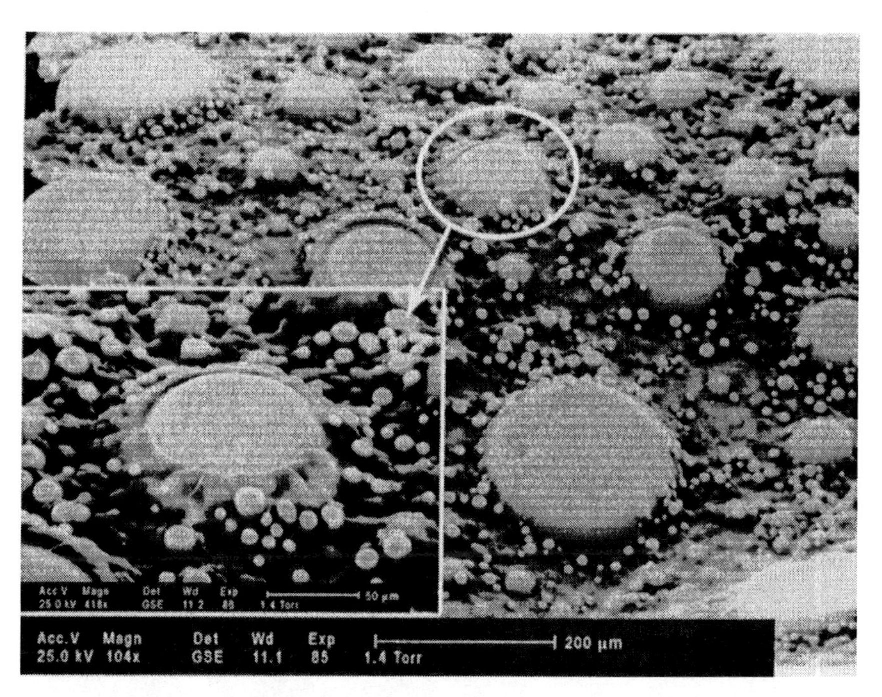

Figure 6d. SEM micrograph of surface morphology and surface detail of PC/PBrS (40/60).

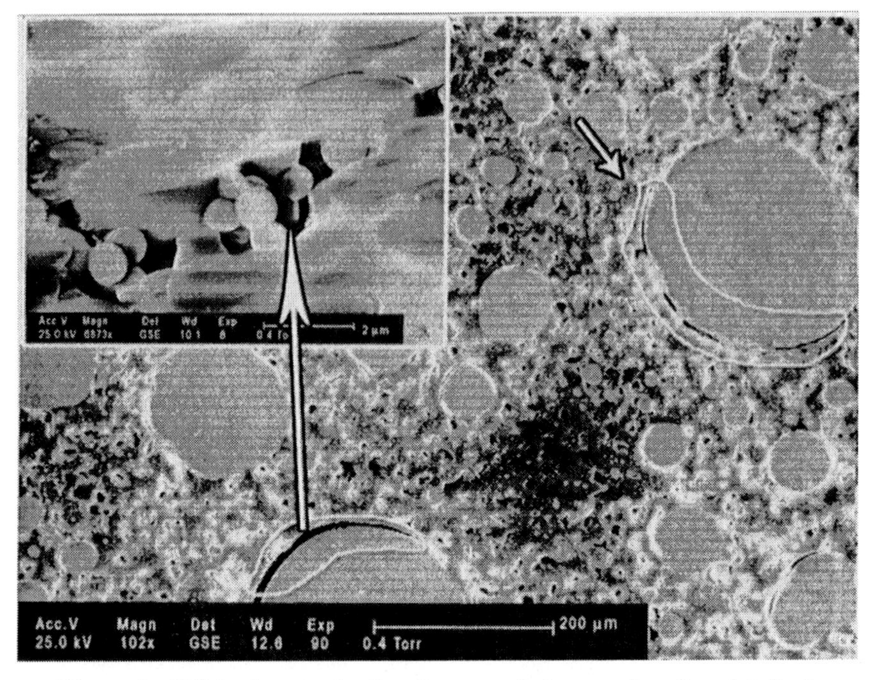

Figure 6e. SEM micrograph of surface morphology and surface detail of PC/PBrS (30/70).

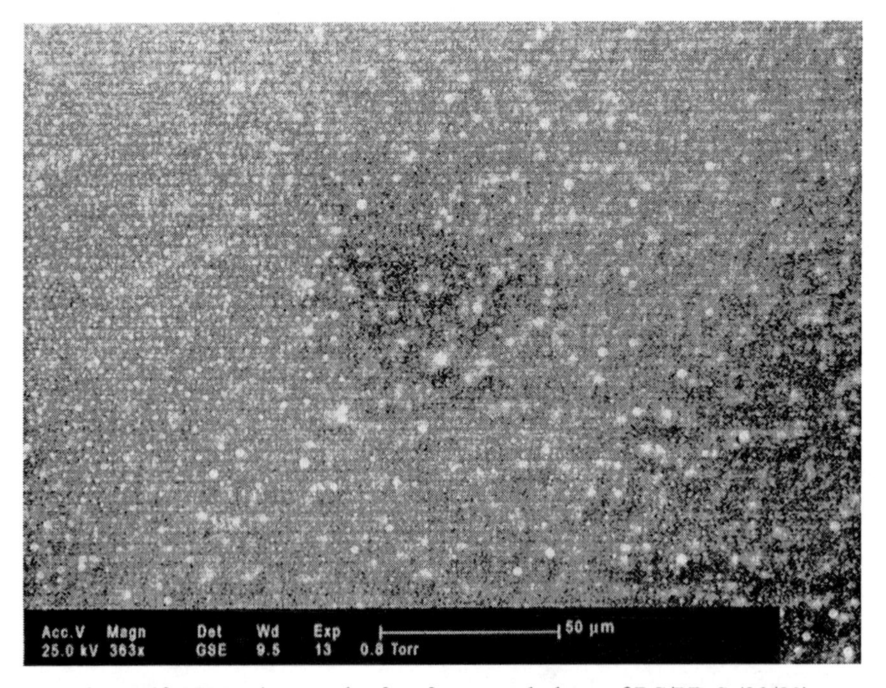

Figure 6f. SEM micrograph of surface morphology of PC/PBrS (20/80).

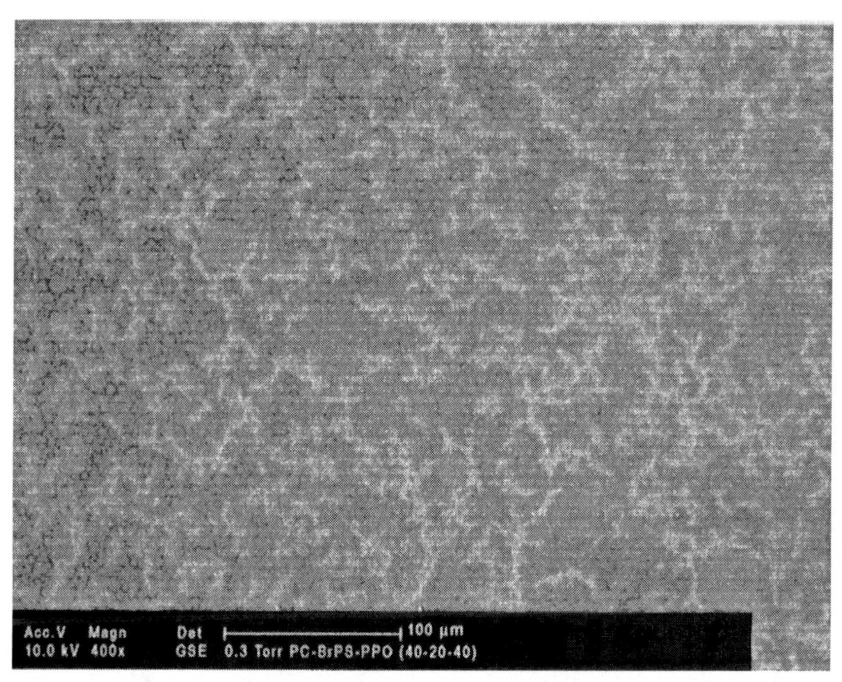

Figure 6g. SEM micrograph of surface morphology of PC/PBrS/PPO (40/20/40).

Figure 6h. SEM micrograph of surface morphology of PC/PBrS/PPO (40/40/20).

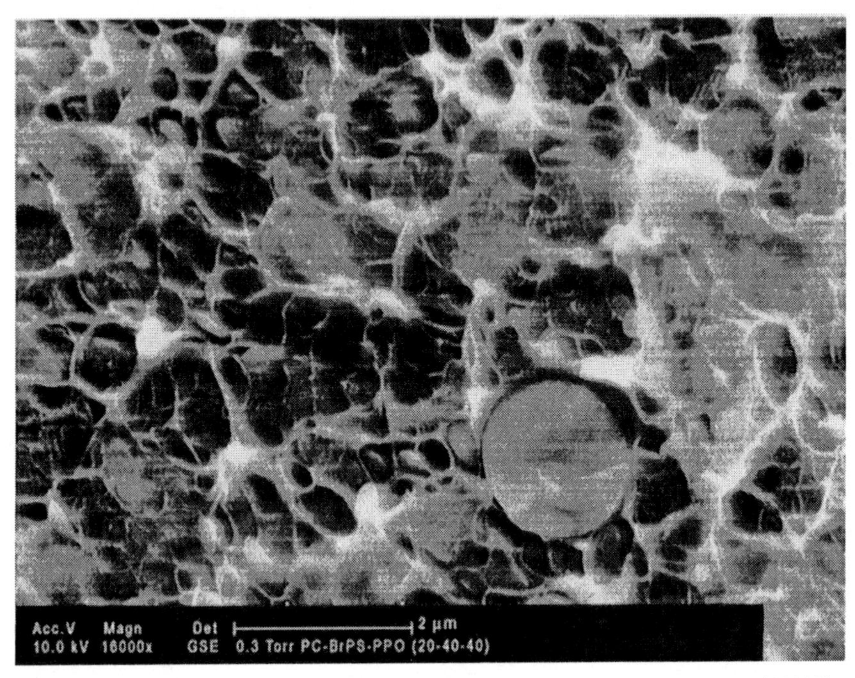

Figure 6i. SEM micrograph of surface morphology of PC/PBrS/PPO (20/40/40).

DSC results in which a single T_g was obtained for this composition. Figure 6c shows the surface morphology of PC/PBrS (50/50) blend. The particle size becomes larger over the surface. Surface and surface-detail were investigated for another blend, PC/PBrS (40/60) (Figure 6d). Excessively big particles can be observed. These migrographs show that the domain-sizes of the dispersed phase becomes larger and the dispersed phase is pulled inside of the matrix (wetting). Increasing the PBrS content in the blend causes the nonwetting behavior, PC/PBrS (30/70) (Figure 6e). There are holes around of the spherical particles. These holes indicate that interfacial adhesion between two polymers was poor. This is due to the limited compatibility between two polymers. Gel structures were observed inside the holes (insert micrograph in Figure 6e). Figure 6f shows the surface morphology of PC/PBrS (20/80) blend. The big particles are pulled inside the matrix. Homogenous morphology was found. DSC results also verify that this blend composition is miscible.

Figures 6g-6i show PC/PBrS/PPO ternary blends migrographs in different compositions. As seen from Figure 6g, a homogenous structure of PC/PBrS/PPO (40/20/40) is present. PBrS and PPO particles are distributed in the PC matrix. The size of the spherical particles in the ternary blend of PC/PBrS PPO (40/40/20) is heterogeneous but the distribution of these particles is homogenous (Figure 6h). Figure 6i shows PC/PBrS/PPO (20/40/40) ternary blend morphology. Again a homogenous structure can be seen. Extremely big particles are mostly eliminated. Big particle-dimensions decreased from 500 μm to 104 μm. A dimpling effect can be seen. If the binary and ternary blends morphologies are compared, it can be observed that the spheres are pulled out of matrix to the surface in the binary blends. However, in the ternary blends the spheres are inside of the matrix.

Conclusions

DSC studies show that the binary blends of PC/PBrS are incompatible and show two T_g values. However, blends containing small amounts of second components indicate a single T_g. DSC results of ternary blends studied in this work indicate a single T_g value for each composition. Morphological studies by scanning electron microscope support the above observations and provide additional information on the structures of the blends studied in this work.

Acknowledgment

This work was supported by *Research Foundation of the University of Istanbul, Contract Grant Number :1081/031297,* B.M.Baysal acknowledges support from *TUBA-Turkish Academy of Sciences.*

REFERENCES

1. Stoelting, J.; Karasz, F.E.; MacKnight, W. *J. o f Polym. Eng. Sci.* **1970**, 10, 133.
2. Prest, W.M.J.; Porter, R.S. *J. Polym. Sci. Polym. Phys. Ed.* **1973**, 10, 1639.
3. Pochan, J.M.; Beatty, C. L.; Pochan, D.F. *Polymer* **1979**, 20, 879.
4. Macoumanchia, A.; Kambour, R.P.; White, D.M.; Rostani, S. Walsh,; D. J. *Macromolecules* **1984**, 17, 2645.
5. Hseih, D.T.; Peiffer, D.G. *Polymer* **1992**, 33, 1210
6. Bazuin, C. G.; Rancourt, L.; Villenneuve, S.; Soldera, A. *J Polym Sci Polym Phys Ed* **1993**, 31, 1431
7. Kambour R.P.; Bendler, J. T.; Bopp R.C. *Macromolecules* **1983**, 16, 753.
8. Brinke, G.T.; Karasz F.E. ;MacKnight, W.J *Macromolecules* **1983**,16, 1827.
9. *In compatibility Studies of Poly(2,6-dimethyl-1,4-phenylene oxide) Blends in polymer Compatibility and Incompatibility-Principles and Practices*; Fried J.R.; Hanna, G.A.; Kalkanoglu, H. Solc, K., Ed,; Harwood . New York, 1980.
10. Kambour, R.P.; Bendler, J.T. *Macromolecules* **1986**, 19, 2679.
11. Strobl, G. R.; Bendler, J.T.; Kambour R.P.; Shultz, A.R. *Macromolecules*, **1986**, 19, 2683.
12. Aroguz A.Z.; Baysal B.M. *J of App Polym Sci* **2000**, 75, 225.
13.*High Performance Polymers: Their Origin and Development;* J.M. Heuschen, R.B. Seymour and G.S. Kirschenbaum, 1986.

Polymers for Environmentally Sustainable Applications

Chapter 27

Polymers for the Conservation of Cultural Heritage

M. Cocca[1], L. D'Arienzo[1], G. Gentile[1], E. Martuscelli[1,*], and L. D'Orazio[2]

[1]CAMPEC, Consortium for the Application of Polymeric Materials, P.le E. Fermi c/o CRIAI, 80055 Portici (NA), Italy
[2]Institute of Chemistry and Technology of Polymers, National Research Council of Italy, Via Campi Flegrei 34, 80078 Pozzuoli (NA), Italy

In this paper the results of a research aimed at developing radically innovative process tailored for the restoration of both stone and textiles are reported. For stone, a new polymerisation procedure of poly(urethane-urea) by in situ polymerisation inside stone is described. Through the method set up a good penetration depth of the polymer into the stone pores is achieved, and the material obtained is characterized by high aggregative and consolidating efficiency. For textile materials, a new consolidating procedure, based on the grafting copolymerisation of acrylic monomers on the cellulose substrate has been set up and a high volume grafting chamber has been designed and realized, in order to carry out the grafting procedure onto large textile items of historical interest.

Introduction

Objects of historical and artistic value undergo to an inevitable degradation due to physical, chemical, mechanical and biological deterioration of the constituent materials.

The most important purpose of restorers and conservators is to slow down these degradation processes through conservative interventions, consisting of restoration, protection and periodical treatments of maintenance *(1)*. After evaluating the status of an artwork and its final placement, and after pointing out the necessary operations to restore it (cleaning, integration of lost fragments, consolidation, protection), the restorer usually faces the problem of the choice of the most appropriate materials useful for the restoration intervention.

Since the late sixties polymeric materials have been considered by restorers as the answer to many problems. Polymer based products have been applied, with more or less satisfactory results, on different substrata (stone, wood, textiles, paper, paintings on wall and on canvas) and they are still widely used as consolidating or pre-consolidating agents, as adhesives, as integrating and supporting materials, as co-adjuvant for cleaning (ionic exchange resins, anti-redeposition agents on textiles) or as protective agents *(2, 3, 4, 5, 6)*.

Nowadays chemists' and restorers' common experience have pointed out the main characteristics of synthetic materials to be used for restoration.

First requirement is the mechanical compatibility between synthetic materials and materials constituting the object: for example adhesive and consolidating agents should "reply" to dimensional changes undergone by artworks under environmental variations.

Another requirement is reversibility, i.e. the possibility of removing the material used for the restoration, even long time after the intervention and with the chance of coming back to the status before the intervention.

Nowadays the main requirement of synthetic materials for conservation purposes is durability; all the materials, both natural and synthetic, undergo to degradation processes that induce chemical modifications. The decreasing in their flexibility (with loss of mechanical compatibility), and the loss of solubility (with loss of reversibility) are phenomena often strictly related each other. An example is the behaviour of polyacrylates, a class of polymers widely used in restoration, that undergo to a photo-oxidative degradation which leads to a strong yellowing, a partial cross-linking and a subsequent decreasing of their solubility *(7, 8)*. The requirement of durability also includes that conservative treatments have to be carried out by using materials that, in principle, will not preclude in the future further restoration interventions *(9)*.

In general, macromolecular substances are suitable for application in conservation field for the relevant diversification of their properties.

Unfortunately, most of the polymers do not fulfil the above-described prerequisites. The use of these substances in conservation field is due to the need of the restorers to find materials with high consolidating or protective properties compared with old natural resins and waxes. Furthermore, the application of synthetic substances is fast and easy. It can be stated that until few decades ago the application of polymeric materials on artworks was made without taking into account their characteristics and thus without foreseeing their eventual negative consequences. Recently, much has been done to explain the mechanisms of action of the products used and the efficacy of the treatments. Nevertheless a methodical study on their intrinsic chemical properties is still needed. In fact, there are many ambiguous situations and it is not always possible to associate the trade name of a product to its chemical composition. The technical sheets often seem to be drawn up for exclusive use of manual operators, because, apart from the mandatory safety data, they only report applicative information. Information about chemical characteristics and composition of products are often missed.

A very interesting example is the case of the commercial polyacrylate dispersions Primal AC33, a copolymer ethylacrylate-methylmetacrylate made and commercialised by Rohm and Haas, widely used as consolidating agent for stones and mortar *(10, 11)*, for parchments *(12)*, for fossilized wood *(13)*, as adhesive of paint layers on different type of substrates (wood, wall, ceramic) *(14)* and as adhesive agent for lining processes on textiles *(15)*.

Primal AC33 was introduced on the U.S.A. market by Rhom and Haas in 1953 with the trade name Rhoplex AC33. Nowadays this commercial dispersion has been discontinued by Rohm and Haas, but several suppliers of products for conservators commercialise different materials in order to replace the original Primal AC33. In one of the most recent research *(16)* the results of the comparison among the chemico-physical characteristics of Primal AC33 and its substitutes are reported. The investigated materials showed remarkable differences in durability: aging treatments under xenon-arc lamp induce structural modifications evidenced by progressive broadening of the carbonyl peak at 1731 cm^{-1}, with simultaneous increasing of new components at about 1780 cm^{-1} and 1665 cm^{-1}. In Figure 1 FT-IR spectra (in the range 1600-1900 cm^{-1}) are showed for thin films of different commercial products, both un-aged and aged for exposure under xenon-arc lamp for up to 250 hours. It can be observed that the relative intensities of the convoluted absorption bands showed by aged materials indicate that these products, all used by the restorers, show very different extents of photooxidative degradation phenomena.

Nevertheless, in conservation, the trend is that both the choice and the final evaluation of the materials applied is carried out by restorers; in general a strict scientific support is missed. These considerations brought us to start a research programme to find and test new polymers and application technologies with

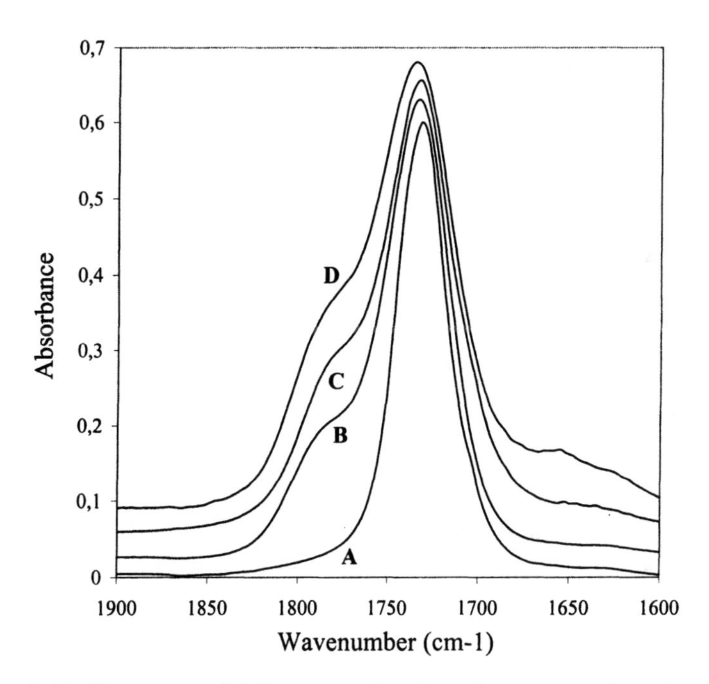

Figure 1. FT-IR spectra of different acrylate-based commercial products, un-aged and aged by exposure to Xenon-arc lamp for 250 hours; A) Primal AC33 un-aged; B) Acrilem IC79 aged; C) Primal AC33 aged; D) Primal B60A aged.

advanced properties tailored for specific restoration treatments. Our goal was to fill the gap between the high level of specialization reached in the researches in the macromolecular field and the comparatively low level of specificity and performance of the materials usually applied in conservation. The ideal aim is to promote new routes for qualified restoration interventions that, starting from the real needs of an artwork, will be carried out by using the most efficient and modern technology.

A New Route for the Consolidation of Stone

Among the materials not usually applied for conservation of stone, polyurethanes show very satisfactory results *(17)*. In fact most of the polymers used in conservation are characterised by Tg values close to or slightly higher than room temperature. This can be easily explained by considering that a polymer coating with a Tg value considerably higher than room temperature is unable to respond to dimensional changes of the treated items and could thus damage them. On the contrary, polymers with Tg value lower than room temperature is too soft to act as a consolidating agent, tending moreover to pick up dirt.

Polyurethane, characterised by hard and soft domains, often show two glass transition, the first one at temperature lower than the room temperature, and the second one at higher values. At room temperature soft segments are above their Tg, while hard micro-domains are below their Tg. Specific Tg values and other chemico-physical properties can be easily obtained by a rational choice of the monomers and their ratio during the synthesis. Furthermore polyurethanes are in general materials with excellent tensile properties, abrasion resistance, good transparency and high photo-oxidative resistance.

Unfortunately, polyurethanes show the same problems of poor penetration inside the stone pores, already shown by other polymeric materials. Usually, polymers are applied on stone in solution of organic solvents or in water dispersion; then solvent or water evaporate through the external surface of the treated item.

For polymeric solutions, the high viscosity strongly reduces the penetration of the material inside stone; moreover part of the material applied often rises up to the surface during the evaporation process of the solvent. For water dispersions the most relevant problem is the size of the dispersed particles. In Figure 2 an example of bad application of polyurethane from water dispersion is reported. The average size of the water dispersed particles applied is too big to let the material penetrate inside the stone pores. The result is an external coating that has not consolidating effect on the inner layer of the degraded stone.

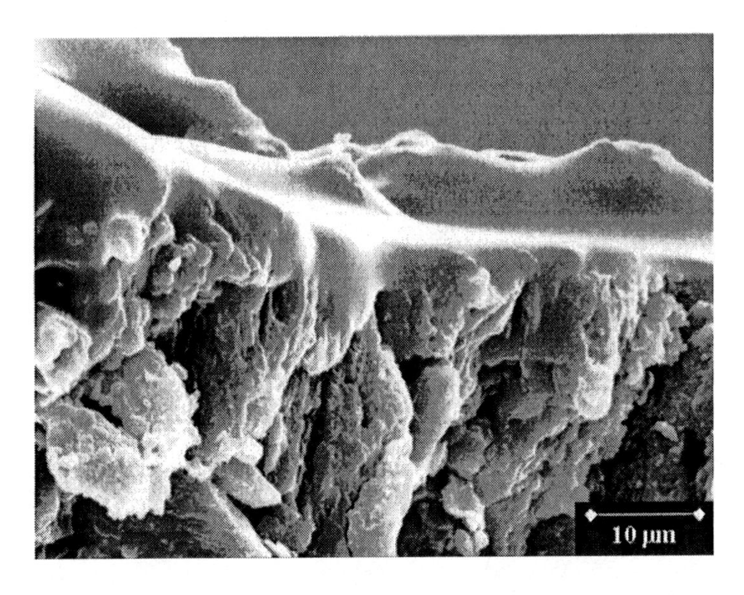

Figure 2. SEM micrograph of sandstone treated with a commercial polyurethane-based water dispersion.

To solve these problems, in the work here reported *(18)* we focused our attention on the in situ polymerisation, consisting of the application of monomers that react during their penetration inside the stone pores. The consolidating effect is obtained through the formation of a layer more or less continuous of polymer, coating the inner walls of the pores within the stone and linking up the disaggregated grains with each other.

A typical reaction scheme of the synthesis of polyurethanes is reported as follows:

Scheme 1

$$\mathrm{R{-}N{=}C{=}O} + \mathrm{HO{-}R_1} \longrightarrow \mathrm{R{-}NH{-}\overset{\overset{\displaystyle O}{\|}}{C}{-}O{-}R_1}$$

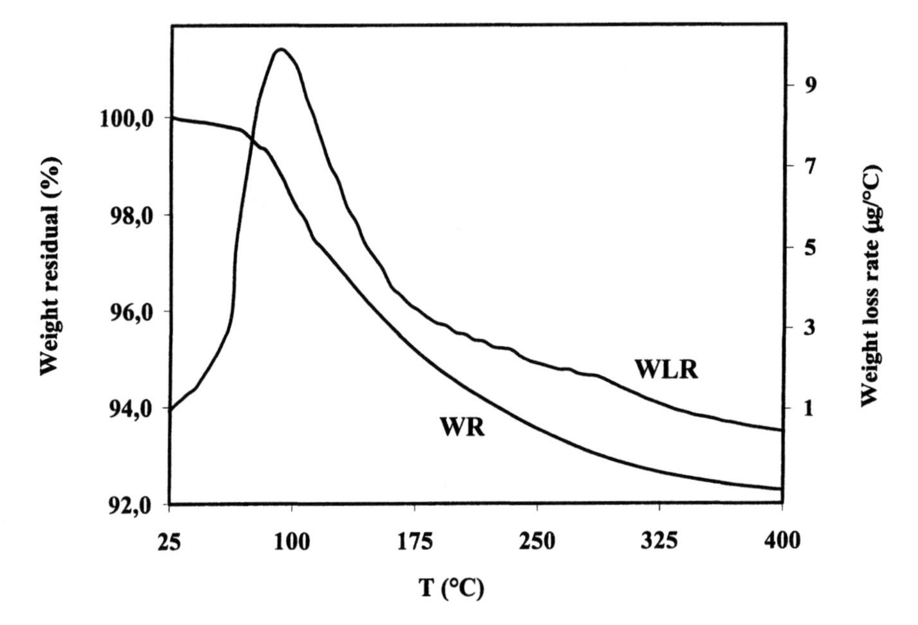

Figure 3. Thermogravimetric curve of Neapolitan yellow tuff; WR = weight residual; WLR = weight loss rate.

The most important parameter in this in situ polymerisation procedure was the room temperature, to reproduce the operative conditions of restorers, and, overall, the presence of bonded water into the stone. In Figure 3 a typical thermogravimetric trace of Neapolitan yellow tuff is reported: it can be observed the large amount of water, evidenced by the weight loss during heating.

This water cannot be removed from stone and it has to be considered as a reactive component of our system, leading to the proceeding of the side reaction giving urea bonds, as reported as follows:

Scheme 2

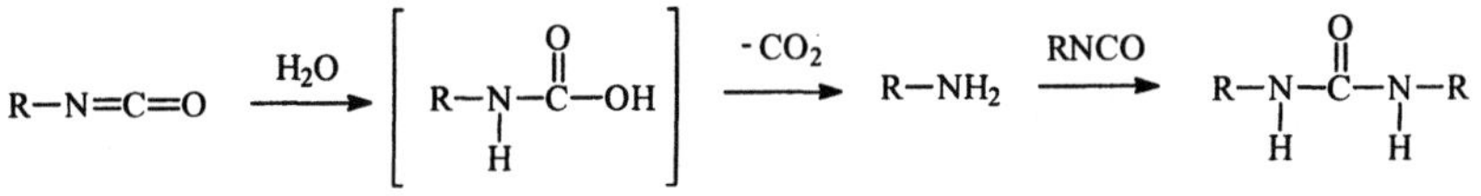

$$R{-}N{=}C{=}O \xrightarrow{H_2O} \left[R{-}\underset{\substack{|\\H}}{N}{-}\overset{\substack{O\\ \|}}{C}{-}OH \right] \xrightarrow{-CO_2} R{-}NH_2 \xrightarrow{RNCO} R{-}\underset{\substack{|\\H}}{N}{-}\overset{\substack{O\\ \|}}{C}{-}\underset{\substack{|\\H}}{N}{-}R$$

Despite to the side reaction, the use of a catalyst selective for the reaction isocyanate-alcohol in respect to the side reaction isocyanate-water, and the right choice of the monomers and their ratio, let us to obtain a poly(urethane-urea) characterised by high depth of penetration and homogeneous distribution into the stone pores. Moreover treated samples showed optimal values of water vapour permeability and a strong decrease in the rate of water capillary absorption (from 32 to 14 mg cm^{-2} $s^{-1/2}$), thus indicating a good protective effect *(19)*.

Furthermore poly(urethane-urea) in situ polymerised inside stone forms a regular homogeneous thin film covering the grains of the stone without modifying their morphological features.

Also mechanical properties of treated stone were very promising. The aggregative efficiency (AE) of this treatment was calculated following the equation:

$$AE\ (\%) = 100^*(\Delta W - \Delta W_{pol})/\ \Delta W \qquad (1)$$

where:

ΔW = average weight loss of 10 untreated tuff samples after 300 abrasion cycles; ΔW_{pol} = average weight loss of 10 tuff samples treated by in situ polymerisation, after 300 abrasion cycles.

Tuff samples treated by in situ polymerisation showed aggregative efficiency values up to 83%.

Moreover compression tests carried out on tuff samples, previously undergone to accelerated weathering by freeze-thaw cycles and then treated by in situ polymerisation procedure, showed high recover in the compression strength.

The recovery in the compression strength (RCS) was calculated following the equation:

$$RCS\ (\%) = 100*(CS_{pol} - CS_{ft})/(CS - CS_{ft}) \qquad (2)$$

where:
CS_{ft} = average compression strength of 15 tuff samples aged by 50 freeze-thaw cycles; CS_{pol} = average compression strength of 15 aged tuff samples, consolidated by the in situ polymerisation technique; CS = average compression strength of 15 unaged, untreated tuff samples.

Freeze-thaw aged tuff samples treated by in situ polymerisation showed recovery in the compression strength up to 85%.

The research is still in progress to evaluate durability properties of this consolidating procedure and efficiency on different stone. Moreover, preliminary tests carried out on different stone substrata gave us other very promising results. As an example SEM micrographs of sandstone, both untreated and after in situ polymerisations procedure are reported in Figure 4. In Figure 4A the fractured surface of untreated sandstone is characterised by a high roughness; furthermore small disaggregated grains (diameter $< 10\ \mu m$) are deposited on the stone surface. In Figure 4B it is well evident the film of polymeric material homogeneously covering the stone: the stone surfaces appears smoothed by the presence of the polymeric film in which disaggregated small grains are embedded.

Finally, in Figure 5, a comparison between the treatment developed in the present work and a traditional consolidating method based on the application of ethyl silicate derivatives is reported. In Figure 5A the external surface of untreated sandstone is showed. In Figure 5B the micrograph of sandstone treated by silicate-based material is reported, showing highly extended cracking phenomena on the treated surface. In Figure 5C the morphology of sandstone treated by in situ polymerisation of poly(urethane-urea) is reported: the polymeric film is regular, homogeneous, and does not drastically modify the morphological features of the substratum on which it is applied.

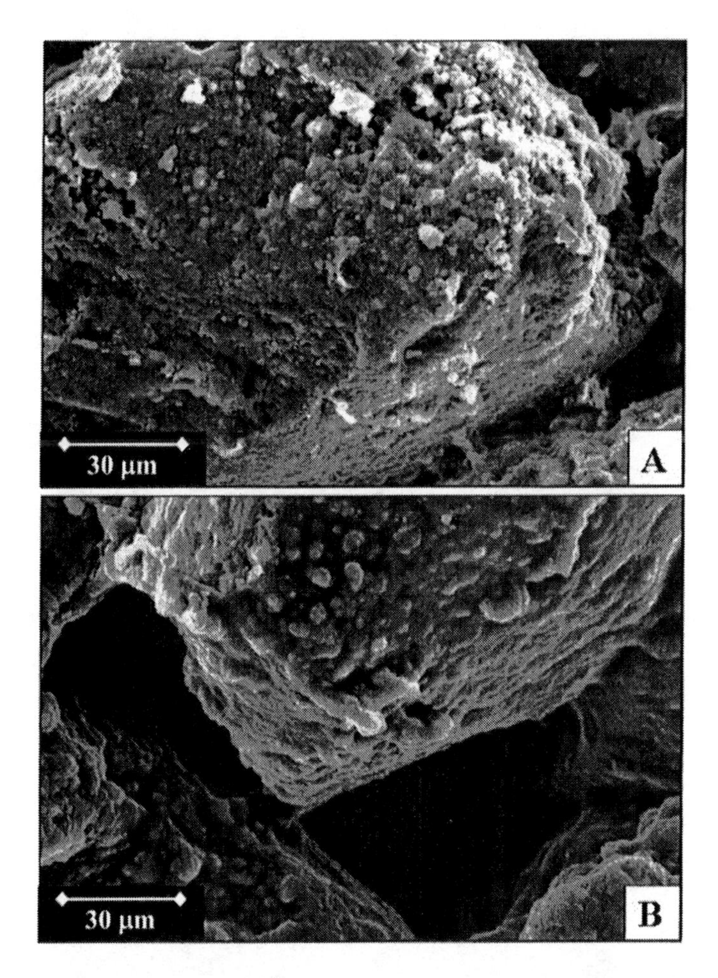

Figure 4. SEM micrographs of fracture surface of: (A) untreated sandstone; (B) sandstone treated by in situ polymerisation of poly(urethane-urea).

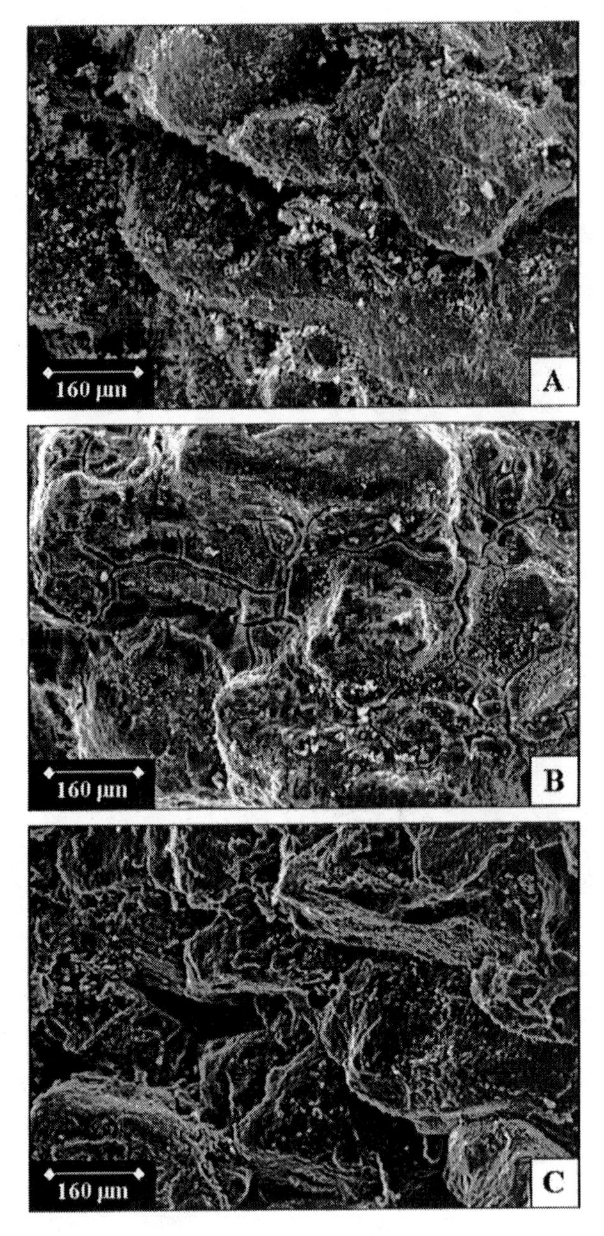

Figure 5. SEM micrographs of external surfaces of untreated sandstone (A), sandstone treated by application of silicate-based material (B) and sandstone treated by in situ polymerisation of poly(urethane-urea) (C).

New Polymer-based Processes for the Restoration of Textiles

Polymeric materials have been subjected to many studies to evaluate their usefulness in replying to specific requirements for conservation of artworks constituted by stone. Their efficacy in restoration treatments of textiles materials is still an open field for investigations.

Acrylic and vinylacetate dispersions have been used as consolidating and adhesive agents on textiles for more than 20 years. There is much anecdotal information about unsuccessful treatments but there are few quantitative data on the effects of the applications of these materials for consolidating and adhesion purposes *(2, 4, 5, 8, 20)*. The aim of consolidation is to hold together degraded fibres improving physical strength of yarns and fabrics. Nowadays a large number of polymers are applied on ancient textiles by impregnation to increase their tearing and breaking strength *(5)*. In adhesive techniques a support fabric is coated with an adhesive material and put in contact with the verso of the textile object to be conserved. Two adhesive techniques are currently used by restorers: heat sealing, for textiles, and cold vacuum table, for the lining of paintings on canvas *(21, 22)*.

The work started from the needs of better understanding the phenomena related to the application of polymeric materials on textiles, with particular attention to the extent of absorption phenomena, to the morphological analysis of the coatings and to the correlation among the physico-chemical properties of the material applied and the effects obtained on the treated items *(23)*.

All the materials tested, both acrylic and acetovinylic, applied on cotton textiles, are absorbed on the textile materials with amounts mainly related to the dilution of the commercial water dispersions used for impregnation. The relative amount A of polymeric materials impregnating the yarns is well fitted by a sigmoid curve, whose equation is here reported:

$$A = \frac{A_0 - A_{max}}{1 + e^{(x - x_{center})/dx}} + A_{max} \qquad (3)$$

where:

A = percent relative amount of dried polymers impregnating the yarns; x = dry content (content of polymeric material) of the water dispersion, at different dilutions, used for impregnation; A_0 = A value for x = 0; in our case: A_0 = 0; A_{max} = asymptote of the impregnation curve at high values of the dry content in water dispersions; in our case: A_{max} = value of A corresponding to impregnation with the as-supplied water dispersions; x_{center} = x value for A = $0.5*A_{max}$; dx = parameter related to shape of sigmoid curve.

In figure 6 the sigmoid curve of the percent relative amount of dried polymer impregnating the yarns as a function of the polymer content in the water dispersion used for the impregnation (RH = 50%, T = 25°C) is reported for the commercial vinylacetate-co-butylacrilate water dispersion Mowilith SDM5.

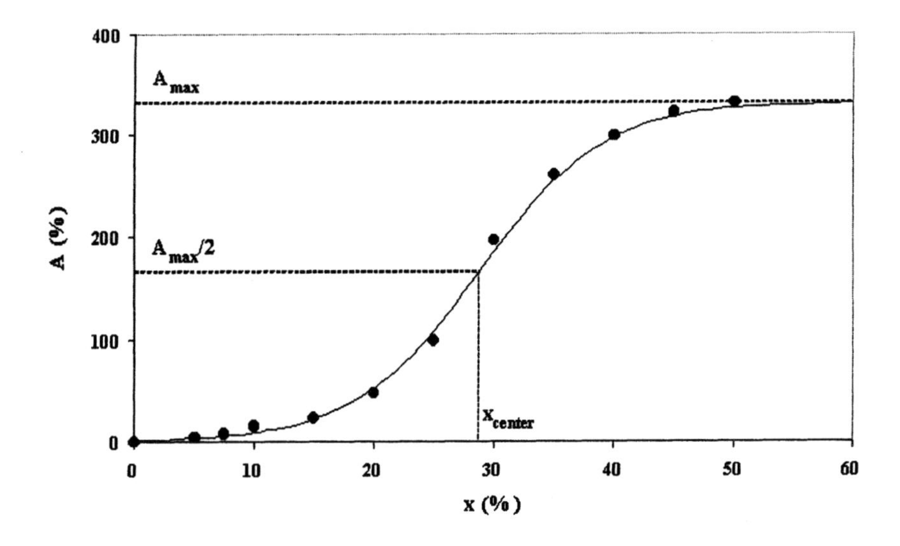

Figure 6. Mowilith SDM5: sigmoid curve of percent relative amount of dried polymers impregnating the cotton yarns (A, %) versus the dry content of the water dispersion used for the impregnation (x, %).

In figure 7 typical SEM micrographs of cotton fibres, before (Figure 7A) and after immersion in two vinylacetate polymeric dispersions (dry content 10% wt/wt) are reported. In the first case (Mowilith SDM5, Figure 7B) it can be underlined that, after the treatments, the morphological features characteristic of cotton fibres (e.g. convolutions around the fibre axis), are unaffected by the presence of the polymer, that forms an homogeneous, regular and smooth coating. Conversely Mowilith DMC2 precipitated irregularly on the cotton fiber surfaces, giving rise to the formation of drop-like domains (see Figures 7C). Such a morphology suggests that formation of lumps occurs during the water casting process.

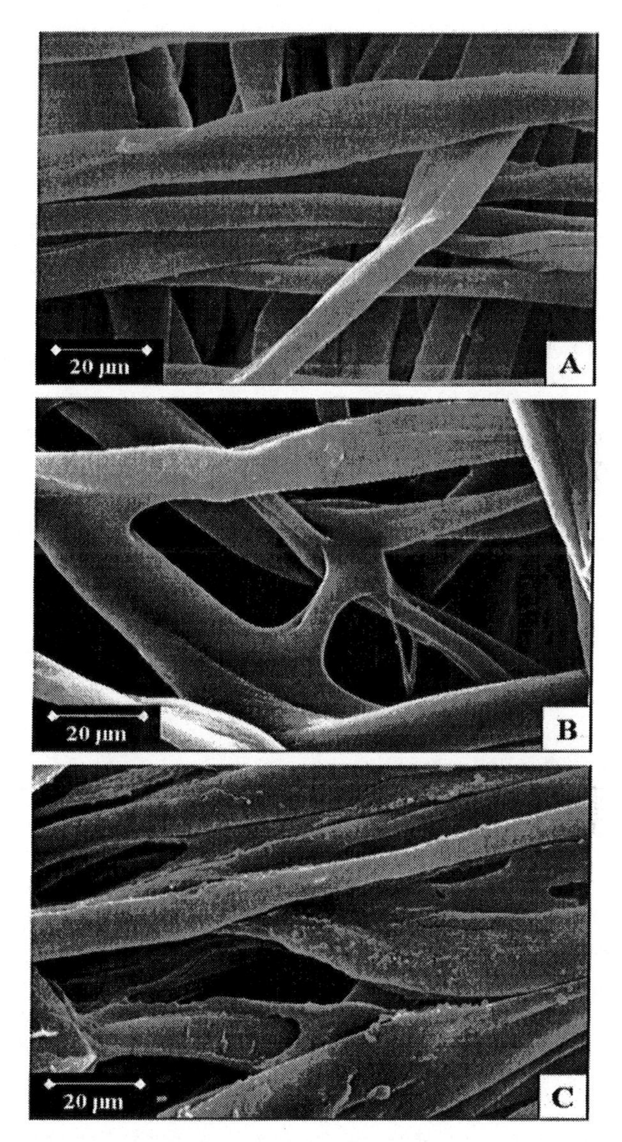

*Figure 7. SEM micrographs of cotton fibres untreated and treated by immersion in polymeric dispersions at dry content of 10%:
A) untreated cotton fibres; B) cotton fibres impregnated by Mowilith SDM5;
C) cotton fibres impregnated by Mowilith DMC2.*

Similar phenomena are observed also for acrylate-based commercial products and seems to be independent from the treated substrata. Smoothness and regularity of the polymer coating result because of the size and size distribution of the polymer particles in the dispersion and/or their intrinsic film-forming properties. Moreover the most interesting result shown by the SEM micrographs is formation of bridges constituted by polymeric thin films between adjacent fibres for impregnations in polymeric water dispersions characterised by good film-forming properties. The formation of these bridges explains the consolidating effect (i.e. an increasing of the mechanical properties of the yarns) of the treatments caused by an adhesion phenomenon among the fibres constituting a yarn and, for higher amount of material applied, among the yarns constituting the fabric. A typical bridge formed between adjacent fibres is well evidenced in Figure 7B.

Tensile recovery percent of the stress at maximum load (RS(%)) of cotton yarns, previously undergone to accelerated aging treatments by immersion in HCl solution, and therefore treated by different commercial products, was determined by using the following equation:

$$RS\,(\%) = \frac{(S_{treat} - S_{degr})}{(SP_{unag} - S_{degr})} * 100 \qquad (4)$$

where:

RS (%) = recovery percent of the stress at maximum load of aged cotton yarns; S_{treat} = stress at maximum load of cotton yarns undergone to accelerated aging and then treated with polymeric materials; S_{degr} = stress at maximum load of cotton yarns undergone to accelerated aging; S_{unag} = stress at maximum load of un-aged, untreated cotton yarns.

In Figure 8 RS (%) is reported for cotton yarns HCl-degraded and then treated with an acrylate commercial dispersion (Acrilem 674, ICAP SIRA, Parabiago, Milano), as a function of the amount of polymeric materials impregnating the yarns.

The shape of the curves is the same also for other materials, both acrylate and vinylacetate. The curve can be subdivided to three parts. The first part (I, relative amount of polymeric materials ranging from 0 to 11.5%) shows that the increase of recovery in the stress at maximum load is roughly linear with the amount of polymeric material impregnating the yarns; the second part (II, relative amount of polymeric materials ranging from 11.5 to 25-35%) is a plateau in which RS(%) is approximately constant; in the third part (III, relative amount of polymeric materials > 25-35%) the recovery in the specific stress at maximum load starts again to increase to higher values.

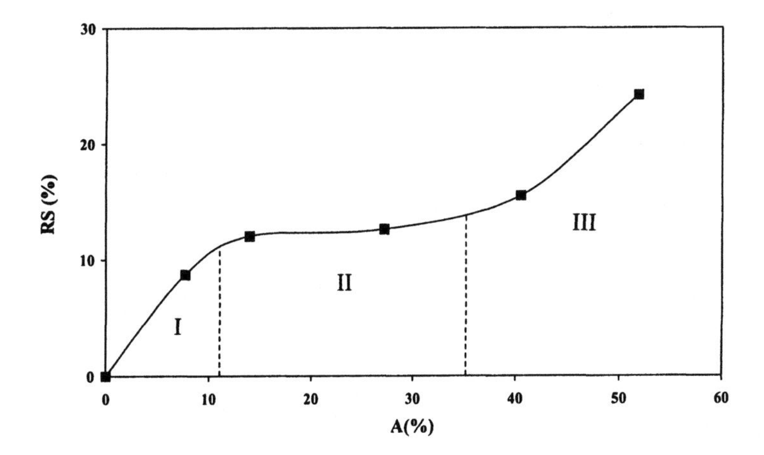

Figure 8. Recovery in the specific stress at maximum load (Rstress, %) for HCl-degraded cotton yarns impregnated with different relative amount (A, %) of the acrylic water dispersion Acrilem 674.

This behaviour can be explained by considering that the first part of the curves is related to amounts of polymer which allow the formation of films homogeneously coating the single fibres together with the formation of significant number of polymer connections (bridges) among adjacent fibres and yarns. In the second part of the curve the maximum effect in the recovery of the stress at maximum load is gained and with further increases in amount of polymer applied, there is no corresponding increase in the tensile properties of the treated yarns: the polymer film presumably coated the whole substratum, moreover filling the free volume. In the third part of the curve, the amount of polymer applied is so high that we pass from a system in which the single fibres of the cotton yarns are coated and interconnected to a different system in which polymeric material becomes the matrix encapsulating the fibres as dispersed phase. The system tends to be similar to a composite material textile-polymer and the increase in the tensile properties cannot be useful for conservative and restorative purposes because the appearance and the hand of such a system are not acceptable.

Finally, the adhesive properties of commercial products were evaluated by preparing cotton fabric swatches laminated together and by measuring the peel adhesion, defined as the force required to delaminate a structure or to separate two stitched layers. In our case, peel adhesion can be defined as the force required to separate two fabrics preliminarily laminated by means of a thin layer of adhesive coating.

It has been found that peel strength values are not dependent from the chemical nature (acrylate and vinylacetate) of the materials used *(23)*. The detachment of the cotton fabric from each other can be better described by a cohesive failure of the adhesive layer. This failure depends upon the properties of the polymer, such as the glass transition temperature, rather than on the interactions between the polymer and the fabric. Polymers with a Tg significatively lower than the temperature of the test (20°C) exhibit a viscous flow behaviour and are weak adhesives *(20)*; polymers with a Tg similar or higher than the temperature of the test are in a glassy state and are not useful as adhesive materials. Better values of peel strength are therefore shown by samples stitched with materials characterized by a Tg slightly lower than the temperature of use.

Starting from this comprehension of several phenomena occurring during the application of polymeric materials on textile items for conservation purposes, our attention was focused on a different applicative techniques, to evaluate the possibility of obtaining high consolidating efficiency also at low amount of materials applied.

The consolidating treatments here briefly described *(24, 25)* consist in a grafting polymerisation process carried out on cellulose textiles. The graft copolymerization method, first proposed by the British Library *(26)*, involves treating papers with gaseous monomeric ethyl acrylate and methyl methacrylate. In the British Library method the initiation of the polymerisation process is induced by irradiating cellulose material with low intensity gamma rays. Conversely, the method here proposed starts from the presence of oxidized groups, always found on degraded cellulose items. These oxidized groups can be used as active sites for the insertion of acrylic monomers from which the polymerization reaction proceeds. Oxidized groups are activated by an UV treatment, thus noticeably simplifying the British Library procedure based on the use of gamma rays. Therefore long-chained polymers are deposited on and between the cellulose fibers thereby strengthening the treated items, without. affecting the crystallinity of cellulose and introducing only a very modest amount of oligomers. Physical and chemical characteristics of grafted materials have been tailored following the results of the researches carried out on commercial products.

By the right chose and balance of acrylate and methacrylate monomers, high yield values were obtained (from 60 to 85%) for the grafting polymerisation, to which high recovery in the flexural resistance (up to 60%) and in the tensile strength (up to 35%) of aged materials grafted with acrylate/methacrylate monomers corresponded. These results confirmed that the grafting process is suitable for the consolidation of precious cellulose based materials of historical interest.

Following the laboratory results, carried out both on cellulose materials previously undergone to accelerated aging processes and on real samples, a scale up of our system has been carried out. A high volume grafting chamber (1 m^3) has been designed and realised, to expand the consolidating methods to larger textile items of historical interest. In Figure 9 a schematic representation of this grafting chamber is reported.

The object to be treated is placed in the quartz chamber (A). Monomers are poured into the tank (B). Through the control panel (C) the UV lamp (D) is switched on, the quartz chamber is evacuated and the temperature cycle of the system is programmed. The vacuum-refrigerator system is placed in (E). A venting system is positioned above the lamp (F). Finally, through the control panel, the insertion of the monomer into the quartz chamber is carry out with a programmed flow rate and the grafting reaction starts.

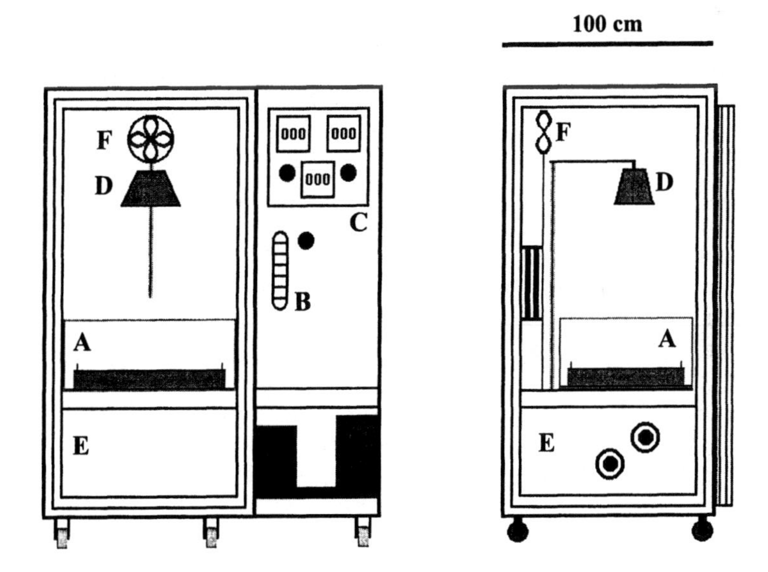

Figure 9. Scheme of the grafting chamber for the consolidation of textile items. (A) quartz chamber; (B) tank for the monomers; (C) control panel; (D) UV lamp (E) vacuum-refrigerator system zone; (F) venting system.

Concluding remarks

With the ideal aim of setting up new methods for qualified restoration interventions that, starting from the real needs of an artwork, are carried out by using the most efficient and modern technology today accessible, we started a research to find and test polymeric materials obtained through advanced technologies, in comparison with materials usually applied in the restoration field. The work was firstly focused on the comprehension of the limits of the materials and technologies today applied, and therefore aimed at developing radically innovative processes tailored for the restoration of both stone and textiles.

For consolidating treatments of stone, a new polymerisation procedure of poly(urethane-urea) by in situ polymerisation inside stone at room temperature has been set up. The relative amount of the reaction components as well as different catalysts have been tested in order to optimise the ratio urethane/urea. High amount of urethane bonds have been obtained setting the ratio [-NCO]/[-OH] = 1 mol/mol and using ZrAcAc as catalyst. Moreover through the method set up a good penetration depth of the poly(urethane-urea) is achieved. The poly(urethane-urea) synthesised inside stone form a regular homogeneous film covering minerals without modifying their peculiar morphological features. Poly(urethane-urea) in situ polymerised strongly decreases the stone water capillary absorption without occluding its porosity, thus leaving stone still permeable to water vapour. Moreover, performed treatments are characterised by high consolidating and aggregative efficiency, thus confirming that they are very promising for conservative purposes.

For textile materials, a new consolidating procedure, based on the activation of the oxidised groups of cellulose by UV light and on the grafting copolymerisation of acrylic monomers on the cellulose substrate, has been set up. Moreover a scale up of the laboratory system has been carried out, through the design and the realisation of a high volume grafting chamber, in order to perform the proposed grafting procedure onto large textile items of historical interest.

Acknowledgement

This work has been partially supported by the Italian MIUR in the framework of the National Research Programme on Cultural and Environmental Heritage (D.M. 14.03.2000, n. 167/Ric, Pract. n. 66413).

References

1. *Il Restauro della Pietra*, Lazzarini, L., Tabasso, M.L., CEDAM, Padova, **1986**.
2. *Materials for conservation, organic consolidants, adhesives and coatings*, Horie, C.V., Butterworths, Repr., London and Boston, **1990**.
3. *I tessili. Degrado e restauro*, Pertegato, F., Nardini Editore, Fiesole, Firenze, Italy, **1993**.
4. *The Textile's Conservator Manual, second edition*, Landi, S., Butterworth-Heinemann Ltd, Oxford, 1992.
5. *Chemical principles of textile conservation*, Timár-Balázsy, A., Eastop, D., Butterworth-Heinemann, Oxford, **1998**.
6. Cocca, M., D'Arienzo, L., D'Orazio, L., Gentile, G., Mancarella, C., Martuscelli, E., Polcaro, C., *J. Cult. Heritage*, **2004**, in press.
7. *Investigation of the cross-linking of thermoplastic resins effected by UV radiation*, Ciabach, J., in: Proceedings of the symposium resins in conservation, held at Edinburgh, 21st-22nd 1982 - Scottish society for conservation and restoration, Edinburgh, 5.1-5.8, **1983**.
8. *Adhesives for the consolidation of textiles*, Verdu, J., Bellenger, V., Kleitz, M.O., in Preprints of the contributions to the IIC Paris congress, 2-8 september 1984, Norman, S., Pye, E.M., Smith, P., Thomson, G., eds., IIC, **1984**, London, 64.
9. Carta del restauro, 1972.
10. *Introduction to Danish Wall Paintings-Conservation Ethics and Methods of Treatment*, Trampedach, K., National Museum of Denmark-Conservation Department, July **2001**.
11. Fernandez de Castro, V., *WAAC Newsletter*, **1990**, 12, 5.
12. *An Evaluation of selected applied polymers for the treatment of parchment*, Gomaa, A.M., Proceedings of the 15th World Conference on Non-destructive Testings, Rome, 15-18 October, 2000.
13. Grattan, G., *CCI Newsletter*, **1999**, 23, 3.
14. Plummer, P., *Conservator News*, **2002**, 77, 44.
15. Hartman, V., *I Beni Culturali*, **1996**, 4-5, 17.
16. Cocca, M., D'Arienzo, L., D'Orazio, L., Gentile, G., Martuscelli, E., *Polymer Testing*, **2004**, 23, 333.
17. D'Orazio, L., Gentile, G., Mancarella, C., Martuscelli, E., *Polymer Testing*, **2001**, 20, 227.
18. Italian Patent NA **2004** A000021.
19. Cocca, M., D'Arienzo, L., D'Orazio, L., Gentile, G., Martuscelli, E., *Luournal Applied Polymer Science*, in press.
20. Down, J.L., MacDonald, M.A., Tetreault, J., Williams, R.S., *Studies in Conservation*, **1996**, 41, 19.
21. Duffy, M.C., *Journal of the American Institute for Conservation*, **1989**, 28, 67.

22. Lewis, G., Muir, N., Yates, N.S., *The link between the treatments for paintings and the treatments for painted textiles*, in: Conservation and restoration of church textiles and painted flags; investigation of museum objects and materials used in conservation and restoration, Fourth International Restorer Seminar, Veszprém, Hungary, 2-10 July 1983.
23. D'Arienzo, L. Gentile, G., Martuscelli, E., Polcaro, C., D'Orazio, L., *Textile Research Journal*, **2004**, 74, 281.
24. Italian Patent NA **2003** A000065.
25. Margutti, S., Vicini, S., Proietti, N., Capitani, D., Conio, G., Pedemonte, E., Segre, A.L., *Polymer*, **2002**, 43, 6183.
26. Carter, H.A., *Journal Chemical Education*, **1996**, 73, 1160.

Chapter 28

Matrix Free Ultra-High Molecular Weight Polyethylene Fiber-Reinforced Composites: Process, Structure, Properties and Applications

Tao Xu and Richard J. Farris*

Silvio O. Conte National Center for Polymer Research, Polymer Science and Engineering Department, University of Massachusetts at Amherst, Amherst, MA 01003

A novel process called high-temperature high-pressure sintering coupled with thermoforming was studied and explored to make shapeable matrix free composites by consolidating and molding layers of ultra high molecular weight polyethylene (UHMWPE) fabric at an elevated temperature under pressure. The optimal processing temperature was found to be near the fiber's melting point (150°C). The Process-Structure-Property relationship was investigated to give an in-depth and better understanding of this creative process. Compared to other UHMWPE fiber materials, Spectra® cloth is shown to be the best material to use this process and make ballistic shields and other high performance products.

Oriented Polyethylene

Historically, polyethylene (PE), a material typically found in items such as "milk jug" plastics, was known as a low strength and stiffness material. Yet theory has long predicted that if the flexible polyethylene chains could be fully extended and "frozen" in a highly dense packing state to form a fully aligned crystalline structure, the resulting oriented material would be extraordinarily strong because of the intrinsic high strength of the carbon-carbon bond. Based on numerous experimental and theoretical studies (*1*), it is widely accepted that the crystalline modulus of PE fiber is estimated to be 300±20GPa. Superdrawn single-crystal mat of ultra high molecular weight PE (UHMWPE) has achieved a tensile modulus as high as 220GPa, approaching the theoretical maximum (*2*). As for the theoretical strength of PE fiber, there is less agreement among researchers and the reported values range from 20GPa to 50GPa (*3, 4*). The actual strength of oriented PE produced by various researchers is around 3-6GPa (*5-8*).

The basic principles to make high modulus PE fibers are (i) use high molecular weight to give high draw ratios, (ii) ultradrawn to remove chain entanglements and induce high molecular orientation, and (iii) induce high crystallinity and near-perfect crystal alignment through flow-induced orientation. Various draw techniques have been developed to produce highly oriented PE structures. Porter et al. (*9*) pioneered solid-state (co)extrusion or rolling as the initial deformation step followed by a uniaxial tensile draw. In this technique, PE polymer is (co)extruded through a conical die at a temperature below its melting point. Due to flow-induced orientation and pressure effects, the extruded material is effectively oriented and well consolidated. The extrudate is then tensile-drawn at a temperature below its melting point to give a fibrous material with high modulus and strength. The combined draw ratio from extrusion and stretch can be as high as 350 (*10*). Ward et al. (*11*) developed a special drawing technique that utilizes a melt spinning method combined with a tensile drawing process at a temperature close to the polymer melting point to yield a high modulus fiber. Farris et al. (*12*) invented a novel radial-compression method to prepare ultradrawn PE. They used highly stretched elastomeric (Spandex) filaments wound around a PE cylinder to create radial pressure causing the polymer to neck in a controlled manner at an elevated temperature well below the melting point of PE. The deformed material has comparable high orientation and mechanical properties to that produced by the solid-state extrusion method. More recently, Lesser et al. (*13*) performed the drawing of commercial UHMWPE fiber in supercritical CO_2. The high pressure CO_2 enhances the drawability of the fiber by permeating the amorphous region to improve the segmental mobility. The drawn fiber shows a 50% increase in tensile modulus and 8% increase in strength compared to the undrawn virgin fiber. Oriented PE morphologies can also be created by other methods, such as direct extrusion (*14*), hydrostatic extrusion (*15*), and zone drawing and annealing (*16*). The most notable process is the solution/gel spinning developed by Pennings (*17*), Smith and Lemstra (*18*). This technique involves a solvent or second phase combined with UHMWPE to facilitate deformation by extrusion and tensile draw from a gel state. DSM and Honeywell are utilizing this process to commercially produce ultra high modulus PE fibers under the trade names of Dyneema® and Spectra®, respectively.

One-polymer Composites

UHMWPE fibers possess extraordinary physical and mechanical properties: low specific gravity, high modulus, high strength, high impact resistance, high cut and abrasion tolerance, excellent chemical resistance, low dielectric constant, good UV resistance, low moisture absorption, excellent vibration damping capability, low coefficient of friction and self-lubricating properties, etc (*19, 20*). Despite the impressive list of properties, UHMWPE fibers also have limitations. Due to their chemical inertness and lack of functional groups, UHMWPE fibers are difficult to bond to most materials, which makes it difficult to produce UHMWPE fiber reinforced polymeric composites. There are several ways to overcome the obstacle by means of various fiber pre-treatments. UHMWPE fibers can be chemically etched by chromic acid (*21*) or oxidized by polypyrrole (*22*) to increase the surface roughness. Plasma and corona treatments (*23, 24*) in O_2 or CO_2 introduce chemical groups onto the fiber surface through chain scission and substitution, as well as etch and roughen the fiber surface. All these fiber pre-treatments have been proven to significantly improve the bonding strength of UHMWPE fibers to matrices, however, the fiber properties generally degrade. Even with good bonding to the fibers, the transverse properties are limited by the very poor lateral adhesion within the fiber. This is also true with Kevlar® and other highly oriented polymer fibers and it is this characteristic that gives these fibers their high cut resistance and good ballistic performance.

Yet, there is another way to circumvent the problem. Since polyethylene is a thermoplastic, surface melting can be used to fuse the fibers together by careful heating and pressure. As the majority of the fiber is intact, the resulting product upon cooling is a UHMWPE fiber reinforced composite without a matrix; this is an example of a homocomposite, a composite in which the matrix and reinforcement have the same chemical composition. The interfacial adhesion is greatly improved because the matrix and reinforcement are essentially the same material, and more importantly, there is molecular continuity throughout the composite. Homocomposites produced from thermoplastics are "recycle-friendly", compared to two-component systems such as glass fiber and epoxy resin. The idea of homocomposites is not new; Capiati and Porter (*25*) presented the first "one-polymer composite" example by embedding a high modulus PE filament in a block of HDPE in 1975. Later, other researchers have extended this technique and developed the film stacking (*26, 27*), powder impregnation (*28*) and solvent impregnation (*29*). In the abovementioned cases, a secondary polymer, besides polymeric fibers, is used as binder or matrix to form the continuous phase. Ward et al. at University of Leeds has introduced a concept called "hot compaction". The method utilizes only one starting material, a highly oriented fiber or tape, to make homocomposites in a two-stage process: soaking and compacting. They have studied the systems of melt-spun PE, gel-spun PE, PET, PP, and Vectran® LCP fibers (*30-39*). Meanwhile, a parallel study conducted by Farris et al. at University of Massachusetts has developed a similar process. They use a one-stage press-and-heat approach to make protective coatings for optics on military airplanes using Spectra® woven cloth. This research has led to several patents regarding the making of high strength and high modulus polymeric materials for impact resistant applications (*40-44*).

One major interest in UHMWPE fiber reinforced composites is their potential as ballistic protective materials. Today's state-of-the-art products use Kevlar® fiber reinforcement. UHMWPE fibers have one major advantage over Kevlar® fibers and that is their lower density, 0.97g/cm^3 compared to 1.44g/cm^3 for Kevlar®. UHMWPE fiber based armor has light weight without sacrificing the high level of protection; composite armor produced from Spectra® fiber weighs 30%-50% less than Kevlar® fiber composite armor for the same performance (*45*). Light weight allows more combat equipment to be mounted on the helmet while maintaining a low weight. Nowadays fiber reinforced composite armor is made by laying up patterned woven cloth or prepreg to produce doubly curved structures via compression molding.

Matrix Free Spectra® Fiber Reinforced Composites

Since UHMWPE fibers are thermally formable, it is possible to directly shape and consolidate UHMWPE fiber materials at elevated temperature without cutting patterns. A novel processing method called high-temperature high-pressure sintering has been developed (*46-48*). In this process, layers of Spectra® cloth are properly confined to prevent shrinkage and exposed to an appropriate heating and stretching sequence followed by a consolidation and cool-down step. Coupled with thermoforming, it is feasible to produce a thick and rigid ballistic shield with desired curvatures and shapes.

Optimization of Processing Conditions

There is a certain amount of fiber melting associated with the high-temperature high-pressure sintering process and the processing temperature plays a major role in determining the extent of the melting. A small amount of surface melting is needed to bond the fibers and cloth to develop the lateral strength of the structure. Excessive melting destroys the original highly oriented crystals and forms less oriented crystals that offer much less in the longitudinal strength of the structure. In the high-temperature high-pressure sintering, polyethylene molecules tend to relax towards their preferential random coil conformation at high temperatures even though they are under substantial constraint. The outcome is the loss of crystalline and chain orientation, resulting in a decrease of longitudinal strength. The goal was to find the proper sintering conditions at which a compromise between longitudinal and lateral strength was achieved. Differential scanning calorimetry (DSC) suggests that the temperature range of 150°C-152°C would be an ideal processing temperature; at these temperatures a minimum of 13% of the original crystals are melted and a 95% overall crystallinity is maintained after consolidation. Wide angle X-ray diffraction (WAXD) and the calculation of Hermans orientation function confirm that the optimal processing temperature should not be higher than 152°C; above this temperature the decrease in crystalline orientation is rather significant. It is interesting to find that crystallinity and orientation is linearly correlated as a function of processing temperature due to the extended chain conformation and parallel crystalline alignment of Spectra® fiber.

As a candidate for ballistic shields, impact properties of matrix free Spectra® fiber composites are of great interest. A series of single-layer samples were produced at different processing temperatures and times under a constant pressure (7.6MPa) and their impact properties, obtained by puncture tests, are compared. Consolidation at high temperature and pressure produces interfiber adhesion and greatly improves the impact properties of the cloth. In the best scenarios, the impact resistance is increased by nearly 6 times that of the as-received cloth. The maximum impact properties are achieved when the Spectra® cloth is sintered between 150°C-152°C depending on the processing time. There is a time-temperature superposition; as time becomes shorter a slightly higher temperature is needed to produce a specimen with the same level of impact resistance as the one prepared at longer time and a lower temperature. The impact property study confirms that the optimal processing temperature range suggested by the previous DSC and WAXD results is valid. Under a pressure of 7.6MPa, a single layer of Spectra® cloth sintered at 150°C for 20-30 minutes, or at 152°C for 10 minutes has the best impact resistance (Figure 1). Impact resistance is not very sensitive to processing pressure, as opposed to time and temperature. Pressure is used to provide a constraint on the fibers and prevent them from shrinking, to eliminate voids, and to consolidate the structure. A minimum pressure is necessary for sintering but an excessive pressure does not help.

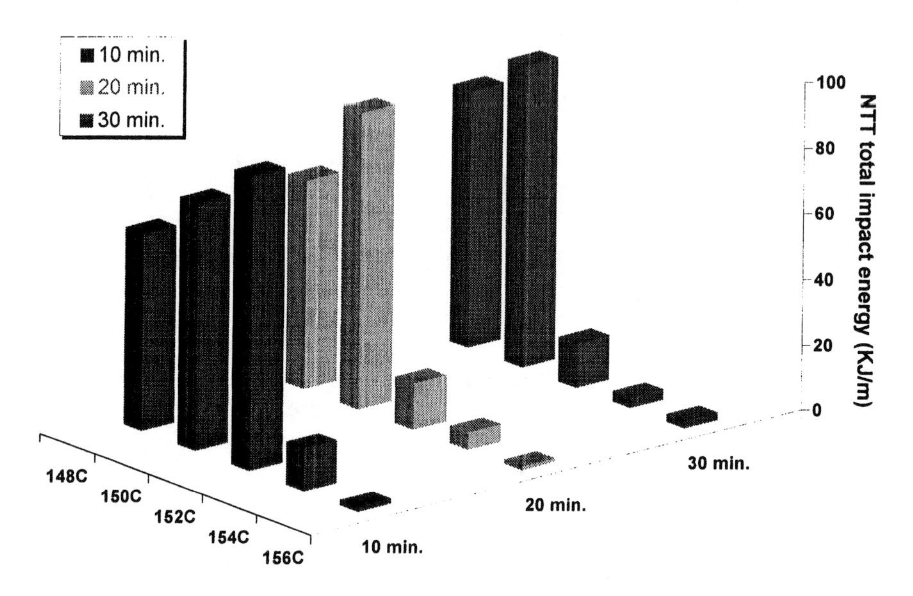

Figure 1. Normalized to thickness total impact energy vs. sintering temperature and time for single layer of consolidated Spectra® cloth under 7.6MPa pressure

Process-Structure-Property Relationship

The crystallinity and orientation of the sintered Spectra® cloth determine the thermo-mechanical properties of the consolidated structures and these two aspects are greatly influenced by the processing conditions. The consolidated Spectra® cloth is mechanically, morphologically, and ballistically characterized. Various characterization methods are used including mechanical testing, scanning electron microscopy (SEM), DSC, and WAXD.

Interlayer adhesion facilitates stress transfer between fibers and matrix material, and provides lateral strength to the composites. The T-peel test is a simple and effective way to evaluate adhesion. T-peel strengths for Spectra® cloth bilayers sintered at different processing temperatures increases monotonically over the temperature range since higher temperatures results in increased melting and recrystallization on cooling, hence better bonding. One would anticipate that consolidation at higher temperatures is required to make a good composite as better adhesion results in better load transfer. However, for laminated materials made for ballistic purposes, the stronger the adhesion the poorer the ballistic performance. When a high speed projectile hits a target, its kinetic energy needs to be consumed in order to stop it and delamination / debonding between layers and fibers provides a good channel for the dissipation of energy. Excellent interface adhesion between layers and fibers worsens the ballistic protection and comes at the expense of the longitudinal strength since more crystalline melting and orientation loss happens at higher temperatures. It has been shown that the tensile failure of the fibers during ballistic impact absorbs a large portion of the kinetic energy (*49*), therefore the preservation of the fibers' longitudinal strength is of great importance.

Flexural properties of multilayer sintered cloth were measured by three-point bend tests to characterize the rigidity and stiffness of the materials. No specimens exhibited any failure or yield within the 5% strain limit. At a much higher strain than 5%, the specimens sintered at lower temperatures showed delamination between the layers while the specimens sintered at higher temperatures showed no visible damage. This echoes the finding that specimens sintered at lower temperature have lower interlayer adhesion, therefore fail in shear mode. The flexural modulus reaches a maximum of 6GPa in the processing temperature range of 150°C - 160°C. Consolidated multilayer Spectra® cloth panels are very stiff and rigid. Strain wave propagation speed is proportional to the square root of the ratio of the material modulus and density ($\sqrt{E/\rho}$). The high modulus and low density of Spectra® material allows for fast wave propagation; the impact energy generated by a projectile can spread quickly in the plane rather than through the thickness and more material will be involved in the energy dissipation.

Ballistic tests were performed using the V_{50} standard test: V_{50} ballistic limit is the velocity at which the probability of penetration of an armor material is 50%. V_{50} ballistic limits of matrix free Spectra® fiber composite, and Spectra® fiber and vinyl ester matrix composites against different projectiles are compared. In Figure 2, the ordinate is V_{50} velocity and the abscissa is the ratio of the presented areal density of the armor system to the areal density of the projectile. A_d refers to the area density of the specimen, A_p is the presented area density of the projectile, and m_p is the projectile's mass. The filled symbols correspond to the matrix free Spectra® cloth composites and the unfilled symbols correspond to the composites using Spectra® cloth with the vinyl ester matrix. All the data are plotted using appropriate symbols representing different projectiles. RCC stands for right circular cylinder type projectile (1 Grain = 64.799 milligrams). Also, plotted in the figure are today's state of the art Kevlar® KM2 (PVB-Phenolic) armor and the expected PBO fiber composite armor performance, the best armor system that the Army can offer to date.

Higher V_{50} values at a given areal density level indicate higher ballistic resistance. Matrix free Spectra® composites perform as well as, or slightly better than the system containing matrix, and better than the Kevlar® system at the same areal density level. Matrix free Spectra® composites are lighter in weight than the Spectra® with matrix or the Kevlar® system due to the elimination of the matrix and the low density of the fiber. Also the manufacturing process is simpler for the matrix free Spectra® composites than for the other two systems. Ballistic test results confirm that matrix free Spectra® fiber reinforced composite made by high-temperature high-pressure sintering is a promising candidate for high performance light weight body armor and helmets.

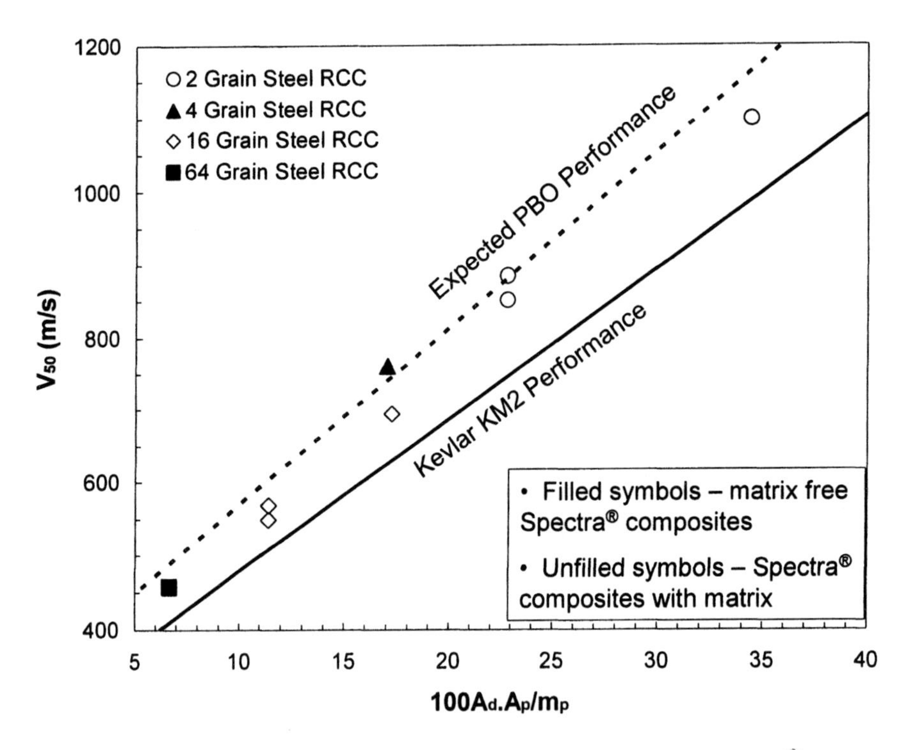

Figure 2. Ballistic performance of the multilayer consolidated Spectra® cloth compared with other composites (from the US Army)

SEM images reveal the cross-section of twelve layers of consolidated Spectra® cloth. These samples were etched with permanganate sulfuric acid solution. Overall the multilayer cloths are well consolidated, although some crevices and voids can be seen due to knife damage or etching. The voids mostly appear at the junction regions between the warp and weft yarns, which indicate weak adhesion in these regions. The junction regions are more susceptible to knife damage and etching reagent. There are fewer visible voids with increasing sintering temperature of the multilayer cloths. There is no significant fiber melting to fill the voids. Instead, the voids are filled by fiber deformation and impingement; there are signs of local welding at fiber boundaries. The consolidation mechanism of fiber lateral deformation to fill voids and spot welding by the recrystallization of melted fiber surfaces is confirmed by SEM. The topology is quite different for the specimens sintered at lower (150°C-152°C) and higher (154°C-156°C) temperatures as a result of the etching process prior to investigation. The etchant removes the density deficient regions: the amorphous regions and the recrystallized phases, but not the original highly oriented crystals. Melting and recrystallization that produces the less dense phases increases at higher processing temperatures at the fiber surface and core. Hill and valley like features are present in the cross section of the samples processed at lower temperatures. In addition, numerous conical pits are seen inside individual fibers since there are density deficient region within Spectra® fiber (*50*). On the contrary, relatively smooth surfaces are seen in the samples processed at higher temperatures as a result of uniform etching.

Thermoforming of Spectra®

Spectra® fiber is a thermoplastic polyethylene. An advantage of thermoplastics is their ability to be repeatedly heated, softened, and shaped provided the temperatures are not too high. Thermoforming of Spectra® cloth was attempted using a homemade hemispherical mold. In the previous literature, Ward et al. has concluded that “Composites with straight GSPE (Gel-Spun Polyethylene) fibres in 0°/90° fibre arrangement could not be moulded into helmet shells” (*35*). We have succeeded in this task using four different shaping and molding schemes (*46, 47*).

- Consolidate layers of Spectra® cloth to a thick flat panel, clamp the panel onto the mold, and apply heat and pressure to develop the shape
- Consolidate each layer of Spectra® cloth individually to make flat sheets, lay the consolidated sheets one on top of another and clamp them onto the mold, and apply heat and pressure to shape
- Shape individual layers of Spectra® cloth using the mold, stack the shaped layers one on top of another, and sinter them in the mold

- Consolidate and shape layers of Spectra® cloth directly in the mold to form the structure

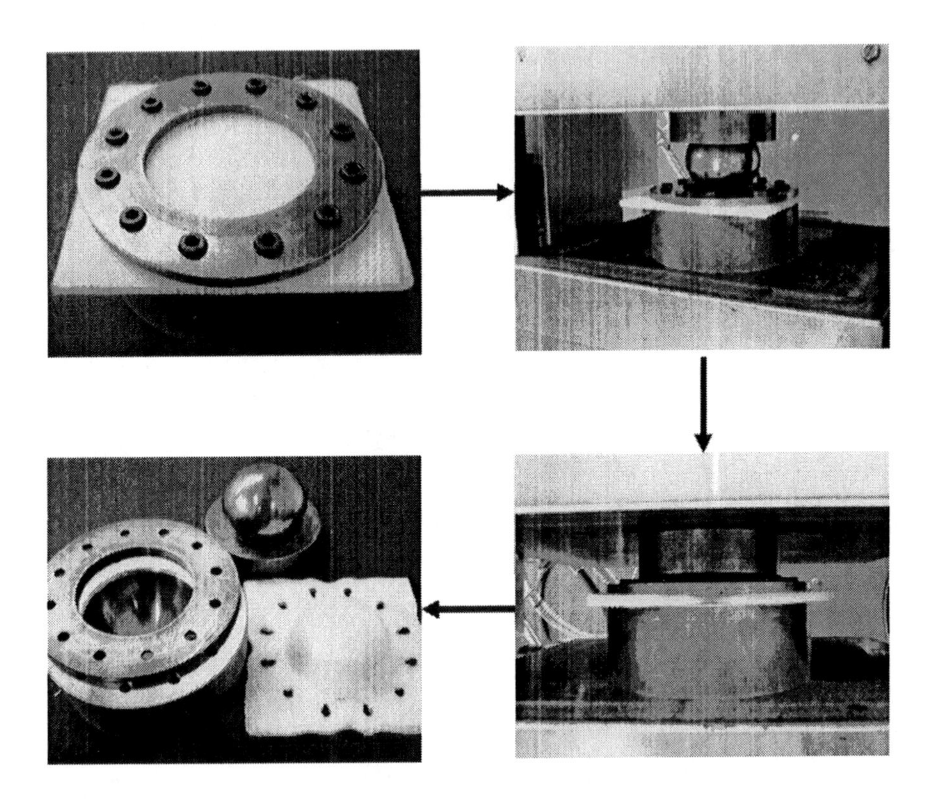

Figure 3. Typical molding procedures using a twenty-five layer consolidated Spectra® cloth flat panel

A typical sequence is illustrated by a set of pictures using the "flat panel first" route as an example in Figure 3. A pre-consolidated flat panel, which consists of 25 layers of cloth, is drilled and bolted tightly between the blank holder and the socket die. Meanwhile, the press is heated to the desired sintering temperature, for example 150°C. The punch is placed on top of the panel and carefully aligned in the press. A small initial force is applied to secure the position of the mold. Once the mold and the cloth reach the desired sintering temperature (about 10 minutes) the force is gradually increased up to 10 tons over a 5-minute period. The cloth fully conforms to the hemispherical shape of the mold and after 5 minutes is consolidated. The mold assembly is rapidly cooled to ambient temperature, by running tap water through the press's platens, while maintaining pressure. The hemispherical dome can be easily removed after pressure is released and the blank holder is lifted. By shaping Spectra® cloth into doubly curved structures, we have demonstrated that the feasibility of using Spectra® cloth to make ballistic shields without adding matrix or cutting patterns.

Spectra® fiber does not stretch significantly at room temperature with only about a 4% strain at break. At elevated temperatures, Spectra® fiber is quite "rubbery" and stretchy. Spectra® fiber has the unique extended chain morphology with minimum chain folds and entanglements. The polymer is chemically nonpolar and there are no strong secondary forces between molecules. At higher temperatures, when chain mobility increases, the molecules slide and pass each other relatively easily under a large load. Spectra® fiber pressurized at an elevated temperature induces a polyethylene crystal transformation, an orthorhombic to hexagonal phase transition (o-h transition) as reported by many researchers (*51-55*). Hexagonal crystals possess a relatively low viscosity compared to orthorhombic crystals and chain mobility is rather high in this so-called "mobile phase". Due to the o-h transition at high pressure and temperature, Spectra® fiber is capable of undergoing large axial strains with relative ease via this metastable, transient hexagonal mesophase. Shown on the macroscopic scale, Spectra® fiber can accommodate high deformation without failing such as those produced in the formation of a hemispherical structure.

Other UHMWPE Fiber Materials

To complete the study, Dyneema Fraglight® nonwoven felt and Spectra Shield® Plus PCR prepreg, two commercially available materials containing the extended chain UHMWPE fibers and designed specifically for ballistic applications, were chosen for parallel comparisons. The high-temperature high-pressure sintering process was applied, and the physical, thermo-mechanical and microstructural properties of the consolidated products were determined and compared to those of the sintered Spectra® cloth. Consolidated Spectra® cloth achieves the highest crystallinity and crystalline orientation. Consolidated Spectra® cloth also shows the best impact resistance and the highest flexural modulus. In terms of interlayer adhesion, Dyneema Fraglight® has an advantage which prompts the use of alternative layers of Spectra® cloth and Dyneema® felt to make composites having both high longitudinal and lateral strengths. The properties of consolidated Spectra® cloth are very sensitive to the processing temperature which is not the case for Dyneema Fraglight® and Spectra Shield® Plus PCR. All the studied materials are thermoformable; Dyneema Fraglight® is the easiest to shape while Spectra Shield® Plus PCR has difficulty in molding due to the flow of matrix at high temperatures. Matrix free Spectra® fiber reinforced composite made of Spectra® cloth is proven to be the best material to make body armor and helmets using our sintering and shaping process.

Applications and Extensions

Besides the potential for ballistic protective shields, there are many other prospective applications for matrix free Spectra® fiber reinforced composites such as: orthopedic implants such as artificial joints, pressure vessels, high strength tubes, automobile parts, sporting goods such as kayaks, tennis racquets, tear and cut resistant protective coatings similar to Tyvek®, sailcloth, and radomes. Orthopedic implants are used to replace fractured bones as well as to reconstruct various degenerated joints in the human body. The most commonly used implant today is the artificial hip joint. Polymers distribute the load and stress more evenly and they are chemically and biologically more compatible to bones than metals. Currently, people are using powder sintering of UHMWPE to make joints and there are problems associated with wear, fatigue and creep, etc. Using carefully designed molds and processing techniques, it should be possible to make better bone replacement by consolidating Spectra® cloth. In conclusion, matrix free Spectra® fiber reinforced composites made by high-temperature high-pressure sintering process offer a novel and promising approach to make products with light weight and extraordinary performance.

Apart from Spectra® fibers and Dyneema® fibers, other materials may be also consolidated by this high-temperature high-pressure sintering technique such as: PET fibers, PP fibers, Vectran® LCP fibers, Nomex® papers, Lycra® fibers, and Nylon fibers, etc. Theoretically, any thermoplastic fibrous material should work, especially high molecular weight materials. Mixed compatible materials can also be sintered to prepare products with combined properties and multiple functions. In fact, BP has commercialized the consolidated PP tapes using the hot compaction process developed by Ward et al. (*56*). The product, *Curv*, mainly targets the market of automobile parts as an environment and recycle friendly replacement compared to glass fiber reinforced products. There is a bright future for this novel process and its unique products.

Acknowledgement

We would like to thank the US Army Soldier and Biological Chemical Command at Natick Massachusetts for funding this research and supplying the

Dyneema® felt. Honeywell kindly provided us the Spectra® materials used in this study.

References

1. Nakamae, K.; Nishino, T.; Ohkubo, H. *J. Macromol. Sci. Phys.* **1991**, 30, 1-23.
2. Southern, J. H.; Capiati, N. J.; Kanamoto, T.; Porter, R. S.; Zachariades, A. E. in *Encylopedia of Polymer Science and Engineering*; Mark, H., Bikales, N.; Overberger, C.; Menges, G., Eds.; Wiley: New York, 1989; Vol. 15, pp 346.
3. Termonia, Y.; Meakin, P.; Smith, P. *Macromolecules* **1985**, 18, 2246-2252.
4. Prevorsek, D. C. in *Polymers for Advanced Technologies*; Levine, M., Ed.; VCH Publishers: New York, 1988; p 557.
5. Kalb, B.; Pennings, A. J. *J. Mater. Sci.* **1980**, 15, 2584-2590.
6. Smith, P.; Lemstra, P. J. *J. Mater. Sci.* **1980**, 15, 505-514.
7. Hoogsteen, W.; van der Hooft, R. J.; Postema, A. R.; ten Brinke, G.; Pennings, A. J. *J. Mater. Sci.* **1988**, 23, 3459-3466.
8. Porter, R. S.; Wang, L. H. *J. Macromol. Sci., Rev. Macromol. Chem.* **1995**, C35, 63-115.
9. Southern, J. H.; Porter, R. S. *J. Appl. Polym. Sci.* **1970**, 14, 2305-2317.
10. Kanamoto, T.; Sherman, E. S.; Porter, R. S. *Polym. J.* **1979**, 11, 497-502.
11. Capaccio, G.; Ward, I. M. *Polym. Eng. Sci.* **1975**, 15, 219-224
12. Griswold, P. D.; Porter, R. S.; Desper, C. R.; Farris, R. J. *Polym. Eng. Sci.* **1978**, 18, 537-545.
13. Hobbs, T.; Lesser, A. J. *Polym. Eng. Sci.* **2001**, 41, 135-144.
14. Imada, K.; Yamamoto, T.; Shigemat, K.; Takayana, M. *J. Mater. Sci.* **1971**, 6, 537.
15. Predecki, P.; Statton, W. O. *J. Polym. Sci. Part B Polym. Lett.* **1972**, 10, 87.
16. Kunugi, T.; Aoki, I.; Hashimoto, M. *Kobunshi Ronbunshu* **1981**, 38, 301-306.
17. Smith, P.; Lemstra, P. J.; Kalb, B.; Pennings, A. J. *Polymer Bulletin* **1979**, 1, 733-736.
18. Smith, P.; Lemstra, P. J. *J. Mater. Sci.* **1980**, 15, 505-514.
19. Kumar, S.; Wang, Y. J. in *Composites Engineering Handbook*; Mallick, P. K., Ed.; Marcel Dekker: New York, 1997; pp 51-61.
20. Kavesh, S.; Prevorsek, D. C. *Int. J. Polym. Mater.* **1995**, 30, 15-56.
21. Silverstein, M. S.; Breuer, O.; Dodiuk, H. *J. Appl. Polym. Sci.* **1994**, 52, 1785-1795.
22. Chiu, H. T.; Wang, J. H. *J. Appl. Polym. Sci.* **1998**, 68, 1387-1395.

23. Kaplan, S. L.; Rose, P. W.; Nguyen, H. X.; Chang, H. W. *SAMPE Quarterly,* **1988**, 19, 55-59.
24. Tissington, B.; Pollard, G.; Ward, I. M. *J. Mater. Sci.* **1991**, 26, 82-92.
25. Capiati, N. J.; Porter, R. S. *J. Mater. Sci.* **1975**, 10, 1671-1677.
26. Teishev, A.; Incardona, S.; Migliaresi, C.; Marom, G. *J. Appl. Polym. Sci.* **1993**, 50, 503-512.
27. Marais, C.; Feillard, P. *Compos. Sci. Technol.* **1992**, 45, 247-255.
28. Hinrichsen, G.; Kreuzberger, S.; Pan, Q.; Rath, M. *Mech. Compos. Mater.* **1996**, 32, 497-503.
29. von Lacroix, F.; Werwer, M.; Schulte, K. *Composites Part A* **1998**, 29, 371-376.
30. Ward, I. M.; Hine, P. J. *Polym. Eng. Sci.* **1997,** 37, 1809-1814.
31. Hine, P. J.; Ward, I. M.; Olley, R. H.; Bassett, D. C. *J. Mater. Sci.* **1993**, 28, 316-324.
32. Yan, R. J.; Hine, P. J.; Ward, I. M.; Olley, R. H.; Bassett, D.C. *J. Mater. Sci.* **1997**, 32, 4821-4832.
33. Rasburn, J.; Hine, P. J.; Ward, I. M.; Olley, R. H.; Bassett, D. C.; Kabeel, M. A. *J. Mater. Sci.* **1995**, 30, 615-622.
34. El-Maaty, M. I. A.; Bassett, D. C.; Olley, R. H.; Hine, P. J.; Ward, I. M. *J. Mater. Sci.* **1996,** 31, 1157-1163.
35. Morye, S. S.; Hine, P. J.; Duckett, R. A.; Carr, D. J.; Ward, I. M. *Composites Part A* **1999**, 30, 649-660.
36. Hine, P. J.; Ward, I. M.; El-Maaty, M. I. A.; Olley, R. H.; Bassett, D. C., *J. Mater. Sci.* **2000**, 35, 5091-5099.
37. Kabeel, M. A.; Bassett, D. C.; Olley, R. H.; Hine, P. J.; Ward, I. M. *J. Mater. Sci.* **1994**, 29, 4694-4699.
38. Hine, P. J.; Ward, I. M.; Jordan, N. D.; Olley, R. H.; Bassett, D. C. *J. Macromol. Sci. Phys.* **2001**, B40, 959-989.
39. Loos, J.; Schimanski, T.; Hofman, J.; Peijs, T.; Lemstra, P. J. *Polymer* **2001**, 42, 3827-3834.
40. Klocek, P.; MacKnight, W. J.; Farris, R. J.; Lietzau, C. U.S. Patent 5,573,824, 1996.
41. Klocek, P.; MacKnight, W. J.; Farris, R. J.; Lietzau, C. U.S. Patent 5,879,607, 1999.
42. Klocek, P.; MacKnight, W. J.; Farris, R. J.; Lietzau, C. U.S. Patent 5,935,651, 1999.
43. Klocek, P.; MacKnight, W. J.; Farris, R. J.; Lietzau, C. U.S. Patent 6,077,381, 2000.
44. Klocek, P.; MacKnight, W. J.; Farris, R. J.; Lietzau, C. U.S. Patent 6,083,583, 2000.
45. Turbak, A. F.; Vigo, T. L. *ACS Sym. Ser.* **1991**, 457, 1-19.

46. Xu, T.; Farris, R. J. *Abstr. Pap. Am. Chem. Soc.* **2003**, 226, 266-PMSE Part 2.
47. Xu, T. Ph.D. thesis, University of Massachusetts, Amherst, MA, 2004.
48. Xu, T.; Farris, R. J. *unpublished.*
49. Morye, S. S.; Hine, P. J.; Duckett, R. A.; Carr, D. J.; Ward, I. M. *Compos. Sci. Technol.* **2000**, 60, 2631-2642.
50. El-Maaty, M. I. A.; Olley, R. H.; Bassett, D. C. *J. Mater. Sci.* **1999,** 34, 1975-1989.
51. Pennings, A. J.; Zwijnenburg, A. *J. Polym. Phys.* **1979**, 17, 1011-1032.
52. van Aerle, N. A. J. M.; Lemstra, P. J.; Braam, A. W. M. *Polym. Commun.* **1989**, 30, 7-10.
53. Kurelec, L.; Rastogi, S.; Meier, R. J.; Lemstra, P. J. *Macromolecules* **2000**, 33, 5593-5601.
54. Waddon, A. J.; Keller, A. *J. Polym. Sci. B Polym. Phys.* **1990**, 28, 1063-1073.
55. Rastogi, S.; Kurelec, L.; Lemstra, P. J. *Macromolecules* **1998**, 31, 5022-5031.
56. Nathan, S. *Process Eng.* **2003**, 84(6), 19-20.

Indexes

Author Index

Aroguz, A. Z., 351
Bair, Harvey E., 339
Bakocevic, Ivana, 215
Bantchev, Grigor B., 106
Baysal, Bahattin M., 119, 351
Baysal, Kemal, 119
Bleha, Tomas, 238
Blume, A., 92
Bogdanić, Grozdana, 229
Bozic, Branislav, 201
Busse, K., 92
Chee, M. J. K., 282
Ciardelli, Francesco, 18
Cifra, Peter, 238
Cocca, M., 370
Cvetkovska, Maja, 119
D'Arienzo, L., 370
D'Orazio, L., 370
Daniel, Christophe, 171
Ding, Liming, 76
Dunjic, Branko, 201
Edwin, Nadia J., 106
Farris, Richard J., 391
Fekete, Zoltán A., 63
Fleš, Dragutin, 229
Freeman, B. D., 187
Gang, Oleg, 158
Gentile, G., 370
Gido, Samuel P., 309
Guerra, Gaetano, 171
Hammer, Robert P., 106
Holcomb, Brandi D., 339
Hussain, H., 92
Ikkala, Olli, 34
Jovanovic, Slobodan, 215
Kammer, H. W., 282
Karasz, Frank E., 76
Kayaman-Apohan, Nilhan, 119
Kerth, A., 92
Kim, Seunghyun, 158
Korugic-Karasz, Ljiljana S., 1, 215
Kressler, J., 92
Kukureka, Stephen N., 138
Kummerlöwe, C., 282
Kuzmić, Ana Erceg, 229
Lee, Kenneth M., 187
Liao, Liang, 76
Liuzzo, Vincenzo, 18
Lu, Ying, 187
MacKnight, William J., 1, 138
Martuscelli, Ezio, 1, 370
Matovic, Radomir, 201
Matsushima, Shigeru, 47
Matusinović, Zvonimir, 325
McCarley, Robin L., 106
Melik-Nubarov, N. S., 92
Mensitieri, Giuseppe, 171, 296
Milano, Giuseppe, 171
Misirli, Z., 351
Misner, Matthew J., 158
Musto, Pellegrino, 171, 296
Nakamura, Naotake, 47
Ocko, Ben, 158
Pajic-Lijakovic, Ivana, 252
Pang, Yi, 76
Panic, Davor, 215
Pilic, Branka, 215
Plavsic, Milenko, 252

Porjazoska, Aleksandra, 119
Prasad, Paras N., 6
Pucci, Andrea, 18
Putanov, Paula, 252
Ragosta, G., 296
Rizzo, Paola, 171
Rogošić, Marko, 325
Ruggeri, Giacomo, 18
Russell, Thomas P., 158
Russo, Paul S., 106
Saicic, Radomir N., 201
Sanchez, I. C., 187
Scarinzi, G., 296
Šegudović, Nikola, 325
Sievert, James D., 158
Stoiljkovic, Dragoslav, 215
Stone, Matthew T., 187
Tasic, Srba, 201
ten Brinke, Gerrit, 34
Teramoto, Akio, 47
Tomić, Tatjana, 325
Tripathy, Amiya R., 138
Venditto, Vincenzo, 171
Vuković, Radivoje, 229
Wang, Xiao-Yan, 187
Williams, Graham, 268
Xu, Tao, 391
Xu, Ting, 158
Yilmaz, Oksan Karal, 119
Yoshiba, Kazuto, 47

Subject Index

A

Absorption
poly(*m*-phenylenevinylene) derivatives, 80, 81, 82*f*, 85*t*
sorption time of ammonia in neat polyimide, 305*f*
volatile organic compounds (VOCs) from water, 176–177
See also Polyimides (PIs)
Acrylated polyesters, hyperbranched in UV curable systems, 202
Acrylates. *See* Elastomer blends
Acrylic dispersions, textile restoration, 381
Active center ensembles, Ziegler–Natta polymerization, 219–221
Adhesion
matrix free Spectra® fiber composite, 396–397
products for textile restoration, 385–386
Adsorption profiles
compensation of size exclusion and adsorption, 245–246
confined macromolecules, 244–245
reduced, for long chains and five concentrations, 245*f*
See also Confined macromolecular systems
Aggregative efficiency (AE)
calculation for stone, 377
See also Stone consolidation
Aliphatic polyesters, copolymerization or blending, 120
Alzheimer's disease, β-amyloid peptide, 107
Ammonia
conversion of imide groups vs. time for polyimide and silica hybrid, 306*f*, 307
correlation of diffusion coefficient with van der Waals volume in polyimide, 306*f*
diffusion behavior in polyimide, 297
gravimetric analysis of diffusion in Kapton® polyimide, 304
mechanism likely for interaction with polyimide, 304
normalized absorbance for sorption test of, in polyimide, 304, 305*f*
sorption in polyimide and hybrid, 302–307
time-resolved FTIR spectroscopy, 302, 303*f*
See also Polyimides (PIs)
Amorphous cells, model building, 193
Amorphous polymers
α-relaxation, 270–273
dielectric and dynamic mechanical behavior, 269
energy landscape, 277–278
mode-mode coupling theory, 270
stretched exponential behavior, 270
Amphiphilic block copolymers
penetration into lipid monolayers, 101–102
See also Block copolymers with poly(ethylene oxide) and poly(perfluorohexylethyl methacrylate)
β-Amyloid fibril formation
aggregation kinetics of solution, 110
5-carboxyfluorescein-labeled peptide at different pH, 112–113

concentration during aggregation, 111
diffusion as function of dyed to undyed Aβ, 112*f*
diffusion coefficient values and theoretical predictions, 112–113
dynamic light scattering (DLS) method, 109
experimental, 109
fluorescence microscopy image of aggregated peptide, 113–114
fluorescence photobleaching recovery (FPR) method, 109
growth kinetics of, 110*f*, 111
log-log curves showing FPR effectiveness, 114–115
materials and methods, 107–109
sample preparation for DLS, 108
sample preparation for FPR, 108–109
signal intensity in FPR and aggregation, 111–112

Anisotropic proton conductivity
poly(4-vinylpyridine):pentadecylphenol (P4VP:PDP), 37*f*
polystyrene-*b*-P4VP(PDP), 40*f*

Anomalous group velocity dispersion, photonic crystals, 13

Anomalous refractive index dispersion, photonic crystals, 13

Architectures
constructing self-assembled supramolecules, 35
See also Comb-shaped supramolecules

B

Ballistic shields
matrix free Spectra® fiber reinforced composites as candidate, 395
testing matrix free Spectra® fiber composite, 397–398

Biological systems. *See* Block copolymers with poly(ethylene oxide) and poly(perfluorohexylethyl methacrylate)

Biomedical applications
biodegradable and bioabsorbable polyesters, 120
See also Triblock copolymer of poly(DL-lactide) and poly(ethylene glycol)

Biophotonics
new frontier, 7
polymeric nanoparticles for, and nanomedicine, 13–15, 16*f*
See also Photonics

Bis(salicylaldiminate) metal(II) complexes
dispersion into polyethylene, 19, 22
structure, 23*f*
synthesis, 20–21
See also Metal complexes in ultra high molecular weight polyethylene (UHMWPE)

2,2-Bis(trifluoromethyl)-4,5-difluoro-1,3-dioxole (BDD). *See* Tetrafluoroethylene/2,2-bis(trifluoromethyl)-4,5-difluoro-1,3-dioxole (TFE/BDD) random copolymers

Bis-phenol-A polycarbonates (BPACY)-based nanocomposites
effects of ductility with domain size, 144–145
effects of nanophase separation on mechanical properties, 143–145
in situ blending with styrene-acrylonitrile (SAN), 140–141
kinetics of phase coarsening, 141, 143
thermodynamics and phase behavior of BPACY-SAN system, 140–141

topological effects of cyclic oligomers on blend miscibility, 145–146
transmission electron microscopy (TEM) of 50/50 BPACY/SAN blend, 141, 142*f*
See also Cyclic ester oligomers (CEO)-based nanocomposites
Blend miscibility, effects of cyclic oligomers on, 145–146
Blends. *See* Elastomer blends; Triblock copolymer of poly(DL-lactide) and poly(ethylene glycol)
Block copolymers
lamellar morphology, 311*f*
self-assembly, 44
See also Comb-shaped supramolecules; Triblock copolymer of poly(DL-lactide) and poly(ethylene glycol)
Block copolymers with poly(ethylene oxide) and poly(perfluorohexylethyl methacrylate)
biotechnology applications, 93
chemical structure of lipid 1,2-diphytanoyl-*sn*-glycero-3-phosphocholine (DPhPC), 95*f*
chemical structure of PFMA-*b*-PEO-*b*-PFMA copolymer, 94*f*
collapse pressure of monolayer, 99–100
concentration dependent cytotoxicity for human erythroleucemia cells, 103*f*
cytotoxicity measurements, 96–97
cytotoxicity results, 102–103
DLS (dynamic light scattering) measurement for copolymers, 97*f*
DLS method, 95
evidence for presence of micelles in solution, 97–98
experimental, 93–97
inner structure of clusters, 98
interfacial properties, 99–100
molecular characteristics of copolymers, 94*t*
penetration of, into lipid monolayers, 101–102
SAXS and SANS traces of three copolymers, 98*f*
small angle neutron scattering (SANS) method, 95
small angle X-ray scattering (SAXS) method, 95
surface pressure and infrared reflection absorption spectroscopy, 95–96
surface pressure-area isotherms of, 99*f*
synthesis, 93, 95
Broadband dielectric spectroscopy (BDS)
amorphous polymers, 270–273
Arrhenius relation for thermal activation, 277
broadening α relaxation, 270–275
dipolar group motions within polymer materials, 268
dynamic heterogeneity, 271
effects of temperature and applied pressure, 275–279
effects of temperature and volume on α-process in amorphous polymers, 276
energy landscape approach, 271–272
energy landscape for amorphous polymer, 277–278
ensembled-averaged time correlation function (TCF) and time-averaged TCF, 272–273
merging dielectric α and β processes forming aβ process at high temperatures, 278–279
molecular dynamics (MD) and Monte Carlo (MC) simulations, 273

orientational trajectories of fluorescent dye-moieties in polymer chain, 273
partially crystalline polymers, 274–275
polymer blends, 274
pressure effects, 275, 276
real-time studies, 268–269
time correlation functions (TCFs), 271–273
volume effects, 276–277
Brominated polystyrene (PBrS)
characterization, 352
formula, 353
See also Polymer blends

C

Camphorsulfonic acid (CSA), protonation of polyaniline, 37–38
ε-Caprolactone, copolymer nanocomposites, 147
5-Carboxyfluorescein β-amyloid
fluorescence microscopy image of aggregated peptide, 113–114
fluorescence photobleaching recovery (FPR) of, and β-amyloid, 111–112
See also β-Amyloid fibril formation
Catalyst productivity, propene polymerization in $TiCl_4/AlEt_2Cl$ on graphite, 226, 227*f*
Catalysts. *See* Ziegler-Natta (ZN) polymerization
Catalysts productivity, Ziegler–Natta polymerization by charge percolation mechanism, 223
Cavities, shape and volume of syndiotactic polystyrene, 173–176
Cavity size distribution
poly(1-trimethylsilyl-1-propyne) (PTMSP), 195, 196*f*
simulation, 193–194
tetrafluoroethylene/2,2-bis(trifluoromethyl)-4,5-difluoro-1,3-dioxole (TFE/BDD), 195, 196*f*
See also Transport properties
CdS quantum dots, dispersion in poly(methyl methacrylate) (PMMA), 10*f*
Cell growth experiments
composite graph for cell attachment and growth on copolymer and blend, 134*f*
culture of cells and experiments, 123
mouse fibroblasts, 123
viability of cells on ABA triblock copolymer and blend films, 131, 133
See also Triblock copolymer of poly(DL-lactide) and poly(ethylene glycol)
Cell viability
block copolymers and human erythroleucemia cells, 102–103
mouse fibroblasts and triblock copolymer and blends, 131, 133, 134*f*
Chain conformation. *See* Elastomer blends
Charge percolation mechanism (CPM)
computer simulation of Ziegler–Natta (ZN) polymerization by, 223–226
ZN polymerization, 221–222
See also Ziegler–Natta (ZN) polymerization
Chemical composition distribution, dual detection size exclusion chromatography, 335–336
Chien, James, Karasz and, 2
1-Chloropropane (CP)
host-guest interaction energies, 183*t*
See also 1,2-Dichloroethane (DCE)

Cluster connectivity, spectral dimension, 255
Cobalt bis(salicylaldiminate) complex
preparation, 21
structure, 23*f*
See also Metal complexes in ultra high molecular weight polyethylene (UHMWPE)
Colloids
depletion interaction, 247–250
See also Confined macromolecular systems
Comb-shaped supramolecules
diblock copolymer-based, 38–44
diblock copolymer of polystyrene (PS) and poly(4-vinylpyridine) (P4VP), 38–41
hierarchical structure, 39
homopolymer-based, 36–38
lamellar-within-lamellar structure, 39–40
microphase separated morphology, 38–39
polarized luminance using poly(2,5-pyridinediyl) (PPY), 36–37
poly(aniline) (PANI), 37–38
polyisoprene (PI) and poly(2-vinylpyridine) (P2VP), 42, 43*f*
preparation of nano-objects, 44
protonation of P4VP, 36
self-assembly for preparing nano-objects, 44
self-assembly in block copolymer systems, 40–44
Compatibility, synthetic materials for restoration, 371
Complex dynamic viscosity
hyperbranched polyols, 207*t*
hyperbranched urethane-acrylates, 209*t*, 210
Composites
one-polymer, 393–394
See also Matrix free Spectra® fiber reinforced composites; Nanocomposites; Ultra high molecular weight polyethylene (UHMWPE)
Confined macromolecular systems
adsorption profiles, 244–245
characteristic behavior, 241
coil-to-pore size ratio, 239–240
compensation of size exclusion and adsorption, 245–246
concentration profiles and compensation effect, 241–246
confinement force calculation, 248
deletion potential, 249–250
depletion attraction, 241
depletion concentration profiles for good solvent and theta solvent, 242*f*
depletion forces in polymer-colloid systems, 241
depletion interaction in colloidal systems, 247–250
depletion profiles, 241–244
effect of depletion interaction, 241
free energy of confinement, 247–248
general features, 239
mixtures of colloid particles and polymers, 240–241
partition coefficient K, 239
partitioning of polymer into porous medium, 239–240
polymer partitioning, 240
ratio of intra-pore and bulk pressure at full equilibrium, 248*f*, 249
reduced adsorption profiles for long chains and five concentrations, 245*f*
reduced depletion profile for long chain and five concentrations, 243*f*
separation modes in liquid chromatography of polymers, 246*f*

steric exclusion/adsorption mechanism, 240
typical concentration profiles for weak adsorption close to compensation point, 246*f*
variation of partitioning with concentration for attractive pores in good solvent, 245*f*
weak-to-strong penetration transition for good solvent and theta solvent, 243*f*, 244
Conformations. *See* Helical conformations of conjugating polymers
π-Conjugated conducting polymers, Karasz and Chien, 2
Conjugating polymers. *See* Helical conformations of conjugating polymers
Connectivity of domains
block copolymer morphology, 310
dark field (DF) field emission gun (FEG) scanning transmission electron microscopy (STEM) of silicon hydrogel, 317
disordered bicontinuous structure, microtomed thin section, and projected image, 314*f*
lamellar bicrystal, 318*f*
lamellar block copolymer morphology, 311*f*
modeling block copolymer nanocomposite, 320
permeabilities of poly(styrene-*b*-butadiene) materials, 320, 321*f*
permeability data for three gases in PS-polydiene block copolymers, 322
phase continuity, 318–319
phase continuity and transport properties, 319–322
projection geometry, 312–313
randomly ordered, microphase separated materials, 313–318
ray tracing of lamellar morphology, 312*f*
scale of phase continuity, 320*f*
silicone hydrogels, 315, 316*f*
simple continuity classification for two component materials, 310*f*
STEM, 315–316
TEM and FEGSTEM projecting electrons through thin sections, 317–318
TEM images of region of spherical morphology at 60° relative tilt, 313*f*
TEM of ordered morphologies, 310–313
transport paths in lamellar materials with small isotropic grains and large anisotropic grains, 321–322
Conservation. *See* Restoration
Contact angle
measurement method, 122
triblock copolymers and blends with poly(caprolactone) (PCL), 130
Contrast
electron microscopy of polymers, 311, 312
transfer function for randomly ordered bicontinuous structure, 314–315
Copolymers. *See* Bis-phenol-A polycarbonates (BPACY)-based nanocomposites; Comb-shaped supramolecules; Crosslinked copolymers; Cyclic ester oligomers (CEO)-based nanocomposites; Dual detection high performance size exclusion chromatography (HPSEC); Reversible surface reconstruction of thin films; Tetrafluoroethylene/2,2-bis(trifluoromethyl)-4,5-difluoro-1,3-dioxole (TFE/BDD) random copolymers; Triblock copolymer of

poly(DL-lactide) and poly(ethylene glycol)
Copper bis(salicylaldiminate) complex
preparation, 21
structure, 23*f*
See also Metal complexes in ultra high molecular weight polyethylene (UHMWPE)
Cotton fibers. *See* Textile restoration
Crosslinked copolymers
copolymerization of methacryl-diisopropylurea (MA-DiPrU) with ethylene glycol dimethacrylate (EDMA) at 0.5:0.5 molar ratio in feed, 231–232
copolymerization of methacryl-isopropylamide (MA-iPrA) with EDMA at 0.5:0.5 ratio in feed, 233–236
mechanism of thermal degradation of, of MA-DiPrU with EDMA, 235
preparation of porous, by thermal degradation of poly(MA-DiPrU-*co*-EDMA), 232–233
removal of volatile isopropylisocyanate (iPrNCO), 235–236
synthesis of MA-iPrA, 233
synthesis of *N*-methacryl-*N,N'*–diisopropylurea (MA-DiPrU), 231
thermogram of, of MA-DiPrU with EDMA, 232*f*
thermogram of model copolymer of MA-iPrA with EDMA, 234*f*
thermogram of porous residue after iPrNCO removal from copolymer, 234*f*
Crystalline polymers, partially, broadband dielectric spectroscopy (BDS), 274–275
Crystallinity, matrix free Spectra® fiber composite, 396
Crystals, photonic
alternating domains of polystyrene and air, 11, 13
transmission and reflection spectra of closely packed polystyrene spheres, 12*f*
Cubic packing of spheres, transmission electron microscopy, 312
Cultural heritage. *See* Restoration
Cured siloxane networks, dynamic moduli, 260
Cyano-substituted polymers
comparing poly[(2-methoxy-5-ethylhexyloxy-1,4-phenylene)vinylene] (MEH-PPV) with cyano-substituted MEH-PPV, 77
ground-state resonance forms of β- and α-cyano-substituted compounds, 78
improving electroluminescence, 78–79
See also Poly(*m*-phenylenevinylene) derivatives
Cyclic carbonate oligomers. *See* Bis-phenol-A polycarbonates (BPACY)-based nanocomposites
Cyclic ester oligomers (CEO)-based nanocomposites
aggregation of 15A clay in matrix of c-PBT/CL copolymer, 152
clay layer movement during polymerization, 149–150
copolymer preparation of c-PBT/ε-caprolactone (CL), 150
cyclic poly(butylene terephthalate) (c-PBT)/clay nanocomposite preparation, 146–147
gel permeation chromatography (GPC) traces of c-PBT/CL copolymer and composites, 152–153
general approach to synthesis, 138–139

macrocyclic polyester oligomers, 146
mechanical testing of c-PBT/CL copolymer composites, 154–155
Na-montmorillonite clay modifications, 150
stress strain curves of c-PBT/CL composites, 155*f*
thermogravimetric analysis (TGA) of c-PBT/CL nanocomposites, 154
thermogravimetric analysis of nanocomposites, 149–150
transmission electron micrograph of PBT/clay before and after polymerization, 148–149
wide angle X-ray diffraction (WAXD) of 15A and C16-MMT nanoclays c-PBT/CL copolymer nanocomposites, 151*f*
X-ray diffraction curves of organoclay/CEO mixtures before and after polymerization, 147–148
X-ray diffraction of c-PBT/CL based nanocomposites, 150–152
See also Bis-phenol-A polycarbonates (BPACY)-based nanocomposites
Cyclic poly(butylene terephthalate) (c-PBT). *See* Cyclic ester oligomers (CEO)-based nanocomposites
Cylinders, transmission electron microscopy (TEM), 312
Cytotoxicity, block copolymers and human erythroleucemia cells, 102–103

D

Dark field (DF) field emission gun (FEG) scanning transmission electron microscopy (STEM), silicone hydrogel, 317
Degradation. *See* Restoration
Dense wavelength division multiplexing (DWDM) function, random glass/polymer nanocomposite, 8, 9*f*
Depletion attraction, colloidal suspension, 241
Depletion interaction
colloidal systems, 247–250
See also Confined macromolecular systems
Depletion profiles
confined macromolecules, 241–244
good and theta solvents, 242*f*
reduced, for long chain and five concentrations, 243*f*
Detection systems, size exclusion chromatography (SEC), 326–327
Deterioration. *See* Restoration
Diagnostics, polymeric nanoparticles, 14–15, 16*f*
Diblock copolymers. *See* Comb-shaped supramolecules; Reversible surface reconstruction of thin films
1,2-Dichloroethane (DCE)
conformational selectivity of DCE, 1,2-dichloropropane (DCP), and 1-chloropropane (CP), 177–181
desorption kinetics from amorphous and clathrate phases of syndiotactic polystyrene (s-PS), 180
electrostatic potential maps, 182*f*
host–guest interaction energies, 183*t*
partitioning in amorphous and crystalline phases, 180–181
potential energy of trans and gauche conformers, 182
role of electrostatic interactions, 181–183
sorption kinetics from solution by s-PS, 176–177
See also Syndiotactic polystyrene (s-PS)

1,2-Dichloropropane (DCP)
host-guest interaction energies, 183*t*
See also 1,2-Dichloroethane (DCE)
Dielectric spectroscopy. *See* Broadband dielectric spectroscopy (BDS)
Differential scanning calorimetry (DSC)
Goretex® fluorocarbon and poly(tetrafluoroethylene) (PTFE) samples, 346–347
hyperbranched polymers, 207*f*
method, 122, 203–204, 354
quenched and annealed poly(ethylene terephthalate) (PET) films, 345*f*
thermal characterization of polymer blends, 357, 359*f*, 360*f*
triblock copolymer and blend with poly(caprolactone) (PCL), 127, 128*f*
See also Polymer blends
Diffusion
5-carboxyfluorescein-labeled and unlabeled β-amyloid, 111–113
modeling methodology, 192
simulation, 194
simulation results, 198
See also Transport properties
Diffusion coefficient
ammonia in polyimide and silica hybrid, 304, 306*f*
water in polyimide and polyimide/silica hybrid, 301–302
1,2-Diphytanoyl-*sn*-glycero-3-phosphocholine. *See* Block copolymers with poly(ethylene oxide) and poly(perfluorohexylethyl methacrylate)
1,4-Divinylbenzene
dynamic reaction coordinate (DRC) potential energy curves, 68*f*, 69*f*
excited-state molecular dynamics, 67
See also Molecular orbital (MO) theory
Dodecyl benzene sulfonic acid (DBSA), protonation of polystyrene-*b*-poly(4-vinylpyridine), 40, 42*f*
Domains. *See* Connectivity of domains
Drug delivery, nanoclinics, 14
Dual detection high performance size exclusion chromatography (HPSEC)
chemical composition distribution (CCD), 335–336
chromatograms of polystyrene (PS) with different concentrations, 331, 332*f*, 333*f*
conditions for homo and copolymerization, 328*t*
copolymers of styrene and methyl methacrylate (SMMA), 331, 332*t*
experimental, 327–329
molecular mass averages and sample composition by SEC, 332*t*
molecular masses for copolymers, 335
refractive index (RI) and ultraviolet (UV) detectors, 335
sample preparation, 327–328
SEC measurements, 328–329
styrene mass fraction values in SMMA copolymer, 336
UV chromatograms of copolymers with different compositions, 331, 333*f*, 334*f*
UV chromatograms of PS and poly(methyl methacrylate) (PMMA), 331
UV spectra of PS and PMMA, 329, 330*f*
variation of apparent composition distribution, 337

Durability
consolidating procedure for stone, 378
synthetic materials for restoration, 371
Dynamical mechanical analysis, hyperbranched urethane–acrylates, 210–212
Dynamic heterogeneity, amorphous polymers, 271
Dynamic light scattering (DLS)
aggregation kinetics of β-amyloid fibril formation, 110–111
diblock and triblock copolymers, 97*f*
method, 95, 108, 109
Dynamic moduli scaling, experimental data, 258–260
Dynamic reaction coordinate (DRC)
potential energy curves for 1,4-divinylbenzene, 68*f*, 69*f*
tracing evolving excited state, 66
Dyneema Fraglight®, ultra high molecular weight polyethylene (UHMWPE), 401, 402

E

Elastomer blends
acrylate and fluorocarbon rubber blend (ACM/FKM), 261–265
causes for synergistic effects, 265
chain conformation, 263–264
cured siloxane networks, 260
excluded volume theory, 254–255
experimental data on contribution of phases, 260–261
experimental data on dynamic moduli scaling, 258–260
FKM/ACM blend properties vs. composition, 263*f*
moduli and conformation, 263–265
modulus and connectivity levels, 255–258
Mooney–Rivlin plots of natural rubber/styrene-butadiene rubber (NR/SBR) blend, components, and phases, 262*f*
phantom network, 257
poly(dimethylsiloxane) (PDMS) gels, 259
poly(ε-caprolactone)diol network with poly(styrene-*co*-acrylonitrile) (PCL/SAN), 258–259
polybutadiene (BR) and ethylene-propylene-ethylidienenorbornene rubber (EPDM), 258
polymer networking theory, 254
polyurethane networks from poly(ethylene oxide) (PEO) and polyisocyanates, 259
shear relaxation, 256
tensile and plateau moduli for FKM/ACM, 261, 262*f*
tensile and plateau moduli for poly(2,6-dimethyl-1,4-phenylene oxide)/polystyrene (PPO/PS), 253, 262*f*
theoretical background, 254–255
Electrical analogy, modulus and connectivity, 255–256
Electroluminescence (EL)
EL spectra for light-emitting diode (LED) devices, 83, 84*f*
light-emitting diode (LED) device, 77
poly(*m*-phenylenevinylene) polymers, 85*t*
voltage dependence of current density and luminance for LED devices, 86, 87*f*
See also Poly(*m*-phenylenevinylene) derivatives
Electronically conducting nanowires, polyaniline (PANI), 37–38
Electron microscopy
contrast, 311, 312

See also Transmission electron microscopy (TEM)
Energy landscape approach, amorphous materials, 271–272
Environmental scanning electron microscopy (ESEM)
binary blends of polycarbonate/brominated polystyrene (PC/PBrS), 362*f*, 363*f*, 364*f*
characterization of microstructural behavior, 354
microstructural examinations of polymer blends, 357, 367
polycarbonate (PC) surface morphology, 362*f*
ternary blends PC/PBrS/poly(2,6-dimethyl-1,4-phenylene oxide (PPO), 365*f*, 366*f*
See also Polymer blends
Ethylene glycol dimethacrylate (EDMA). *See* Crosslinked copolymers
Ethylene-propylene-ethylidienenorbornene rubber (EPDM), dynamic moduli of blends, 258
Excluded volume theory, power law, 254–255
Exfoliation, polymer penetration into silicate, 146–147

F

Field emission gun (FEG) scanning transmission electron microscopy (STEM), projecting electrons through thin section, 317–318
Films. *See* Reversible surface reconstruction of thin films
Flat panel first route, matrix free Spectra® fiber composite, 400–401
Flexibility, hyperbranched urethane-acrylates, 212–213
Flexible spacers, hyperbranched urethane-acrylates, 211
Flexural properties, matrix free Spectra® fiber composite, 397
Flory–Huggins approximation, polymer blend solution, 283
Fluorescence, poly(*m*-phenylenevinylene) derivatives, 80, 81, 82*f*, 83, 85*t*
Fluorescence microscopy, 5-carboxyfluorescein β-amyloid, 113–114
Fluorescence photobleaching recovery (FPR)
aggregation kinetics of peptide, 111–112
effectiveness as tool, 114–115
measuring diffusion coefficients, 107
method, 108–109
See also β-Amyloid fibril formation
Fluorocarbon rubbers. *See* Elastomer blends
Fourier transform infrared (FTIR) spectroscopy
characterization of polymer blends, 354, 357
method, 122, 353–354
triblock copolymer and blend with poly(caprolactone) (PCL), 124, 126*f*, 127
See also Polymer blends; Time-resolved Fourier transform infrared (FTIR) spectroscopy
Fractional free volume (FFV)
polymers with high permeability, 188
See also Transport properties
Free energy of confinement, colloid particles, 247–248
Free volume
simulation results, 194, 195*t*
sorption of water in polyimides, 299

See also Transport properties
Frontiers of polymer research, Karasz, 3

G

Gel permeation chromatography (GPC)
cyclic poly(butylene terephthalate)/caprolactone (c-PBT/CL) copolymer nanocomposites, 152–153
cyclic poly(butylene terephthalate)/caprolactone copolymer, 152–153
molar mass determination, 121
triblock copolymer characterization, 124*t*
Gel point, theory, 254
General Electric (GE) silicone RTV-615. *See* Silicone films
General Electric Research Laboratory, Karasz, 1–2
Geometrical confinement macromolecular chains, 239
See also Confined macromolecular systems
GISAXS. *See* Grazing incidence small angle x-ray scattering (GISAXS)
Glass, sol-gel processing for glass-polymer composite, 7–8
Glass transition temperatures
hyperbranched polyols, 207*t*
hyperbranched urethane-acrylates, 210–212
UV cured coatings, 213*t*
Glassy polymers. *See* Transport properties
Goretex® fluorocarbon material
clothing and footwear for outdoor activity, 346
differential scanning calorimetry (DSC), 346–347
effectiveness in allowing water vapor to pass, 349
permeability, 346–347, 349
scanning electron micrograph, 347, 348*f*
Graft polymerization process, textile restoration, 386–387
Grazing incidence small angle x-ray scattering (GISAXS)
diblock copolymer film after acetic acid dip, 163, 164*f*
diblock copolymer film on passivated surface after acetic acid dip, 164–165
pore penetration through films, 159–160
re-annealing diblock copolymer thin film, 165–167
regeneration of initial nanotemplate by thermal annealing, 160
See also Reversible surface reconstruction of thin films

H

Hardness, hyperbranched urethane-acrylates, 212–213
Havriliak–Negami (HN) function, dielectric α-process, 269, 270
Heeger, Alan, π-conjugated conducting polymers, 2
Helical conformations of conjugating polymers
dilute solution properties of linear polymers, 48
experimental results, 51–53
Kuhn's dissymmetry ratio g_{abs}, 51
Kuhn's dissymmetry ratio g_{abs} for two poly(silylene)s, 52–53
liquid crystal formation, 48
microscopic-macroscopic structure correlation, 56–58
order-disorder transition in aqueous solutions of schizophyllan, 57*f*

partition function of chain, 49
poly(silylene)s, 51–53
polyisocyanates, 52–53
schizophyllan, 56–57
single-stranded helices, 48
solutions of stiff poly(silylene)s, 57–58
statistical weight matrix, 49
theoretical analysis, 54–56
theory of conformational transitions in dilute solution, 49–51
Heritage. *See* Restoration
4-Hexylresorcinol (Hres), hydrogen bonding for polyaniline, 37–38
Hierarchical structures
concept for constructing, 35
proton conductivity, 39
High performance liquid chromatography (HPLC), coupling different detectors, 327
High performance size exclusion chromatography (HPSEC). *See* Dual detection high performance size exclusion chromatography (HPSEC)
Host-guest interactions. *See* Syndiotactic polystyrene (s-PS)
Huggins' equation, viscosity of polymer solutions, 283, 284
Hybrids. *See* Polyimides (PIs)
Hydrogen bonding
comb-shaped supramolecule by, of pentadecylphenol (PDP), 36
4-hexylresorcinol (Hres) for polyaniline, 37–38
octyl gallate (OG) with poly(2,5-pyridinediyl) (PPY), 36–37
OG for polyisoprene-*b*-poly(2-vinylpyridine), 42, 43*f*
PDP for polystyrene-*b*-poly(4-vinylpyridine), 38–41
sorption of water in polyimides, 298–299
Hydrolytic degradation
ABA triblock copolymer and blends with poly(caprolactone), 131, 132*f*
studies, 122–123
Hydrophobic photosensitizer, doped nanoparticles, 14–15, 16*f*
Hyperbranched polymers
acrylated polyesters in UV curable systems, 202
applications, 202
end groups for incorporation of crosslinkable groups, 202
properties, 201–202
Hyperbranched polyols
complex dynamic viscosity, 207*t*
esterification with α-haloacids, 204–205
glass transition temperatures, 207*t*
properties, 207*t*
synthesis, 204
See also Hyperbranched urethane-acrylates
Hyperbranched urethane-acrylates
differential scanning calorimetry (DSC), 203–204, 207
dynamical mechanical analysis, 210–212
experimental, 203–205
general testing, 203–204
glass transition temperatures, 213*t*
hardness and flexibility, 212–213
idealized structure, 209
impact of hydroxyl groups on thermal properties, 207
influence of xanthate groups on properties of crosslinked materials, 212
materials, 203
modification of hyperbranched polyester with xanthate groups, 206
partial modification of hyperbranched polyols and properties, 206, 207*t*

polyalkyleneoxide chains as flexible spacers, 211
synthesis, 204–205
synthesis of, with xanthate moieties, 205, 206*t*
synthesis of hyperbranched polyesters with xanthate moieties, 205
synthesis of hyperbranched polyol, 204
synthetic routes, 208
tan δ curves, 210, 211*f*
unsaturation type and cure properties, 211
xanthate mediated controlled radical polymerization, 202–203

I

Ideal gel, phantom network, 257
Intercalation, polymer penetration into silicate, 146
Interfacial properties, block copolymers, 99–100
Interlayer adhesion, matrix free Spectra® fiber composite, 396–397
Interpenetrating polymer networks (IPN), in situ formation, 140
Interpenetration function, polymer blend solutions, 293–294

K

Karasz, Frank Erwin
biography, 1–3
education, 1
Kleiner–Karasz–MacKnight equation for moduli of blends, 253
optical nanocomposite by sol-gel processing, 7–8
professional associations, 2–3
segmented blue emitting polymers, 64
Kinetics
chemical reaction of ammonia and polyimide, 306*f*, 307
phase coarsening, 141, 143
sorption and desorption of water in pure polyimide and hybrid, 300–301
See also β-Amyloid fibril formation; Syndiotactic polystyrene (s-PS)
Kleiner–Karasz–MacKnight equation, elastomer moduli of blends, 253
Kohlrausch–Williams–Watts (KWW) function
amorphous polymers, 270–273
dielectric α-process, 269

L

Lamellar morphology
block copolymer, 311*f*
comparison of possible transport pathways, 321–322
polystyrene-*b*-poly(4-vinylpyridine)(nonadecylphenol), 39*f*
polystyrene-*b*-poly(methyl methacrylate) films, 162
ray tracing of, 312*f*
See also Connectivity of domains
Lamellar-within-lamellar structure, polystyrene-*b*-poly(4-vinylpyridine), 39–40
Lexan®. *See* Polycarbonate film
Light-emitting diode, polymeric (PLED). *See* Poly(phenylene-vinylene) (PPV) materials
Light scattering
theory, 285–286
See also Polymer blend solutions
Lipid monolayer

penetration of amphiphilic block copolymers, 101–102
See also Block copolymers with poly(ethylene oxide) and poly(perfluorohexylethyl methacrylate)
Liquid chromatography, separation modes in, of polymers, 246
Liquid crystals, helical conformation, 48
Local field enhancement, photonic crystals, 13
Low molar mass (LMM) glass-forming liquids
dielectric properties, 269
dynamic heterogeneity and energy landscapes, 271
predicting structural relaxation, 272
See also Broadband dielectric spectroscopy (BDS)
Luminescence. *See* Poly(*m*-phenylenevinylene) derivatives; Poly(phenylene-vinylene) (PPV) materials

M

MacDiarmid, Alan, π-conjugated conducting polymers, 2
Macrocyclic polyester oligomers
properties, 146–147
See also Cyclic ester oligomers (CEO)-based nanocomposites
Macromolecular chains
geometrical confinement, 239
See also Confined macromolecular systems
Mass-thickness contrast, electron microscopy of polymers, 311
Matrix free Spectra® fiber reinforced composites
applications and extensions, 402
ballistic performance of multilayer consolidated Spectra® cloth, 398*f*
ballistic tests, 397–398
crystallinity and orientation, 396
flat panel first route, 400–401
flexural properties, 397
high-temperature high-pressure sintering, 394
interlayer adhesion, 396–397
molding procedures, 400*f*
normalized to thickness total impact energy vs. sintering temperature, 396*f*
optimization of processing conditions, 395–396
other ultra high molecular weight polyethylene (UHMWPE) materials, 401–402
processing method, 394
process-structure-property relationship, 396–399
scanning electron microscopy (SEM) images, 398–399
thermoforming of Spectra®, 399–401
Mechanical compatibility, synthetic materials for restoration, 371
Mechanical testing
cyclic poly(butylene terephthalate)/caprolactone (c-PBT/CL) copolymer composites, 154–155
effects of nanophase separation of nanocomposites on mechanical properties, 143–145
See also Elastomer blends
Mechanism
charge percolation, of Ziegler–Natta polymerization, 221–222
combined steric exclusion/adsorption, for polymer partitioning, 240
likely reaction of ammonia in polyimide, 304

thermal degradation of poly(methacryl-diisopropylurea-*co*-ethylene glycol dimethacrylate), 235
Metal complexes in ultra high molecular weight polyethylene (UHMWPE)
apparatus and methods, 19–20
bis(salicylaldiminate) metal(II) complex ($NiC_{18}OSalophen$, $CuC_{18}OSalophen$, and $CoC_{18}OSalophen$) preparation, 20–21, 22
cobalt(II) complex, 21
complex structures, 23*f*
copper(II) complex, 21
$d \rightarrow \pi^*$ transition band of Co and Cu bis(salicyladiminate) complexes, 30–31
energy dispersive spectrum (EDS) method, 20
experimental, 19–21
film preparation by solution casting, 21
materials, 20–21
nickel(II) complex, 21
opto-electronic properties, 27–29
$\pi \rightarrow \pi^*$ electronic transitions of nickel complex, 29–30
polymer orientation, 21
scanning electron microscopy (SEM) analysis method, 20
SEM of 3 wt% $CuC_{18}OSalophen$/UHWMPE film, 22, 25*f*
SEM of 3 wt% $CuC_{18}OSalophen$/UHWMPE film near surface and 10μm deeper, 22, 26*f*
SEM of 3 wt% $NiC_{18}OSalophen$/UHWMPE film, 22, 24*f*
UV-vis spectra in polarized light of UHMWPE oriented film, 27–28
UV-vis spectra of $CoC_{18}OSalophen$/UHMWPE oriented film, 30
UV-vis spectra of $CuC_{18}OSalophen$/UHMWPE oriented film, 28–29
UV-vis spectra of $NiC_{18}OSalophen$/UHMWPE oriented film, 28
X-ray EDS of 3 wt% $CuC_{18}OSalophen$/UHWMPE film, 22, 25*f*
X-ray EDS of 3 wt% $NiC_{18}OSalophen$/UHWMPE film, 22, 24*f*
Methacryl-diisopropylurea (MA-DiPrU). *See* Crosslinked copolymers
Methane sulfonic acid (MSA), protonation of polystyrene-*b*-poly(4-vinylpyridine), 39–40, 41*f*
Microcavity effect, photonic crystals, 13
Microphase separated morphology, polystyrene-*b*-poly(4-vinylpyridine), 38–39
Migration regions, active center ensembles, 219–220
Miscibility
effects of cyclic oligomers on blend, 145–146
polymer blend, 283–284
Modeling. *See* Transport properties
Mode-mode coupling theory (MCT), amorphous polymers, 270
Moduli
blend components, 253
Kleiner–Karasz–MacKnight equation, 253
See also Elastomer blends
Molding procedures, matrix free Spectra® fiber composite, 400–401
Molecular dynamics (MD), polymer chain dynamics, 273
Molecularly imprinted polymers

preparation methods, 230
thermal degradation of crosslinked copolymers, 230–231
See also Crosslinked copolymers
Molecular mass, size exclusion chromatography (SEC), 326
Molecular orbital (MO) theory
computational methods, 65–66
conformational mapping of poly(phenylene-vinylene) (PPV)-based polymers, 67, 70*f*
dynamic reaction coordinate (DRC) potential energy curves for 1,4-divinylbenzene, 68*f*, 69*f*
excited-state molecular dynamics, 67, 68*f*, 69*f*
potential energy curves for phenylene ring rotation, 70*f*
schematic of cut through potential energy surfaces, 66*f*
See also Poly(phenylene-vinylene) (PPV) materials
Molecular sieves. *See* Syndiotactic polystyrene (s-PS)
Monte Carlo (MC) simulations, polymer chain dynamics, 273
Morphology
polystyrene-*b*-poly(4-vinylpyridine)(nonadecylphenol), 39*f*
polystyrene-*b*-poly(methyl methacrylate) films, 162
See also Connectivity of domains
Mouse fibroblasts. *See* Cell growth experiments

N

Nanoclinics, targeted delivery approach, 14–15, 16*f*
Nanocomposites
clay layer movement during polymerization, 149–150
dense wavelength division multiplexing (DWDM) function, 8, 9*f*
general approach to synthesis, 138–139
photorefractive, for optoelectronics, 8, 10
schematic of exfoliation of nylon 6 and modified nanoclay, 139*f*
schematic of polymer dispersed liquid crystal (PDLC) system, 10*f*
sol-gel processing, 7–8
See also Bis-phenol-A polycarbonates (BPACY)-based nanocomposites; Composites; Cyclic ester oligomers (CEO)-based nanocomposites
Nanomedicine, polymeric nanoparticles and biophotonics and, 13–15
Nano-objects, self-assembly of block copolymers, 44
Nanoparticles
biophotonics and nanomedicine, 13–15
organically modified silica, encapsulating hydrophobic drugs, 16*f*
Nanophotonics, opportunities for polymeric nanostructures, 7
Nanostructures, photonic crystals, 11–13
Nanowires, polyaniline for electronically conducting, 37–38
Nickel bis(salicylaldiminate) complex
preparation, 21
structure, 23*f*
See also Metal complexes in ultra high molecular weight polyethylene (UHMWPE)
Nitrogen sorption, experiments of syndiotactic polystyrene, 175–176

Nonadecylphenol (NDP), hydrogen bonding of polystyrene-*b*-poly (4-vinylpyridine), 39*f*
Noryl®
development, 2
family of engineering thermoplastics, 253
See also Elastomer blends
Nuclear magnetic resonance (NMR)
^{1}H NMR spectrum of triblock copolymer, 124, 125*f*
method, 121

O

Octylgallate (OG)
hydrogen bonding for poly (2,5-pyridinediyl) (PPY), 36–37
hydrogen bonding for polyisoprene-*b*-poly(2-vinylpyridine), 42, 43*f*
See also Hydrogen bonding
Optical activity
helical conformation, 48
poly(silylene)s, 51–52
Optical properties
photonic crystals, 11, 13
poly(*m*-phenylenevinylene) derivatives, 85*t*
Opto-electronic properties
sensitivity of metal complexes dispersed in oriented polyethylene, 27–30
See also Metal complexes in ultra high molecular weight polyethylene (UHMWPE)
Optoelectronics, photorefractive nanocomposites, 8, 10
Order-disorder transition, aqueous solutions of schizophyllan, 57*f*
O'Reilly, James, Karasz collaborator, 1–2
Organoclays. *See* Cyclic ester oligomers (CEO)–based nanocomposites
Orientation, matrix free Spectra® fiber composite, 396

P

Partially crystalline polymers, broadband dielectric spectroscopy (BDS), 274–275
Partition coefficient K. *See* Confined macromolecular systems
Pentadecylphenol (PDP)
hydrogen bonding for poly(4-vinylpyridine) (P4VP), 36
See also Hydrogen bonding
Peptides. *See* β-Amyloid fibril formation
Percolation theory, distribution of relaxation times, 256
Performance polymers, Noryl® development, 2
Permeability
amorphous poly(ethylene terephthalate) (PET), 344–346
calculation, 342–343
container for water vapor transmission (WVT) studies, 340
cured silicone film, 343, 344*f*
experimental, 340–341
Goretex® fluorocarbon material, 346–347, 349
invention of methods and apparatus, 340–341
partially crystalline PET, 346
poly(tetrafluoroethylene) (PTFE), 346–347, 349
polycarbonate film, 342–343
rubbery silicone film and Goretex® membrane, 349*f*
schematic of WVT capsule, 341*f*
testing small size film, 340

Permeability coefficients
modeling methodology, 190
simulation, 194
simulation results, 198–199
See also Transport properties
pH, diffusion coefficient values for β-amyloid, 112–113
Phantom network, ideal gel with, 257
Phase continuity
CO_2 permeabilities of poly(styrene-*b*-butadiene), 321
microphase separated materials, 318–319
scale, 320*f*
transport properties, 319–322
See also Connectivity of domains
Phase moduli, coupling for blends, 260–261
Phenylene-vinylene (PV) polymers. *See* Poly(phenylene-vinylene) (PPV) materials
Photodynamic therapy, photosensitizers, 14–15
Photoluminescence, poly(*m*-phenylenevinylene) derivatives, 80, 81, 82*f*
Photonic crystals
alternating domains of polystyrene and air, 11
anomalous group velocity dispersion, 13
anomalous refractive index dispersion, 13
local field enhancement, 13
microcavity effect, 13
stop gap, 11
transmission and reflection spectra of, by close packing polystyrene spheres and air, 12*f*
See also Biophotonics
Photonics
applications of polymeric media, 6–7
biophotonics, 7
dense wavelength division multiplexing (DWDM) function, 8, 9*f*
nanophotonics, 7
optoelectronic application using photorefractivity, 8, 10
photonic crystals, 11, 13
polymeric nanoparticles for biophotonics and nanomedicine, 13–15
random nanocomposites, 7–10
transmission and reflection spectra of photonic crystal, 12*f*
See also Biophotonics
Photorefractivity, optoelectronic application using, 8, 10
Photosensitizers, nanoparticle approach, 14–15, 16*f*
Physiological function, β-amyloid peptide, 107
Polarized luminance, poly(2,5-pyridinediyl) (PPY), 36–37
Polyaniline (PANI)
camphorsulfonic acid (CSA) for protonation of, 37–38
4-hexylresorcinol (Hres) for hydrogen bonding to, 37–38
See also Comb-shaped supramolecules
Polybutadiene (BR), dynamic moduli of blends, 258
Poly(butylene terephthalate), cyclic (c-PBT). *See* Cyclic ester oligomers (CEO)-based nanocomposites
Polycarbonate (PC)
blends with brominated polystyrene (PBrS), 352
See also Bis-phenol-A polycarbonates (BPACY)-based nanocomposites; Polymer blends
Polycarbonate film
permeability equation, 342–343

weight loss of water vapor through, 342*f*
Poly(dimethylsiloxane) (PDMS), dynamic moduli of blends, 259
Poly(DL-lactide). *See* Triblock copolymer of poly(DL–lactide) and poly(ethylene glycol)
Poly(ε-caprolactone) (PCL). *See* Cyclic ester oligomers (CEO)–based nanocomposites; Polymer blend solutions; Triblock copolymer of poly(DL–lactide) and poly(ethylene glycol)
Poly(ε-caprolactone)diol (PCL) network, dynamic moduli of blends, 258–259
Poly(2,6-dimethyl-1,4-phenylene oxide) (PPO)
mixtures of PPO and polystyrene (PS), 352
moduli of blend with polystyrene, 253, 261–263
Noryl® family of engineering thermoplastics, 253
structure, 353
See also Elastomer blends; Polymer blends
Poly(ethylene) (PE)
applications, 392
principles to make high modulus PE fibers, 392–393
See also Matrix free Spectra® fiber reinforced composites; Metal complexes in ultra high molecular weight polyethylene (UHMWPE); Ultra high molecular weight polyethylene (UHMWPE)
Poly(ethylene glycol). *See* Triblock copolymer of poly(DL-lactide) and poly(ethylene glycol)
Poly(ethylene oxide) (PEO)
amphiphilic block copolymers of, 93
copolymerization or blending, 120
See also Block copolymers with poly(ethylene oxide) and poly(perfluorohexylethyl methacrylate); Polymer blend solutions
Poly(ethylene terephthalate) (PET)
amorphous PET in water vapor transmission (WVT) capsule, 345, 346*f*
broadband dielectric spectroscopy (BDS), 275
comparative differential scanning calorimetry (DSC) scans of quenched and annealed films, 345*f*
DSC scan, 344–345
partially crystalline PET, 346
rate of water vapor transmission through amorphous PET film, 346*f*
Poly(3-hydroxy butyrate) (PHB) blends
blends with poly(ethylene oxide) (PEO) and poly(ε-caprolactone) (PCL), 284
experimental, 286–287
Huggins coefficient vs. blend composition, 289*f*
interpenetration function, 293–294
intrinsic viscosities and Huggins coefficients for blends with PHB, 288*t*
light scattering experiments with blends, 287
light scattering for blend solutions, 290*t*
molar coil volume vs. blend composition, 291*f*
second virial coefficient times square of molecular mass vs. mass fraction for PEO/PHB and PCL/PHB, 292*f*
viscometry of blends, 287–288
viscosity measurement of blends, 286

See also Polymer blend solutions
Polyimides (PIs)
absorbance for sorption test of ammonia in neat, 305*f*
ammonia and water as probes, 297
applications, 297
comparing polyimide-silica hybrid, 297
correlation of diffusion coefficient vs. van der Waals volume of penetrant, 306*f*
curve fitting analysis of spectrum of water absorbed in neat, 298*f*
curve fitting analysis of spectrum of water absorbed in polyimide/silica hybrid, 299*f*
diffusion coefficient vs. average water concentration for pure, and hybrid, 302*f*
Fourier transform infrared (FTIR) transmission spectra of ammonia sorption in neat, 303*f*
gravimetric analysis of ammonia diffusion of Kapton® polyimide, 304
kinetics of chemical reaction with ammonia, 307
mechanism likely for ammonia sorption, 304
relative conversion of imide vs. time for neat, and hybrid, 306*f*, 307
sorption and desorption kinetics of pure, and hybrid, 301*f*
sorption isotherms as sorbed water per matrix vs. water vapor activity for neat and hybrid sample, 300*f*
sorption of ammonia in, and hybrid, 302–307
sorption of water in, and hybrid, 298–302
time-resolved FTIR spectroscopy, 297, 300, 302, 303*f*
Polyisocyanates
helical conformation, 48
optical rotation, 55, 56*f*
temperature dependence of circular dichroism, 53, 54*f*
See also Helical conformations of conjugating polymers
Polyisoprene-*b*-poly(2-vinylpyridine) (PI-*b*-P2VP)
octyl gallate (OG) as hydrogen bonding amphiphile for comb-shaped , 42, 43*f*
See also Comb-shaped supramolecules
Poly(lactide-*co*-glycolide) (PLGA), biomedical applications, 120
Polymer blends
broadband dielectric spectroscopy (BDS), 274
characterization of microstructural behavior of environmental scanning electron microscopy (ESEM), 354
dispersed spherical particles in binary blends, 361*t*
dispersed spherical particles in ternary blends, 361*t*
DSC thermograms of binary blends of polycarbonate/brominated polystyrene (PC/PBrS), 359*f*
DSC thermograms of ternary blends of PC/PBrS/poly(2,6-dimethyl-1,4-phenylene oxide) (PPO), 360*f*
ESEM of PC/PBrS/PPO surface morphology, 365*f*, 366*f*
ESEM of PC/PBrS surface morphology, 362*f*, 363*f*, 364*f*
ESEM of pure PC surface morphology, 362*f*
experimental, 352–353
formulas of PC, PBrS, and PPO, 353*t*
Fourier transform infrared (FTIR) measurements, 353–354
FTIR characterization, 354, 357

FTIR of pure PC, PBrS, and PPO, 355*f*
FTIR spectra of binary blend PC/PBrS, 356*f*
FTIR spectra of ternary blends of PC/PBrS/PPO, 358*f*
materials, 352–353
microstructural examinations, 357, 367
obtaining desirable properties, 253
preparation of binary and ternary blends of PPO, PBrS, and PC, 353
thermal characterization, 357, 359*f*, 360*f*
thermal properties, 354
See also Elastomer blends; Triblock copolymer of poly(DL-lactide) and poly(ethylene glycol)

Polymer blend solutions
characteristics of poly(3-hydroxybutyrate) (PHB), poly(ethylene oxide) (PEO), and poly(ε-caprolactone) (PCL) and solvent, 287*t*
experimental, 286–287
Huggins coefficients and parameter κ for PEO/PHB and PCL/PHB, 289*t*
Huggins coefficient vs. blend composition, 288, 289*f*
interpenetration function for blends, 293–294
intrinsic viscosity, 284
intrinsic viscosity and Huggins coefficients for blends with PHB, 288*t*
light scattering, 290–294
light scattering experiments, 287
light scattering for PHB/PEO and PHB/PCL blends, 290*t*
light scattering theory, 285–286
miscibility, 283–284
molar coil volume vs. blend composition for PEO/PHB and PCL/PHB, 291*f*
refractive index increments, 287
second virial coefficient of PHB with PEO addition, 292–293
second virial coefficient times square of molar mass vs. mass fraction, 291, 292*f*
specific viscosity, 284
theoretical background, 284–286
viscometry, 287–288
viscosity measurement, 286
viscosity theory, 284–285

Polymer dispersed liquid crystals (PDLC), schematic of photorefractive nanocomposite, 10*f*

Polymer networking, theory, 254

Polymer research, Karasz active at frontiers of, 3

Polymeric light-emitting diode (PLED)
electroluminescence (EL), 77
poly(phenylene-vinylene) (PPV) materials, 64–65
voltage dependence of current density and luminance for LED devices, 86, 87*f*
See also Poly(phenylene-vinylene) (PPV) materials

Polymeric materials. *See* Syndiotactic polystyrene (s-PS); Textile restoration

Polymerization. *See* Ziegler-Natta (ZN) polymerization

Poly(methacryl-*N,N'*-diisopropylurea-*co*-ethylene glycol dimethacrylate). *See* Crosslinked copolymers

Poly(methyl methacrylate) (PMMA)
acetic acid selective solvent for PMMA blocks of diblock, 162
UV chromatograms of PS and PMMA, 331*f*
UV spectra of PS and PMMA, 329, 330*f*

See also Dual detection high performance size exclusion chromatography (HPSEC); Polystyrene-*b*-poly(methyl methacrylate) (PS-*b*-PMMA); Reversible surface reconstruction of thin films
Polyolefins. *See* Metal complexes in ultra high molecular weight polyethylene (UHMWPE)
Polypeptides
helical conformation, 48
order-disorder transition of synthetic, 1–2
See also Helical conformations of conjugating polymers
Poly(perfluorohexylethyl methacrylate) (PFMA). *See* Block copolymers with poly(ethylene oxide) and poly(perfluorohexylethyl methacrylate)
Poly(*m*-phenylenevinylene) derivatives
absorption and photoluminescence (PL) of polymer 8, 80, 82*f*
absorption and PL of polymer 10, 81, 82*f*, 83
chemical structures of polymers 8, 9, and 10, 79
cyano-functional groups for improving electroluminescence, 78–79
electroluminescence (EL) properties, 83, 86
electroluminescence spectra for light-emitting diode (LED) device, 83, 84*f*
fluorescence spectrum of polymer 10 in THF, 81, 83, 84*f*
fluorescence spectrum of polymer 8 in THF, 80, 84*f*
ground-state resonance forms for β- and α-cyano-substituted compounds, 78
molecular weight and optical properties of polymers, 85*t*
polymer synthesis and characterization, 79–86
replacement of *p*-phenylene with *m*-phenylene, 78
schematic of polymer synthesis, 80
thin film properties, 83
voltage dependence of current density and luminance for LED devices, 86, 87*f*
See also Poly(phenylene-vinylene) (PPV) materials
Poly(phenylene-vinylene) (PPV) materials
calculated spectral characteristics of substituted PPV3 molecules, 71*t*
calculated transitions in crosslinked PPV-related polymers, 74*f*
comparing poly[(2-methoxy-5-ethylhexyloxy-1,4-phenylene)vinylene] (MEH-PPV) with cyano-substituted MEH-PPV, 77
computational methods, 65–66
conformational mapping of PPV-based polymers, 67, 70*f*
cyano-substituted derivatives, 77
hole-transport material in light-emitting diodes (LEDs), 77
polymeric light-emitting diode (PLED) materials, 64–65
predicting properties for new, 71–72
schematic of cut through potential energy surfaces, 66*f*
sol-gel processing for glass-polymer composite, 7–8
structures of crosslinked PPV-related polymers and model compounds, 72
structures of model compounds for crosslinked PPV-related polymers, 73

structures of PPV-related polymers and PPV3 model compounds, 64
structures of substituted PPV3 model compounds, 71
substituent effects of PPV3-based blue-emitting chromophores, 71
wavelength multiplexing and narrow bandwidth filtering in sol-gel processed PPV:silica, 9*f*
See also Poly(*m*-phenylenevinylene) derivatives
Poly(2,5-pyridinediyl) (PPY), polarized luminance, 36–37
Poly(silylene)s
Kuhn's dissymmetry ratio, 52–53
optical activity, 51–52
solutions of stiff, 57–58
temperature dependence of cholesteric wavelength of isooctane solution, 58*f*
See also Helical conformations of conjugating polymers
Polystyrene
chromatograms with different concentrations, 331, 332*f*, 333*f*
mixtures with poly(2,6-dimethyl-1,4-phenylene oxide) (PPO), 352
moduli of blend with poly(2,6-dimethyl-1,4-phenylene oxide) (PPO), 253, 261–263
Noryl® family of engineering thermoplastics, 253
photonic crystals, 11–13
UV chromatograms of PS and poly(methyl methacrylate) (PMMA), 331*f*
UV spectra of PS and PMMA, 329, 330*f*
See also Dual detection high performance size exclusion chromatography (HPSEC); Elastomer blends; Polymer blends; Syndiotactic polystyrene (s-PS)
Poly(styrene-*b*-butadiene)
transmission electron microscopy, 311–312
See also Connectivity of domains
Polystyrene-*b*-poly(methyl methacrylate) (PS-*b*-PMMA)
experimental, 160–161
nanoporous films by solvent swelling, 159–160
See also Reversible surface reconstruction of thin films
Poly(styrene)-*b*-poly(4-vinylpyridine) (PS-*b*-P4VP)
complexation of P4VP with methane or toluene sulfonic acid (MSA or TSA), 39–40
hydrogen bonding of nonadecylphenol (NDP) to, 39*f*
hydrogen bonding of pentadecylphenol (PDP) to, 38–39
lamellar-within-lamellar self–assembled morphology, 39*f*
preparation of hairy rod objects from PS-*b*-P4VP(PDP), 44*f*
self-assembly for excess dodecyl benzene sulfonic acid (DBSA), 40, 42*f*
sequence of transitions vs. temperature in PS-*b*-P4VP(MSA)(PDP), 41*f*
See also Comb-shaped supramolecules
Poly(styrene-*co*-acrylonitrile) (SAN), dynamic moduli of blends, 258–259
Poly(tetrafluoroethylene) (PTFE)
differential scanning calorimetry (DSC), 346–347
permeability, 346–347, 349
Poly(1-trimethylsilyl-1-propyne) (PTMSP)
cavity size distribution, 195, 196*f*
diffusivity, 198
free volume, 194, 195*t*

mean-square displacement of CO_2, 197*f*
permeability, 188, 190, 191*t*, 198–199
properties, 189*t*
solubility, 191–192, 198–199
structure, 189*t*
typical packing model, 195*f*
See also Tetrafluoroethylene/2,2-bis(trifluoromethyl)-4,5-difluoro-1,3-dioxole (TFE/BDD) random copolymers; Transport properties
Polyurethanes
characterization, 374
dynamic moduli of blends, 259
reaction scheme for synthesis, 375, 377
selective catalysts, 377
two glass transitions, 374
urea side reaction by water, 377
See also Stone consolidation
Poly(2-vinylpyridine) (P2VP). *See* Polyisoprene-*b*-poly(2-vinylpyridine) (PI-*b*-P2VP)
Poly(4-vinylpyridine) (P4VP). *See* Poly(styrene)-*b*-poly(4-vinylpyridine) (PS-*b*-P4VP)
Porter, Roger, Karasz and, 2
Positron annihilation lifetime spectroscopy, cavity size distribution, 189, 194, 195*t*
Potential energy
curves for 1,4-divinylbenzene, 68*f*, 69*f*
See also Molecular orbital (MO) theory
Predictions for photophysical properties
theoretical calculations allowing, 71–72
See also Poly(phenylene–vinylene) (PPV) materials
Pressure. *See* Broadband dielectric spectroscopy (BDS)
Probability, polymer networking, 254
Process-structure-property relationship, matrix free Spectra® fiber composite, 396–399
Productivity. *See* Catalysts productivity
Propene polymerization, $TiCl_4/AlEt_2Cl$ on graphite, 226, 227*f*
Protonation
camphorsulfonic acid (CSA) for polyaniline, 37–38
CSA for poly(2,5-pyridinediyl) (PPY), 36–37
methane or toluene sulfonic acid (MSA or TSA) for poly(4-vinylpyridine) (P4VP), 39–40
TSA for P4VP, 36
See also Comb-shaped supramolecules
Proton conductivity
poly(4-vinylpyridine):pentadecylphenol (P4VP:PDP), 37*f*
polystyrene-*b*-poly(4-vinylpyridine)(PDP), 40*f*

Q

Quantum dots, dispersion in poly(methyl methacrylate) (PMMA), 10*f*

R

Random copolymers. *See* Tetrafluoroethylene/2,2-bis(trifluoromethyl)-4,5-difluoro-1,3-dioxole (TFE/BDD) random copolymers

Random nanocomposites, sol-gel processing, 7–8
Reconstruction of surfaces. *See* Reversible surface reconstruction of thin films
Recovery in compression strength (RCS)
  equation, 378
  See also Stone consolidation
Regions of free migration, active center ensembles, 219–220
Reinforced composites. *See* Matrix free Spectra® fiber reinforced composites
Relaxation time, percolation theory, 256
Restoration
  characteristics of synthetic materials for, 371
  commercial polyacrylate dispersions (Primal AC33) introduction, 372
  Fourier transform infrared (FTIR) spectra of acrylate-based commercial products, 373*f*
  macromolecular substances for application, 371–372
  material evaluation, 372, 374
  slowing degradation processes, 371
  See also Stone consolidation; Textile restoration
Reversibility, synthetic materials for restoration, 371
Reversible surface reconstruction of thin films
  acetic acid as solvent for generating nanoporous films, 159–160
  approach to nanoporous films of polystyrene-*b*-poly(methyl methacrylate) (PS-*b*-PMMA), 159–160
  atomic force microscopy (AFM) of CSM69k after acetic acid swelling and annealing, 168*f*
  diblock copolymer PS-*b*-PMMA samples (LSM71k and CSM69k), 160–161
  experimental, 160–161
  GISAXS (grazing incidence small angle X-ray scattering), 160, 161
  GISAXS pattern of LSM71k after acetic acid exposure, 163, 164*f*
  GISAXS pattern of LSM71k on passivated surface after acetic acid dip, 164–165
  GISAXS patterns during re-annealing process, 166*f*
  intensity of first order diffraction, 167*f*
  LSM71k thin film preparation on substrates, 164
  morphology of PS-*b*-PMMA LSM71k sample, 162
  pore penetration through entire film, 163–164
  re-annealing process of cylindrical microdomains CSM69k, 165–167
  scanning force microscopy (SFM), 160
  SFM images of swollen diblock copolymer after annealing, 167–168
  transmission electron microscopy (TEM) of annealed LSM71k, 161, 162*f*
Rubber elasticity, theory, 254
Rubbers. *See* Elastomer blends

S

Scanning electron microscopy (SEM)
  bis(salicylaldiminate) metal complexes, 22, 24*f*, 25*f*, 26*f*
  cotton fibers untreated and treated, 382, 383*f*

external surfaces of untreated, silicate-based treated, and poly(urethane-urea) treated sandstone, 380*f*
fracture surface of treated and untreated sandstone, 379*f*
Goretex® fluorocarbon material, 347, 348*f*
matrix free Spectra® fiber composite, 398–399
See also Environmental scanning electron microscopy (ESEM)
Scanning force microscopy (SFM)
regeneration of initial nanotemplate, 160
swollen diblock copolymer thin films after annealing, 167–168
Scanning transmission electron microscopy (STEM), contrast transfer function, 315–316
Schizophyllan
microscopic-macroscopic structure correlation, 56–57
order-disorder transition in aqueous solution, 57*f*
Selective solvent
manipulating microdomain orientation, 159
See also Reversible surface reconstruction of thin films
Self-assembly
block copolymers for preparing nano-objects, 44
concept for constructing supramolecules structures, 35*f*
lamellar-within-lamellar morphology, 39*f*
See also Comb-shaped supramolecules
Separation modes, liquid chromatography of polymers, 246
Silica, sol-gel processing to glass-polymer composite, 7–8, 9*f*
Silica hybrids. *See* Polyimides (PIs)
Silicone films
permeability of cured, vs. temperature, 343, 344*f*
water vapor loss through, 343*f*
water vapor loss through rubbery, 349*f*
Silicone hydrogels
dark field (DF) field emission gun (FEG) scanning transmission electron microscopy (STEM), 317
STEM, 315, 316*f*
TEM, 315, 316*f*
See also Connectivity of domains
Siloxane networks, cured, dynamic moduli of blends, 260
Simulations
building amorphous cells, 193
cavity size distribution, 193–194
computer, of Ziegler–Natta polymerization by charge percolation mechanism, 223–226
diffusion, solubility, and permeability coefficients, 194
See also Transport properties
Sintering
matrix free Spectra® fiber reinforced composites, 394
optimization of conditions, 395–396
Size exclusion, compensation of, and adsorption, 245–246
Size exclusion chromatography (SEC)
molecular mass and molecular mass distribution, 326
See also Dual detection high performance size exclusion chromatography (HPSEC)
Small angle neutron scattering (SANS)
inner structure of block copolymer clusters, 98*f*
method, 95

Smart behavior. *See* Metal complexes in ultra high molecular weight polyethylene (UHMWPE)
Sol-gel processing, glass-polymer composite, 7–8
Solubility
  modeling methodology, 191–192
  simulation, 194
  simulation results, 198–199
  See also Transport properties
Solvents
  manipulating microdomain orientation, 159
  See also Confined macromolecular systems; Reversible surface reconstruction of thin films
Sorption
  kinetics of 1,2-dichloroethane (DCE), by syndiotactic polystyrene, 176–177
  nitrogen experiments, 175–176
  volatile organic compounds (VOCs) from water, 172, 176–177
  See also Syndiotactic polystyrene (s-PS)
Spacers, hyperbranched urethane-acrylates, 211
Spectra® fiber. *See* Matrix free Spectra® fiber reinforced composites
Spectroscopic detectors, size exclusion chromatography (SEC), 327
Spheres, transmission electron microscopy (TEM), 312
Stone consolidation
  aggregative efficiency (AE), 377
  evaluating durability of procedure, 378
  polyurethanes, 374
  recovery in compression strength (RCS), 378
  scanning emission microscopy (SEM) of sandstone with commercial polyurethane-based dispersions, 375*f*
  SEM of external surfaces of untreated, silicate-based application, and in situ polymerization of poly(urethane-urea), 380*f*
  SEM of fracture surface of untreated and poly(urethane-urea) treated sandstone, 379*f*
  thermogravimetric curve of Neopolitan yellow tuff, 376*f*
  urea bonds from water as reaction component, 377
  use of selective catalyst, 377
  See also Restoration; Textile restoration
Stop gap, photonic crystals, 11
Stress strain curves, cyclic poly(butylene terephthalate)/caprolactone (c-PBT/CL) copolymer composites, 154–155
Stretched exponential (SE) behavior, amorphous polymers, 270
Styrene-acrylonitrile copolymer (SAN)
  in situ blending bis-phenol-A carbonate (BPACY) with SAN, 140–141
  See also Bis-phenol-A polycarbonates (BPACY)-based nanocomposites
Styrene-methyl methacrylate (SMMA) copolymers
  homo and copolymerization conditions, 328*t*
  preparation, 327–328
  size exclusion chromatography (SEC) measurements, 328–329
  See also Dual detection high performance size exclusion chromatography (HPSEC)
Superprism phenomenon, photonic crystals, 13

Support nano-particles
distribution of transition metal, 218–219
Ziegler–Natta polymerization, 216–218
See also Ziegler–Natta (ZN) polymerization
Support regions
distribution of transition metal atoms, 220*f*, 221*t*
See also Ziegler–Natta (ZN) polymerization
Supramolecules. *See* Comb-shaped supramolecules
Surface. *See* Reversible surface reconstruction of thin films
Surface pressure-area isotherms, block copolymers, 99*f*
Syndiotactic polystyrene (s-PS)
absorption of volatile organic compounds (VOCs) from water, 176–177
chlorinated compounds 1,2-dichloroethane (DCE), 1,2-dichloropropane (DCP), and 1-chloropropane (CP) as guest molecules, 177–178
comparing conformational equilibria of chlorinated compounds, 178
conformational selectivity, 181
crystalline δ-form, 172
crystal structure parameters, 174*t*
desorption kinetics from amorphous and clathrate phases of, 180
electrostatic potential of trans and gauche conformers of DCE, 182*f*
empty volume calculation, 173–174
fractions of trans conformer (Xt), 178, 179*f*
FTIR study of conformational selectivity, 177–181
guest volume fraction, 174*t*
host-guest interaction energies, 183*t*
host-guest interactions, 177–183
molecular modeling results, 181–183
nitrogen sorption experiments, 175–176
number of N_2 molecules per crystalline cavity, 175–176
partitioning of DCE in amorphous and crystalline phases, 180–181
potential maps, 182*f*
region of empty space calculated for δ-form, 173*f*, 174
role of electrostatic interactions, 181–183
shape and volume of cavities, 173–176
sorption and desorption isotherms of δ- and γ- form samples, 175*f*
sorption kinetics of DCE from saturated or aqueous solutions, 176, 177*f*
sorption studies from liquid and gas phases, 172
synthesis, 172
volume of cavity, 174*t*
Xt values during guest desorption procedures, 180
Synthetic polypeptides, order-disorder transition, 1–2

T

Takayanagi coupling, phase moduli, 260–261
Tan δ curves, hyperbranched urethane-acrylates, 210, 211*f*
Temperature. *See* Broadband dielectric spectroscopy (BDS)
Tetrafluoroethylene/2,2-bis(trifluoromethyl)-4,5-difluoro-1,3-dioxole (TFE/BDD) random copolymers

cavity size distribution, 189–190, 195, 196*f*
commercially available products, 188
diffusivity of CO_2, 198
fractional cavity volume (FCV), 194, 195*t*
mean-square displacement of CO_2 in, 197*f*
permeability, 190, 191*t*, 194, 197*t*
positron annihilation lifetime spectroscopy (PALS), 189, 195*t*
properties, 188, 189*t*
solubility and permeability, 198–199
structure, 189*t*
See also Transport properties
Textile restoration
acrylic and vinylacetate dispersions, 381
adhesive properties of commercial products, 385–386
amount of polymeric materials impregnating yarns, 381
dried polymers impregnating cotton vs. dry content of water dispersion, 382*f*
graft polymerization on cellulose textiles, 386–387
peel strength, 386
scanning electron microscopy (SEM) of cotton fibers before and after immersion in vinylacetate polymeric dispersions, 382, 383*f*
scheme of grafting chamber for consolidation of textile items, 387*f*
stage of polymer interaction with fibers, 385
tensile recovery percent of stress at maximum load (RS), 384, 385*f*
See also Restoration; Stone consolidation
Therapy, polymeric nanoparticles, 14
Thermal characterization. *See* Differential scanning calorimetry (DSC)
Thermal degradation
poly(methacryl-diisopropylurea-*co*-ethylene glycol dimethacrylate), 232–233, 235
See also Crosslinked copolymers
Thermal stability, clay/cyclic ester oligomers (CEO) nanocomposites, 149–150
Thermoforming, matrix free Spectra® fiber composite, 399–401
Thermogravimetric analysis (TGA)
method, 122
nanocomposites based on cyclic ester oligomers, 149–150
nanocomposites of cyclic poly(butylene terephthalate)/caprolactone copolymer, 154
triblock copolymer and blend with poly(caprolactone) (PCL), 127, 129
water vapor transmission (WVT) studies, 340
See also Permeability
Thermoplastic molecular sieves. *See* Syndiotactic polystyrene (s-PS)
Thin films. *See* Reversible surface reconstruction of thin films
Time correlation functions (TCFs)
amorphous polymers, 271–273
ensemble-averaged TCF, 272
time-averaged TCF, 272–273
Time-resolved Fourier transform infrared (FTIR) spectroscopy
diffusion studies, 297
sorption/desorption kinetics of water in polyimides, 300–301
sorption of ammonia in polyimide and hybrid, 302–303
See also Fourier transform infrared (FTIR) spectroscopy

Toluene sulfonic acid (TSA), protonation of polystyrene-*b*-poly(4-vinylpyridine), 38–40
Toxicity, block copolymers and human erythroleucemia cells, 102–103
Transition metal
distribution of, on support, 218–219
See also Ziegler–Natta (ZN) polymerization
Transmission electron microscopy (TEM)
bisphenyl-A-polycarbonate/styrene-acrylonitrile blend progress, 141, 142*f*
contrast between different regions of sample, 311
contrast transfer function, 314–315
lamellar block copolymer morphology, 311*f*
nanocomposites of cyclic ester oligomers (CEO) and clay before and after polymerization, 148–149
nanocomposites of cyclic poly(butylene terephthalate)/caprolactone copolymer, 152*f*
ordered morphologies, 310–313
poly(styrene-*b*-butadiene), 311–312
projection geometry, 312–313
spheres, cylinders and cubic packing, 312
See also Connectivity of domains
Transport properties
building amorphous cells, 193
cavity size distribution, 195, 196*f*
cavity size distribution of material, 189–190
comparing diffusion, solubility and permeability of CO_2 in poly(1-trimethylsilyl-1-propyne) (PTMSP) and random copolymer of 13/87 tetrafluoroethylene/2,2-bis(trifluoromethyl)-4,5-difluoro-1,3-dioxole (TFE/BDD87), 197*t*
comparing fractional cavity volume, average cavity size, PALS of PTMSP and TFE/BDD87, 195*t*
comparing possible transport pathways in lamellar materials, 321–322
diffusion, 192
diffusivity of CO_2 in PTMSP and TFE/BDD87, 198
experimental permeability values of TFE/BDD87 and PTMSP, 191*t*
free volume, 194, 195*f*
mean-square displacement of CO_2 in PTMSP and TFE/BDD87, 197*f*
modeling details, 193–194
modeling methodology, 190–192
permeability, 190
phase continuity and, 319–322
polymers with high fractional free volumes (FFV), 188
positron annihilation lifetime spectroscopy (PALS) measurements, 189
properties of TFE/BDD87 and PTMSP, 189*t*
simulation of cavity size distribution, 193–194
simulation of diffusion, solubility, and permeability coefficients, 194
solubility, 191–192
solubility and permeability, 198–199
structure of TFE/BDD87 and PTMSP, 189*t*
typical packing model for PTMSP, 195*f*

Triblock copolymer of poly(DL-lactide) and poly(ethylene glycol)
ABA triblock copolymer synthesis for A=poly(DL-lactide) (PDLLA) and B=poly(ethylene glycol) (PEG), 121
characteristics of copolymers, 124*t*
characterization, 121–122
composite graph for cell attachment and growth on, and PDLLA-PEG-PDLLA/PCL blends, 134*f*
contact angle measurements, 130*t*
copolymer synthesis and blend preparation with poly(caprolactone) (PCL), 123–127
culture of cells and cell growth experiments, 123
differential scanning calorimetry (DSC) of triblock, PCL, and their blends, 127, 128*f*
experimental, 120–123
Fourier transform infrared (FTIR) spectra of PDLLA-PEG-PDLLA and blends with PCL, 124, 126*f*, 127
^{1}H nuclear magnetic resonance (NMR) spectrum of PDLLA-PEG-PDLLA, 125*f*
hydrolytic degradation for, and its blends, 131, 132*f*
hydrolytic degradation studies, 122–123
mass and molar mass losses for, and its blends, 131, 132*f*
materials, 120–121
porous film fabrication and characterization, 122
preparation of PDLLA-PEG-PDLLA/PCL blends, 121
surface free-energy components for, and blends with PCL, 131*t*
thermogravimetric analysis (TGA) of triblock, PCL, and blends, 127, 129*t*
viability of cells, 131, 133
water absorption and contact angel measurements, 130

U

Ultra high molecular weight polyethylene (UHMWPE)
commercial materials, 401–402
Dyneema Fraglight®, 401, 402
one-polymer composites, 393–394
tensile modulus of superdrawn, 392
See also Matrix free Spectra® fiber reinforced composites; Metal complexes in ultra high molecular weight polyethylene (UHMWPE)
Ultraviolet-visible (UV-vis) spectroscopy
bis(salicylaldiminate) metal complexes, 28*f*, 29*f*, 30*f*
poly(*m*-phenylenevinylene) derivatives, 80, 81, 82*f*, 84*f*
polystyrene-*b*-poly(4-vinylpyridine)(dodecyl benzene sulfonic acid), 42*f*
ultra high molecular weight poly(ethylene) oriented film, 27*f*
University of London, Karasz, Frank Erwin, 1
University of Massachusetts, Karasz, Frank Erwin, 2
Unsaturation, hyperbranched urethane-acrylates and curing, 211
Urethane-acrylates. *See* Hyperbranched urethane-acrylates
Urethane-urea polymer. *See* Stone consolidation

V

Vanadium oxide, sol-gel processing to glass-polymer composite, 7–8

Vinylacetate dispersions
amount impregnating cotton vs. dry content of dispersion, 382*f*
textile restoration, 381
Viscoelastic properties, hyperbranched urethane-acrylates, 210–212
Viscometry
polymer-polymer interactions and viscosity, 283
See also Polymer blend solutions; Viscosity
Viscosity
theory, 284–285
See also Viscometry
Volatile organic compounds (VOCs)
absorption from water, 176–177
See also Syndiotactic polystyrene (s-PS)
Volume. *See* Broadband dielectric spectroscopy (BDS)
Vulcanization chemistry, 254

W

Water
absorption of triblock copolymers and blends with poly(caprolactone) (PCL), 130
curve fitting analysis of water absorbed in neat polyimide, 298*f*
curve fitting analysis of water absorbed in polyimide/silica hybrid, 299*f*
diffusion behavior in polyimide, 297
diffusion coefficient vs. water vapor activity for polyimide and hybrid, 301–302
sorption and desorption kinetics in polyimide and hybrid, 301*f*
sorption in polyimide and polyimide/silica hybrid, 298–302
sorption isotherms in neat, and hybrid, 300*f*
time-resolved Fourier transform infrared spectroscopy, 300–301
See also Polyimides (PIs)
Water vapor transmission (WVT)
apparatus and methods of invention, 340–341
container design for small samples, 340
schematic of WVT capsule, 341*f*
See also Permeability
Widom insertion method, solubility and permeability, 194, 198–199

X

Xanthate groups
controlled radical polymerization, 202–203
hyperbranched urethane-acrylates, 212
modification of hyperbranched polyester with, 206–207
synthesis of hyperbranched polyesters with, 205
synthesis of hyperbranched urethane acrylate with, 205, 206*t*
X-ray diffraction
comparing clays for clay/cyclic ester oligomer (CEO) nanocomposites, 150–152
organoclay/cyclic ester oligomer mixtures, 147–148
X-ray scattering
inner structure of block copolymer clusters, 98*f*
method, 95
See also Grazing incidence small angle X-ray scattering (GISAXS)

Y

Yarns. *See* Textile restoration

Z

Ziegler-Natta (ZN) polymerization
catalysts productivity, 223
charge percolation mechanism (CPM), 221–222
computer simulation of, by CPM, 223–226
definition of ZN catalyst, 216
distribution of transition metal (Mt) atoms by support regions, 220*f*, 221*t*
distribution of transition metal on support, 218–219
effect of number of Mt atoms per region on productivity, 225*f*
effect of surface concentration of active centers on productivity, 224*f*
monomer self-organization and charge percolation mechanism, 222*f*
Monte–Carlo procedure, 223
one- and two-dimensional percolation processes, 223
polymer detachment, 222*f*
productivity of catalyst system in propene polymerization, 226*f*
propene polymerization by $TiCl_4/AlEt_2Cl$ on graphite, 226, 227*f*
series of simulations, 223–224
support nano-particles, 216–218
theory of active center ensembles, 219–221
$TiCl_3$ based systems, 216–217